PC Telephony

Bob Edgar
Parity Software Development Corporation

*Flatiron Publishing, Inc.
New York*

First Edition Copyright © 1992 Bob Edgar.
Second Edition Copyright © 1994 Bob Edgar.
Third Edition Copyright © 1995 Bob Edgar.
Fourth Edition Copyright © 1997 Bob Edgar.
All rights reserved.

No part of this publication may be reproduced, stored in a retrieval system or transmitted, in any form or by any means, electronic, mechanical, photocopying, recording or otherwise, without the prior written permission of the author.
Company and product names used in this book may be trademarks or registered trademarks of their respective authors.

In particular, Dialogic and Dialogic product names are trademarks of Dialogic Corporation. MS-DOS and Microsoft are trademarks of Microsoft Corporation. PC-DOS, IBM, IBM PC and OS/2 are trademarks of International Business Machines Corporation. Touch-tone and UNIX are trademarks of AT&T. BeSTspeech is a trademark of Berkeley Speech Technologies, Inc. dBASE and Paradox are trademarks of Borland International. VOS is a trademark of Parity Software Development Corporation.

While every effort has been made to ensure complete and accurate information in this book, neither the author nor the publisher can be held liable to errors in printing or inaccurate information in this book. This book is sold "as-is" without warranty of any kind, expressed or implied, of merchantability or fitness for any particular purpose. The reader is encouraged to verify all relevant information with product vendors when making important purchase or design decisions. The author and publisher welcome suggestions and corrections.

ISBN# 1-57820-017-2

Table of Contents

Foreword	15
Preface To The First Edition	19
Preface To The Fourth Edition	21

1   Call Processing — 23
    What Is Call Processing? — 23
    Call Processing Features — 25
    Major Applications — 27

2   Computer Telephony — 31
    Introduction — 31
    Computer Telephony Integration (CTI) — 32
    Stand-Alone Analog VRU — 33
    Stand-Alone Digital With CSU — 34
    Stand-alone With Channel Bank — 34
    Behind A PBX — 35
    In Front Of A PBX — 36
    Call Center — 37
    Service Bureau With Dumb Switch — 39
    VRU With PC Database — 41
    Client / Server Configurations — 42

3   Telephony Applications — 47
    Introduction — 47
    Automotive — 48
    Construction — 49
    Corporate — 50
    Corrections — 52
    Education — 53
    Electronic Media — 55
    Entertainment — 57
    Financial — 59
    Government — 62
    Health Care — 64
    Hospitality — 65
    Insurance — 67
    Manufacturing — 68
    Military — 69
    Power Utilities — 70
    Print Media — 71
    Product Distribution — 73

	Retail Services	74
	Social Services	76
	Sport	79
	Telephone Companies	80
	Transportation	83

4 *Telecommunications*   89

The Phone Line	89
Starting An Outgoing Call: Getting Dial Tone	91
Announcing An Incoming Call: Ringing	91
Dialing A Number	93
Completing The Call	94
Call Progress Analysis	95
The Flash-Hook	96
Terminating A Call: Disconnect	96
T-1 Digital Trunks	96
E-1 Digital Trunks	97
ISDN Digital Trunks	97
DNIS And DID Services	98
Automatic Number Identification (ANI) And Caller ID	99
PBXs: Private Telephone Systems	100
The Telephone Network	102
Telephone Numbers	104
Dial Strings	109
Access Codes	111

5 *Telephone Signaling*   115

Introduction	115
Seize	117
Disconnect	118
Ring	119
Flash-Hook	120
Wink	121
Disconnect Supervision	121

6 *Call Progress*   125

Introduction	125
Call Progress Monitoring	126
Loop Current Supervision	127
Cadence	127
The D/xx Call Progress Algorithms	129
Global Tone Detection	132

7 Fax Integration — 135
- Introduction — 135
- A Brief History Of Fax — 136
- Fax Transmission Standards — 138
- Fax Call Protocol — 139
- Compression Schemes — 141
- Printing The Fax — 143
- Computer-Based Fax — 144
- The Future Of Fax — 145
- The Evolution Of Group 3 — 147
- Fax And X.400 — 149
- Fax Boards — 151
- The CAS Programming Interface — 153
- The Dialogic VFX And FAX/Xx — 157

8 Voice Recognition — 159
- Introduction — 159
- Touch-Tone Availability — 161
- Grunt Detection — 163
- Multiple Matches — 163
- The User Interface — 163
- Vocabulary Features — 165
- Sharing Recognizers — 168

9 Pulse Recognition — 171
- Introduction — 171
- Pulse Recognition Vendors — 173
- Pulse Recognition Equipment — 173
- The User Interface — 175

10 Text To Speech — 177
- Introduction — 177
- TTS Vendors — 179
- Berkeley Speech Technologies — 179
- Digital Equipment Corporation — 180
- First Byte — 181
- Infovox — 181
- TTS Methods — 182
- TTS Features — 184

11 International Issues — 187
- Introduction — 187
- Approvals — 188
- Naive Users — 189
- Algorithms — 189
- Telephone Numbers — 190

    Bi- And Multi-Lingual Applications ... 190
    Script Translation ... 190
    Rotary Phones ... 190
    Trunk Types And Signaling Protocols ... 190
    Regulatory Issues ... 191

12 E-1 ... 193
    Introduction ... 193
    Compelled Signaling ... 194
    Types of R2/MF Signaling ... 196
    E-1 Signaling Bits ... 198
    E-1: The Details ... 199
    E-1 And T-1 Trunks Compared ... 199
    R2/MF Tones ... 202
    R2 Dual Tone Frequencies ... 204
    Interpreting R2 Tones ... 204
    Socotel R2 Signaling ... 207
    Socotel Signal Interpretation ... 209
    ABCD Signaling Bits: CCITT Q.421 ... 210
    ABCD Signaling Bits: Socotel ... 212
    DID Example: CCITT ... 213
    DID Example: Socotel ... 214
    Protocol Outlines For Selected Countries ... 215

13 Voice/Data Protocols ... 219
    Introduction ... 219
    Alternating Voice And Data (AVD) ... 221
    Simultaneous Voice And Data (SVD) ... 221
    Available Protocols ... 222
    Dialogic And Voice/Data ... 223
    VoiceView ... 223
    VoiceView Modes ... 224
    DTE Interface ... 226
    Implementing VoiceView ... 226

14 Voice Boards ... 229
    Introduction ... 229
    The Dialogic D/4x Series ... 237

15 Installing Boards ... 243
    Introduction ... 243
    Resources ... 244
    Hardware Interrupts (IRQs) ... 245
    Hardware Interrupt Priorities ... 247
    Software Interrupts ... 247
    I/O Port Addresses ... 248

    Shared Memory   249
    DMA Channels   252
    Trouble-Shooting Checklist   253
    Trouble-Shooting Strategy   253

16 Dialogic Product Line   255

    Introduction   255
    Types Of Trunk   257
    Voice Buses   258
    The D/4x Board   259
    The D/2x Board   261
    The D/42-xx Boards   262
    The AMX/8x Board   262
    The MF/40   264
    The VR/40   265
    The DID/40   265
    The DTI/124   265
    The DTI/100   267
    The LSI/120   267
    The DID/120   268
    The D/12x   268
    The FAX/120   270
    The VR/121   271
    The VR/xxp   272
    The MSI   272
    The MSI/C (MSI Revision 2)   273
    The MSI/SC   273
    The DMX   274
    The DTI/211   276
    The DTI/212   277
    The D/81A   278
    The LSI/80-xx   278
    The DCB (Spider) Board   279
    HD SC Bus Products   280
    The D/160SC-LS   282
    The D/240SC   282
    The D/320SC   283
    The D/240SC-T1   283
    The D/300SC-E1   283
    Typical Configurations   284
    Dialogic Product Line Summary   285

17 Antares — 291
- The Dialogic Antares Series — 291
- Antares Board Models — 292
- Antares Firmware Products — 293
- Programming For Antares — 295

18 PEB Boards — 297
- Introduction — 297
- Example Configurations — 300
- Drop And Insert Configurations — 302
- The DMX Switching Card — 304

19 SCSA And The SC Bus — 307
- What Is SCSA? — 307
- SCSA Components — 309
- SC Bus — 309
- SCx Bus — 311
- SC Bus Hardware — 312
- SC Firmware — 312
- SC Driver — 312
- Configuration Database — 314
- Device Programming Interface — 314
- Server API — 314

20 The SCX Bus — 317
- Extending The SC Bus — 317
- An SCX Example — 319
- Programming The SCX/160 Adapter — 320
- Programming An SCX Bus System — 320

21 Switch Cards — 323
- Introduction — 323
- Third Party Switch Card Vendors — 324
- The Amtelco XDS Switch — 324
- The Dianatel SS96, SS192 And SB — 325
- The Dianatel CO24 — 327
- The Excel PCX512 Series — 327

22 AEB Switching — 331
- Introduction — 331
- One AMX And Four D/4x Boards — 334
- Two D/4x Cards And One External Device — 335
- One AMX And One D/4x — 336
- Four D/4x Boards Sharing Eight Devices — 337
- Making And Breaking Switchpoints — 338
- Switching For The PEB — 339

23 Telephony Sound Files — 341
Introduction — 341
Digitization — 342
Preparing Digital Audio Files — 343
Digitization Methods — 346
Dialogic Standard ADPCM — 348
VOX Files — 351
Cutting And Pasting ADPCM Data — 356
Indexed Prompt Files (IPFs) — 356
IPF File Format — 357
Data Compression — 358
Should I Use Wave Files? — 358

24 Programming For Dialogic — 361
Dialogic Drivers And APIs — 361
Operating Systems And Driver Types — 362
A Sneak Peak — 365
MS-DOS API — 366
Windows And UNIX Proprietary API — 367
OS/2 API — 368
TAPI — 369
Dialogic API Pros And Cons — 371
File Handles And errno In Win32 — 375
User I/O In Win32 And UNIX — 377
Playing And Recording VOX Files — 378
Fast-Forward, Pause and Rewind In Wave Files — 381
Creating EXEs For Both Windows NT And 95 — 382

25 Dialogic Devices — 385
Introduction — 385
Device Types — 388
Transmit And Listen/Receive Channel Devices — 391
Voice Channel Devices — 392
Analog Interface (LSI) Devices — 392
DTI Channel Devices — 394
Fax Channel Devices — 394
MSI Station Devices — 395
Voice Recognition Resource Devices — 395
Device Names And Numbers — 396
Board Numbers — 396
Voice Channel Numbers — 399
Fax Channel Numbers — 400
DTI Channel Numbers — 400

26 Programming Models — 403
- What Is A Programming Model? — 403
- Multi-Tasking, Multi-Threading and Fibers — 405
- Synchronous And Asynchronous Modes — 406
- Event Notification — 407
- Event Notification By Waiting — 407
- Event Notification By Polling — 408
- Event Notification By Callback — 409
- Event Notification By Window Message — 412
- Hybrid Programming Models — 412

27 Dialogic Programming For MS-DOS — 415
- Introduction — 415
- Driver Functions — 416
- A First Program — 420

28 State Machines — 431
- Introduction — 431
- The State Machine Engine — 434
- A Complete State Machine Program — 436
- Real Life — 441
- States And Super States — 443
- There's More To Life Than Voice — 447

29 PEB Programming — 449
- Introduction — 449
- PEB Routing Functions In The MS-DOS API — 459
- Routing Example: DMX — 460
- Time-Slot Routing For Drop And Insert — 464
- Signaling — 467
- DTI Transmit Signaling — 469
- Idling A DTI Time-Slot — 472
- DTI Signaling Detection — 474
- DTI Hard-Wired Drop And Insert Example — 475
- The DTI/124 — 479
- Controlling The MSI Board — 481

30 SC Bus Programming — 483
- Introduction — 483
- How The SCbus Works — 484
- SC Bus Devices — 486
- SC Bus Routing — 487
- SC Bus Routing API — 490
- Intra- And Inter- Board Connections — 493

31 CTI And TSAPI — 495
- Introduction — 495
- Missing: VRU Media Control — 498
- CSTA Interface — 498
- TSAPI Components — 499
- Application Types — 500
- How To Get TSAPI — 501

32 TAPI — 503
- Introduction — 503
- Desktop Applications — 505
- TAPI Structure — 507
- TAPI Functionality — 509
- Getting The TAPI SDK — 512

33 TAPI Programming — 513
- Introduction — 513
- Line Devices — 514
- Phone Devices — 515
- Calls — 516
- Typical Call — 518
- TAPI Structure — 519
- Assisted Telephony — 520

34 TAPI, Wave And The Dialogic API — 525
- Introduction — 525
- Writing A TAPI Application — 527
- Answering In-Bound Calls — 528
- Monitoring For Hang-Up — 529
- Making Out-Bound Calls And Transfers — 530
- Getting Digits — 532
- Playing And Recording Messages — 533
- Speaking Phrases — 538
- Speed And Volume Control — 539
- Generating And Detecting Custom Tones — 539
- Configuration Information — 540
- C-T and PBX Control — 540
- API Design — 541
- Conclusions: TAPI/Wave Pros And Cons — 541

35 SWV: A Proposed Sound File Format — 543
- Improving On Wave — 543
- Segments — 546
- File Extension — 546
- File Format — 547
- Segment Index Chunk — 547

LIST INFO Chunk	550
Sample Files	550
Future Extension: Phrase Algorithms, RFCs	551

36 The User Interface — 555

Introduction	555
Menus	556
Mnemonic Menus	558
Entering Digit Strings	560
The Overall Structure	563
Damage Control	566
Reporting Information	567
Field Trials	569

37 Speaking Phrases — 571

Introduction	571
Phrase Components	572
Whole Numbers	574
Ordinal Numbers	578
Dates	578
Times	580
Date And Time Stamps	580
Money	581
Digit Strings	582
Phone Numbers	582
Inflection	583
Conclusion	586

38 ActiveX Controls — 589

What Is A Control?	589
Control Properties	591
Control Events	593
Control Methods	594
Invisible Controls	595
Controls: A Summary	595
Custom Controls: VBXs, OCXs And ActiveX	595

39 Telephony Controls — 597

Extending Visual Tools	597
ActiveX For Telephony	598
A Simple Example	600
Handling Caller Hangup	601
Processing Hangup Error Traps	601
Problems With Error Traps	602
The VoiceBocx Programming Interface	604
VoiceBocx Properties	604

VoiceBocx Methods	607
VoiceBocx Events	610
Application Generators	610
CallSuite AppWizard	611
CallSuite Wizard	611
Routines	613
Routine Example	614
Recording Prompts And Testing Routines	616
Using CallSuite Wizard With Development Environments	617

40 VOS 619

How To Avoid C Programming	619
What Is VOS?	620
Why VOS?	621
Greeter Revisited	623
Dialogic Basics	624
Creating A Touch-Tone Menu	625
Answering An Incoming Call	626
Making An Outgoing Call	626
Transferring A Call	627
Terminating A Call	628
Detecting A Disconnect	628
Robust and Fast	630
Best Features Of C And Pascal, Easier To Learn And Use	630
Source Code	631
Values	632
Variables	632
Constants	633
Arithmetic Expressions	634
Logical Conditions	634
Assignments	636
String Concatenation	636
Loops	636
Switch / Case	639
Goto And Jump	640
Calling VOS Functions	641
VOS Language Functions	642
Functions Files And Libraries	644
Include Files	645

41 Sizing Your System — 647
- Introduction — 647
- How Many Lines Do You Need? — 648
- Hard Disk Size — 649
- Hard Disk Speed, Number Of Lines — 650
- Automated Testing — 656

42 Which Operating System? — 659
- Introduction — 659
- Platforms — 660
- MS-DOS — 663
- Windows 3.1, Windows for Workgroups — 664
- Visual Basic — 665
- Windows NT — 666
- Windows 95 — 667
- OS/2 — 667
- UNIX — 668

43 Which Programming Tool? — 669
- Hourglass Of Death: The Sequel — 669
- It's Harder Than You Think — 671
- Ping Pong — 671
- If You Try This At Home... — 672
- Programming Models — 674
- Multiple Lines — 675
- The Tests — 676
- Speed Results — 677
- CPU Load Results — 679
- Memory Usage — 681
- Startup Times — 684
- Interpretation — 686
- Conclusions — 687

Glossary — 691

Index — 759

Foreword

by Harry Newton
Editor-in-Chief
Computer Telephony Magazine

Universities are registering their students and scheduling their classes with PC Telephony. Banks are giving out account information to their customers, who prefer "talking" to the machines than to a live teller. ("They snicker.") Companies are providing information on benefits to their employees and allowing them to change plans on-line. Dealers with broken machines are calling "fax-back" machines and receiving detailed print-outs on how to fix the machines and which parts to order.

Customers don't just want your product. They want information that "surrounds" it. What I call "The Surround Stuff." Your customers want to know the answers to questions like, "Where's your closest dealer?" "Where can I get it repaired?" "Does it work with your older model 265?" "Where can I buy it?" "How much money do I have in the bank? Do I have enough to move some of it to my savings account?" "When does the next plane to Chicago leave?" "Which is the closest movie house to me playing Rambo 37? When does the next show start?"

PC Telephony is superb for giving out Surround Information. Your customer (or prospect) calls, listens to some instructions, punches in his digits, listens to some more instructions, punches in some more digits, hears the information he wants and hangs up, a happy camper. PC Telephony can deliver the information by voice, by fax, by Internet e-mail. That's just for now.

PC Telephony is another term for Computer Telephony, which I define as "adding computer intelligence — hardware and software — to the making, receiving, and managing of phone calls." Today, Computer Telephony is largely PC-based — standalone and local area networked. This means it works like that children's game — Lego, Erector or Mecano (depending on which country you live in). With Lego Set PC Telephony you can build yourself anything your imagination can create and your heart desires.

You can build a system that will ask your callers questions, then, based on the answers they punched in or said (using speech recognition), the machine will transfer them to the right person — the one person who can answer their questions.

You can build an intelligent voice mail system that recognizes your callers and give them sterling or lousy attention — based on how important they are.

You can build a system which lets you dial phone calls from your desktop PC, over your LAN, via your main office telephone system. It's much easier dialing from a PC than from a phone, with its puny screen, upside-down keypad and teeny tiny, poorly-marked buttons.

You can build a system which lets callers from overseas save 50% on the international calls to the U.S. and other countries. This application is called international callback.

You can build a system the gives callers information they want — like the nearest movie theater for their chosen movie and the next time the movie is playing. And you could charge the movie distributor to advertise his movies on your service.

You can build a system which tells callers when your planes are landing and taking off and where their luggage can be picked up. You can build a system to take calls from your utility customers when they call to tell you their lights are out. And you can tell them you already know and are fixing it. All this can save hiring people. On their wages alone, you can pay such PC Telephony systems off very quickly — usually within a few months.

PC Telephony systems are for everyone — from large corporations to small, home-based companies. From governments (the IRS uses them) to small municipalities (who use them to give out bus timetables).

This book is aimed at the developer of PC-based Computer Telephony systems.

That "developer" may be someone who wants to design, build and sell a complete system — a systems integrator, a VAR, a reseller, etc. That "developer" might also be someone who builds part of a system — a voice processing component board, a fax-board, a voice recognition subsystem, a text-to-speech speaker, or programmer, etc. That "developer" might also be an end-user, who deploys his PC telephony system/s in his business.

By reading this book you will learn how to build PC Telephony Systems that are also called Voice Mail, Audiotext, Automated Attendants, Automatic Call Distributors, Interactive Voice Response systems, Predictive Dialers, PBXs, Voice Servers, Voice Clients.

There are two basic reasons companies invest in voice processing equipment — customer nurturing and internal productivity. Companies that provide 24-hour a day information on their products will win more customers than companies that only do it five days a week, eight hours a day. Companies that can get their customers' calls to their people faster — without landing them in Voice Mail Jail —will have happier customers. Companies whose screens "pop" with customer information as the customer calls will attend the customer's problems faster. Companies that can have their computers automatically set up their conference calls will find more time to concentrate on solving problems. This means their employees' "productivity" will go up.

PC Telephony used to be confined to one PC. Now there are standards — like SCSA — which let you build your PC telephony application in one PC and, as it grows, expand it to multiple PCs. This way you get the modular growth economics of the PC and the Client Server LAN architecture.

As a result, PC Telephony now is a technology with no limits. At least one telephone company is now using it as a central office (also called a public exchange). Up till now, central offices have been gigantic, mainframe-like devices. To drop them into the size of a PC is truly the beginnings of a revolution that may be even more dramatic in telecom than what happened in computing when that industry switched from mainframes to PC/Client Server Architecture.

This is the fourth, much-expanded edition of an extraordinarily successful book. Thousands of people have bought this book and built powerful computer telephony systems. While there is a little measure of bias toward Dialogic PC telephony cards (they are the "Intel" of PC Telephony) and VOS, the very successful PC script which Bob Edgar, the author, helped write, the book is jammed with information useful to anyone building a PC Telephony system based on any platform. Bob calls VOS a "Call Processing Application Language." Which is probably a narrow way of looking at it. And it supports, inter alia, voice processing, fax integration, voice recognition, text-to-speech, voice mail and telephony pipes like T-1, E-1, ISDN, ANI, DNIS, and DID. Good news, if you write code in VOS, you can move to several operating systems UNIX, DOS, Windows 3.1 and 3.11, Windows NT and Windows 95.

If you want to learn more about this exploding field of PC Telephony, you should subscribe to Computer Telephony, Call Center, and TELECONNECT Magazines. Fax your requests for subscription forms to 215-355-1068 or send a note to 1265 Industrial Highway, Southampton, PA 18966. You should also attend Computer Telephony Conference and Exposition, held each year in March. At the show, you'll hear Bob Edgar, the author of this book, deliver a one-and-a-half hour tutorial on PC Telephony. You'll also see all the latest hardware and all the latest tools, which you can use to build PC Telephony systems. If you subscribe to one of the above magazines, you'll receive plenty of early warning on the show. I'll see you at the next Computer Telephony Expo. Until then, enjoy this book. It's really good.

Harry Newton
New York City, Summer 1997
harrynewton@mcimail.com

Preface To The First Edition

In 1988 I started my first voice processing project. My experience of telecommunications was limited to picking up a telephone and dialing a number. I had never even heard the term "voice processing", but was vaguely familiar with voice mail. My one advantage was that I was familiar with PCs and had done a fair amount of programming in other areas. I was given a fill-in-the-blanks "application generator" product and asked to develop a rather complex application. The screens of this product presented me with terms like "loop drop." What was a loop, I wondered, and how could it drop?

This is the book I wish that I'd had back in 1988. I've pictured the target reader to be a programmer, engineer, project manager or other decision-maker who understands at least a little programming and wants to get involved in the design and programming of PC-based voice processing systems. The reader is assumed to be technically literate, but perhaps to be a telecommunications novice. This book therefore starts from square one: by explaining how a telephone works.

To cut a long story short, after becoming frustrated with the fill-in-the-blanks method of programming, I started my own company, Parity Software Development Corporation, and became involved in creating an alternative tool for software development — the type of product I would have liked to use in 1988. The result: VOS, an applications-oriented programming language for building voice applications in the tradition of products Visual Basic. VOS is now in use at thousands of sites around the world. Parity Software and myself are therefore active players in the voice processing market, and cannot claim to be impartial observers or commentators on the voice processing scene. We work closely with Dialogic and other vendors to bring the latest technology to our developers, systems integrators and value added resellers.

As you will have gathered, I cannot claim systematic education or training in voice processing or telecommunications. What I have been able to learn has been out in the field, down in the trenches, trying to make stuff work in the real world. Some of the technical explanations may therefore be a little rough at the edges or even incorrect — but I have done my best to go beyond the hardware vendors' documentation and discuss the real issues.

I hope that readers will take the time to comment on possible inaccuracies or make suggestions for future editions of the book. The author and publisher are looking forward to improving and extending this book in future editions. Readers can write:

> c/o Christine Kern
> Telecom Library
> 12 West 21st Street
> New York, NY 10010
> 212-691-8215, Fax 212-691-1191

Finally, many people have been kind enough to devote time and energy to helping me with this book. Harry Newton, publisher of *Teleconnect Magazine* and editor of the Telecom Library series of books, persuaded me that the world needed a voice processing book, and supported me the whole way to publication. Harry kindly put much personal time and energy into reviewing early drafts and contributing material, including many definitions from his indispensable Telecom Dictionary. Dialogic Corporation has also been very supportive. Jim Shinn, Ed Margulies, Eamonn Kearns, Mike Stahl, Bob Pond and others helped in many different ways. An honorable mention should certainly go to Dialogic's indefatigable technical support team, who have had to endure many conversations with me over the past few years going over the fine details of their product architecture. Eamonn Kearns contributed some material for the International chapter. Rich Bernstein of Gammalink generously made some of the information from his white papers on fax technology available. I also thank Michael J. Winsek, Jr., Director of Systems Engineering at Radish Communications Corp. for providing a white paper which was the basis for the technical discussion of the VoiceView protocol.

Errors and omissions are, of course, the responsibility of the author.

Bob Edgar
San Francisco, Fall 1993.

Preface To The Fourth Edition

A great deal has changed in the computer telephony industry since the first edition of this book (then entitled *PC-Based Voice Processing*) appeared. The industry has grown tremendously in size and the technology has grown in capabilities along with the rest of the computer industry.

One of the most significant changes from the application developer's point of view is the increasing popularity of modern operating systems such as Windows NT for developing high-density systems. MS-DOS, while still widely used, is declining as Windows and other platforms gain market share. This new edition of *PC Telephony* reflects this change. I have tried to add more information about developing software for Windows and other non-DOS platforms, concentrating as usual on information which is not available or hard to find in vendor manuals. I have added chapters on using TAPI and Wave to create voice applications, on ActiveX controls including a section on Visual Basic programming, and described benchmarks which make an attempt measure the relative performance of DOS and Windows.

In addition to the people mentioned in the Preface to the first edition who made such valuable contributions to the book, it is my pleasure to thank Laura McCabe and Jennifer Kollmer of Parity Software. Laura encouraged me to start writing the first edition book, Jennifer worked diligently to collect, edit and revise drafts of the fourth edition.

Bob Edgar
Parity Software
Sausalito, CA
June 1997

Chapter 1

Call Processing

What Is Call Processing?

On Sunday, May 3rd 1992, I was in Dallas preparing for Parity Software's participation in the annual Telecom Developer's trade show (now called CT Expo). The US presidential election campaign was in full swing, and the *Dallas Morning News* carried a front-page article featuring the efforts of undeclared Texan candidate H. Ross Perot, under the headline "High-tech tactics by Perot could reshape politics, experts say." The article speculated that Perot could use the latest technology in his campaign:

> "He could speak regularly to supporters, by satellite and audio hookups. Callers to his 800 number would punch buttons to register their opinions on issues, sign up to lend aid, send messages, or receive position papers by fax.
>
> "Meanwhile, the callers' phone numbers could be passed through huge computer databases to help identify similar potential backers, based on details that could be as arcane as frequent-flier mileage and taste in automobiles."

The new technology is *call processing*: the combination of telecommunications and data processing allowing callers to use the telephone almost like a keyboard and screen — as a device for interacting with a computer.

Chances are that you encounter call processing regularly in your daily life. Each time you leave a voice mail message, query your bank balance from a 24-hour automated service line or dial an informational 900 number offering a touch-tone menu selection you are utilizing a voice processing system. The general public tends to lump all this together and call it *voice mail*, since that is currently the most widely used application.

Telecommunications people like to divide the information transmitted over telephone lines into two categories: *voice* and *data*. Voice means *audio*, i.e. sound, which is to be interpreted by the human ear. Data is coded information sent between computers. The main distinction between voice and data from a data transportation point-of-view is the amount of "loss" which can be tolerated. The human ear and brain make a phenomenal pattern-recognition processor very tolerant of transmission errors, which may manifest themselves as clicks, hisses, short gaps, missing high- or low-frequency sound components and so on. Computers, on the other hand, deal in digital data where a single missing or incorrect binary digit is likely to cause a drastic change in the meaning of the transmission. Data transmission must be high

Call Processing 25

fidelity, in other words, while voice transmission may be lower in quality.

Voice processing adds intelligence to the underlying process of transmitting sound from one place to another. A device with voice processing functions is known as a *Voice Response Unit*, or *VRU*.

Call Processing Features

Some of the important functions which VRUs may offer are as follows.

Voice Store And Forward

This is simply industry jargon for record and play-back. A voice processing system with voice store and forward capabilities has the ability to record audio and play it back at a later time. A home telephone answering machine is a primitive VRU with store and forward capability. The archetypal store and forward application is voice mail, which allows a caller to record messages which can be retrieved by the intended recipient at a later time.

Digit Capture

Every telephone allows you to dial digits zero through nine, a voice processing system will often be able to recognize and act on digits dialed by a caller.

Text To Speech

Often abbreviated to *TTS*, this feature refers to the ability of a VRU to synthesize audio messages which were not pre-recorded. A simple example of TTS is speaking a number, 123 can be spoken as the four elements:

   "One" "Hundred" "Twenty" "Three"

With a pre-recorded *vocabulary* of less than 40 elements it is possible to speak any number less than a billion billions. With a few more pre-recorded elements, more elaborate messages such as:

   "You have" "Eighteen" "New messages"

can be strung together. However, Text to Speech more usually refers to the ability to speak almost any text — a newspaper article, for example. TTS technology is only just now becoming widely available, and still has difficulties in some important areas, such as speaking names and addresses.

Voice Recognition

Also known as *Speech to Text*, this allows the VRU to interpret sound as spoken words. Typically, voice recognition systems are only capable of recognizing a limited set of possible words, such as "Yes" or "No", or the digits from "One" to "Nine". Voice recognition systems fall into two major groups: *speaker independent* and *speaker dependent*. Speaker-independent recognition is intended to recognize any caller, regardless of accent and variation in phone line characteristics (background noise, bandwidth etc., which will vary between local, cellular and international calls, for example). Speaker-dependent systems must be trained to recognize a given voice, but will be able to achieve larger vocabularies.

Out-dial

As you would expect, a VRU with out-dial capability is able to dial a telephone number and connect a call. This will usually require that the VRU can perform *call progress analysis*, i.e. determine the result of the attempt to call: a successful connection, a busy signal, ringing through to the dialed number but no answer after a given number of rings, and so on. An important variant on the out-dial theme is a *supervised transfer* on a phone system, where the VRU will attempt to transfer a call to an extension and detect whether the call was picked up.

Fax Store And Forward

Fax data can be stored and re-transmitted in the same way as audio information. Voice processing systems are expanding to encompass fax processing options. Fax documents, for example, may be stored as messages within voice mailboxes.

Fax Synthesis

It often happens that information to be faxed originates in a computer — in a word processing file, a spreadsheet or database, as graphics files scanned in from photographs, and so on. Fax synthesis is the ability of a

VRU to convert data on a computer directly to fax format and transmit without the intermediate step of printing a document.

Major Applications

At the time of writing, the following are some of the most common applications of voice processing technology.

Automated Attendant

An automated attendant (often abbreviated to auto-attendant) is a VRU which answers the phone of a business and performs the basic functions of a receptionist: transferring calls to extensions within the company phone system. A typical auto-attendant will answer the phone with a pre-recorded greeting message like:

> "Hello, this is XYZ company. If you have a touch-tone phone and know the extension of the person you would like to reach, please dial it now. For a company directory, press the star key. If you have a rotary phone, please stay on the line and your call will be transferred to an operator."

Voice Mail

As mentioned earlier, the primary function of a voice mail system is to store messages for later retrieval by the intended recipient. The caller will be greeted with a message like:

> "Hello, this is Bob Edgar. I'm sorry, but I'm on another line or away from my desk at the moment. If you would like to leave me a message, please begin speaking after the tone."

The messages belonging to a single owner are referred to as a *mailbox*. A mailbox owner will be able to call the voice mail system and, usually by dialing a mailbox number and password, listen to his or her messages. Often, the voice mail system will be able to speak the date and time when the message was recorded, re-wind, fast-forward or skip through messages, forward messages to other mailboxes or even distribute copies of messages to groups of mailboxes (e.g. all salespeople).

Many voice mail systems include an integrated auto-attendant, though many businesses prefer to have a live person answer the phone, who can offer the caller the option of being transferred to voice mail if the desired individual is not available to take the call. Auto-attendants are often perceived as providing a lower grade of service than a live person.

Voice mail is sometimes called voice messaging.

Message Delivery

Message delivery applications are designed to store and forward messages to recipients who are *not* subscribers to the system. The classic example is the harried executive at the airport, trying to place a call when the "final boarding" call is heard over the loudspeaker — but the number is busy. By pressing the pound key over the busy signal, the executive is able to record a voice message which the carrier will store and later attempt to deliver to the number which earlier was busy. The delivery attempts might, for example, be once every fifteen minutes until the call is completed and the message delivered.

Audiotex

Sometimes spelled *audiotext*, this refers to systems which are able to provide spoken information through an automated telephone system. An example familiar to many people in the US is automated bank account query systems. By dialing a number, entering an account number and a social security or tax ID number, the caller can listen to a spoken response to a query, which could be the current balance in the account, the most recent checks written, most recent deposits made, and so on. Some writers refer to this type of system as an *Interactive Voice Response*, or *IVR* system, which they use for system which retrieve information from a database. Other writers use IVR system as an alternative term for VRU. I'll avoid the term IVR in this book. An excellent audiotex system in my home city of San Francisco (similar systems are found in some other metropolitan areas) is 415-777-FILM, which provides up-to-date information on movies playing in theaters around town. Movies can be selected by category, and theaters by area. Once a movie and theater selection has been made, the system informs the caller of current show times — dependent, of course, on the time of day, so that performances which are already over or in progress are not mentioned. Audiotex systems are built on *menus* such as:

"If you know the name of the movie you'd like to see, press 1, if you'd like to know what's playing at the theater nearest you, press 2, if you like to choose from a list of movies, press 3."

Pay-Per-Call

In the US, local calls made to the 976 prefix or long-distance calls with the 900 area code are billed at a premium rate. The additional revenues generated by these calls are (after charges deducted by the phone company) passed to the *information provider* owning the number. These services are generically termed *pay-per-call*. While some numbers are answered by live people — for example, a line offering legal advice might route a call to an attorney to discuss a problem, many pay-per-call services are audiotex systems offering pre-recorded information or automated database access. Classic examples are weather forecasting and sports results lines. Pay-per-call is now available in many countries around the world.

Automated Order Entry

Many businesses devote significant resources to accepting re-orders from experienced purchasers — distributors, agents and large repeat customers. A computerized VRU will be able to take a call from a customer, prompt for an account number, and then begin one or more order transactions. The number required, delivery and payment terms can all be selected from menus.

Call Center

Both in-bound and out-bound telemarketing call centers can derive considerable benefits from the use of voice response technology. On the in-bound side, a VRU can answer the call if no agent is available, providing entertainment, information on the current queue (how soon the caller can expect live service), or menu options for obtaining product information. A VRU may also be able to capture information from the phone company such as the *ANI* (the telephone number of the caller), for matching against a customer database.

On the out-bound side, the VRU could be a *predictive dialer*, which continually attempts to complete a call from a database of numbers, transferring to a live agent when a call is answered. Since the process is statistical, there will be times when no agent is available, and the VRU may ask the caller to hold, playing an informational message. This may be used, for example, by collection agencies.

Chapter 2

Computer Telephony

Introduction

Computer Telephony applies computer processing to a telephone call. This can be done in a number of different ways. The two most common types of computer telephony system involve a computer with a *voice card*, and a *CTI link*.

A *voice card* is a computer expansion card which can process a voice telephone call. The prototypical example is the Dialogic D/4x series, which started and defined an entire industry. A D/4x board has connectors for four analog phone lines, and looks like a modem which can handle four lines. A D/4x can make and answer calls, play and record sound from the phone line, detect touch-tones dialed by the caller, and detect when the caller hangs up. A computer with one or more voice cards is called a *voice response unit*, or *VRU*, so-called because it can respond to a caller by pre-recorded voice prompts.

Computer Telephony Integration (CTI)

Computer control can also be applied to telephone switches. Modern switches, including PBXs, are often built with *Computer Telephony Integration*, or *CTI*, links. The CTI link transmits messages from the switch to the computer with notifications of changes in the call state: a new call is detected, a caller hangs up, and so on. In the other direction, the CTI link transmits commands from the computer to the switch to perform operations such as answering and transferring calls. The CTI link may be an RS-232 serial line, Ethernet or other type of connection, and the hardware and software used by the link is proprietary for a given type of switch. Note that while the term "computer telephony" is broadly used for any system which combines computers and a telephone call, "computer telephony integration" or CTI is generally used for this specific type of connection.

Computer Telephony

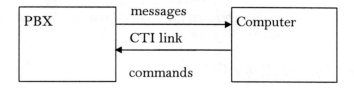

A CTI link is a two-way connection between a telephone switch and a computer. The switch sends messages to the computer with information regarding changes in a call state, such as a new incoming call, the computer sends commands to the PBX to perform operations such as answering or transferring calls.

CTI links are most commonly used in call center applications, where the computer is responsible for tracking how long each caller has been waiting to speak with an agent, logging agents in and out, and so on. A call center may often use both CTI and VRU computer telephony – if no agent is available to answer a new incoming call, the computer may use the CTI link to route the call to a VRU, which might provide interactive services to the caller offering pre-recorded product or account information, messages informing the caller of the average wait time, offering the opportunity to record a message and receive a call back, and so on.

The following sections review some of the most common configurations for VRUs.

Stand-Alone Analog VRU

In a stand-alone analog configuration, regular business or domestic service POTS (Plain Old Telephone Service) lines are connected directly to a VRU, which may be used to answer or originate calls:

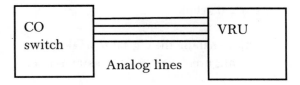

Stand-alone analog. The simplest configuration for a VRU: regular analog phone lines from the phone company are connected directly to the VRU. Appropriate, for example, for simple information-providing audiotex applications.

Stand-Alone Digital With CSU

For larger numbers of lines, a digital trunk may be ordered from the phone company; in the US, this will probably be a T-1 (DS-1) service carrying 24 conversations on two twisted-pair lines. This can be achieved in two variations: using a *Channel Bank*, or a *CSU (Channel Service Unit)*. The CSU is required as Customer Premise Equipment (CPE) by most carriers, and provides loop-back testing and integrity features. The output signal from the CSU is exactly as the input, so this configuration requires a digital telephone interface card, such as a Dialogic DTI/xx or D/xxSC-T1, in the VRU.

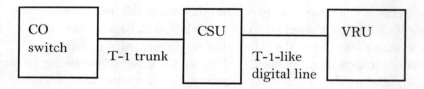

Stand-alone digital with CSU. For a stand-alone connection to a T-1 or similar digital connection, an additional item of CPE will be required: the Channel Service Unit (CSU) which provides testing and integrity features for the carrier. Otherwise, similar to the stand-alone analog configuration except that more lines will typically be supported. Often used for higher volume 800 or 900 applications.

Stand-alone With Channel Bank

The alternative to the CSU is a Channel Bank, which incorporates a CSU, and in addition converts the digital signals to a number of analog lines, 24 in the case of a single T-1 trunk.

With the use of a channel bank to handle the digital to analog conversion, the VRU will be configured exactly as in a stand-alone analog situation.

Computer Telephony

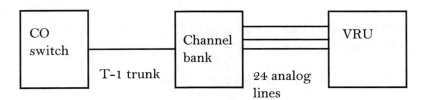

Stand-alone digital with Channel Bank. The Channel Bank is required if the VRU hardware requires an analog rather than a digital interface.

Behind A PBX

The traditional configuration for corporate voice mail environments has the VRU as two-wire analog extensions on the corporate phone system (PBX).

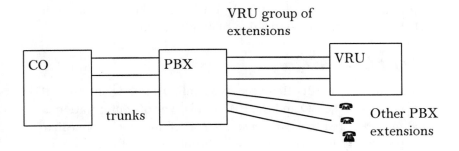

VRU behind a PBX. The most usual configuration for a VRU in a business environment. The VRU will be connected as one or more extensions in a group attached to the company phone system. Using the flash-hook transfer feature of the PBX, the VRU will be able to transfer a call anywhere within the phone system if required.

If the VRU is to perform as an auto-attendant, the PBX will be programmed to route incoming calls to a hunt group of extensions terminating at the VRU. The VRU will be able to answer the incoming call and transfer to the desired extension by using the flash-hook and dial call transfer capability of the PBX.

With an auto-attendant installed, the typical call will proceed along the following steps:

1. In-coming call arrives from CO trunk.

2. PBX hunts for free extension in VRU group, routes call through to that extension.

3. VRU answers call, prompts caller to enter an extension.

4. VRU transfers call to extension by sending flash-hook, waiting for dial tone from PBX, dialing new extension.

An important consideration when selecting a PBX for this type of environment is disconnect supervision. Some PBX designs do not transmit a disconnect signal. In other words, if a caller hangs up, the PBX does not carry the drop in loop current to the extension which was carrying the call. If you are stuck with such a PBX, it may be possible to use the call accounting feature of the phone system by capturing call record information from the PBX's serial port. This may enable the VRU to catch a definite indication of a caller hang-up.

In Front Of A PBX

Less common, but still often used, is a configuration where the VRU is placed between the CO and the PBX. This type of configuration is sometimes referred to as *drop and insert*: a call may be "dropped" to a voice processing board or "inserted" through to the PBX.

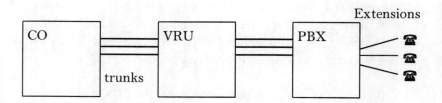

VRU in front of a PBX. This situation allows the VRU to intercept a call before it is passed through to a conventional PBX. The usual reason for this configuration is to take advantage of a trunk type or trunk service which is not supported by the PBX. For example, ANI capture or ANI blocking might be used by the VRU if this capability were not available in the PBX. A CTI link might also be required between the VRU and PBX

so that the VRU can monitor the state of the PBX extensions. In this way, Automatic Call Distribution (ACD) features may be added through the VRU.

Call Center

The traditional PBX configuration can be extended to support a full-blown inbound/outbound telemarketing call center with *ACD* (*Automatic Call Distribution*) and/or predictive dialing features, using PC platforms for both the voice response unit, ACD/predictive dialer, and agent data entry stations, as shown in the following diagram.

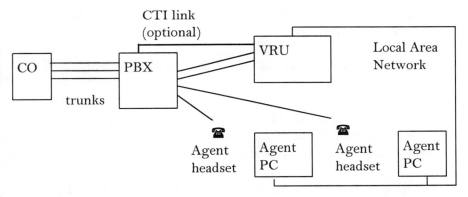

Call Center. Using the type of configuration shown here, complete in-bound and out-bound call centers can be designed using PC hardware rather than proprietary equipment, often with significant cost savings and improved flexibility in programming.

When this type of configuration is used for in-bound telemarketing or a similar application, it is important that the computer with the overall control of the call center (probably the VRU, or another PC connected to the VRU through a LAN) receives an indication of when an agent has finished dealing with a caller so that the agent can be added to the list of those available. The VRU is "out of the loop" once it has accomplished the transfer to the agent, and therefore receives no direct signal when the caller or agent hangs up. The solution may be to capture call accounting information transmitted by the PBX through an RS-232 serial port. Another use for the call accounting data may be to capture the ANI or DNIS information captured by the PBX when the incoming call is answered.

The PC LAN enables the VRU to communicate with the data-entry PCs for each agent station. The LAN can be used, for example, to transmit the ANI and database information captured by the PBX. This allows the account information to be shown to the agent at the time the call is transferred.

With an in-bound call center configured as shown in the diagram, a typical call would proceed as follows:

1. In-coming call arrives on CO trunk.

2. PBX answers call, capturing ANI and/or DNIS information from the phone carrier.

3. PBX hunts for free extension in VRU group, routes call through to that extension.

4. PBX sends initial call information over RS-232 serial connection to VRU, including port number where call was routed, together with ANI and DNIS digits.

5. VRU greets caller with a menu; caller selects options which determine which type of agent will handle call.

6. VRU examines agent queues to see if there is an available agent. If there is no agent, the VRU plays a message to the caller, including perhaps the expected wait and the caller's position in the queue.

7. When a free agent has been identified, the VRU transfers the call to the agent by sending a flash-hook, waiting for dial tone from the PBX, and dialing the new extension. When the agent picks up the call, the VRU will hang up, that VRU extension will now be available to process another call.

8. Using a shared database or LAN communication protocol such as TCP/IP, IPX/SPX or NetBIOS, the VRU sends ANI information to the data-entry PC at the agent station. This

Computer Telephony 39

enables the customer account information to be displayed on the agent's screen at the time the call is transferred.

9. When the caller or agent hangs up to terminate the call, the PBX will detect the disconnect signal and send a call accounting record to the VRU on the RS-232 serial connection. This enables the VRU to update its queue information, noting the fact that the agent who took the call is now free again.

A similar configuration and call sequence applies to out-bound call center applications. The main difference is that the VRU initiates the call using a predictive dialing algorithm designed to complete enough calls to keep agents occupied. The steps in routing a call to the agent are similar to the above. In this case, of course, the VRU knows the dialed number and has the account information available without ANI capture.

Service Bureau With Dumb Switch

A service bureau is required to service a large number of different phone numbers, most of which will be "terminate only" where the VRU handles the entire call. Once the traffic exceeds the capacity of a single PC, a switch will be required to distribute calls to different VRUs. While some conventional PBXs may have enough features that they can be programmed for this application, an increasingly popular solution is to use a *dumb switch*. This is a special type of PBX which delegates its programmable features to an external computer known as the *host*, which allows the system integrator considerable flexibility at the expense of some often exacting programming. In other words, a dumb switch is designed to delegate most of its programmability to the CTI link.

This type of service bureau can be created with the following configuration.

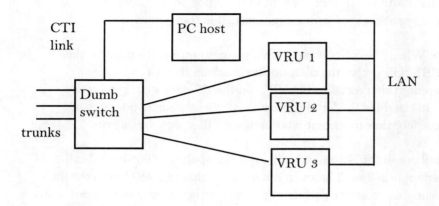

Service Bureau With Dumb Switch. The dumb switch becomes useful when the call volume significantly exceeds the capability of a single PC. The switch will be able to keep track of the load on each VRU and route calls to VRUs which have free capacity and required resources to process any given call.

A typical incoming call will be processed with the following sequence:

1. Incoming call arrives on a trunk from one of the carriers.

2. Dumb switch answers call, following a pre-programmed sequence which may depend on the trunk and/or port number where the call arrives, and which captures DNIS and possibly ANI digits.

3. When the call has been successfully answered and digits captured, a message is sent by the switch to the host on the RS-232 serial connection with a notification of the incoming call and the captured digits.

4. The host examines the current state of the VRUs, and selects a VRU and port number which is free and which is capable of processing the call.

5. The host sends a command to the switch with instructions to route the call to the selected VRU port.

Computer Telephony

6. The VRU processes the call.

7. When the VRU or the caller hangs up, the switch detects the disconnect signal and sends a message to the host with the appropriate port number.

8. The host program receives the message, and updates its tables of active VRU ports.

Additional applications, such as conferencing to other in-bound calls, or to calls originated by a VRU, may also be implemented by taking advantage of the LAN connection between the VRUs and the host and conferencing features in the switch.

VRU With PC Database

Applications which access and update databases in order to respond to caller queries are sometimes referred to as *interactive voice response* or *IVR* systems. This term is avoided in this book since some writers use interactive voice response for any type of application where a computer answers the phone and responds with a recorded or generated voice.

The prototypical IVR applications are banking by phone and automated order entry.

The PC database might be located either on the local hard drive of the VRU or on a LAN server drive, as shown in the following diagrams.

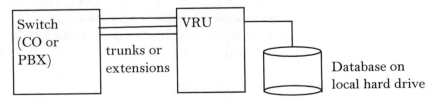

Simple IVR system: database on local hard drive. An IVR application allows the caller to access and perhaps update a database stored on the computer. The simplest way to achieve this is to store the database on the local hard drive mounted in the VRU PC.

Client/Server Configurations

There are many types of client/server configuration which are used for computer telephony systems. The VRU may itself be considered a server (of call processing functions), and it may be connected to other types of server: file servers, database servers and so on.

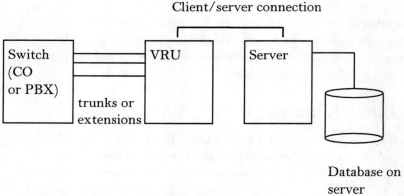

Client/server system with database server.
The database may also be stored on a server and accessed via a LAN.

The link to the server may be of many different types. Some of the more common are:

LAN.
The host computer could be a LAN file server just as in a PC-based network. The database could be accessed through a programming interface provided by a database server vendor or by directly reading the data files through a network drive letter.

RS-232 Serial.
A COM port on the VRU could be used to establish a connection via a serial communications line to a host computer. Most UNIX computers, for example, support serial communications for terminal users. There are two main approaches to serial connectivity:

1. Terminal emulation.

This technique requires the VRU to "pretend" to be a serial terminal. Keystrokes are sent to the host as if typed by a terminal user, responses from the host are analyzed for escape or control codes used to address the serial terminal cursor, change display attributes etc. In this way, the VRU can determine at all times what the screen of the serial terminal would be showing. The main advantage of using this technique is that software probably already exists on the host computer for accessing database information and new software will therefore not be required on the host to allow access through the VRU. The main disadvantages are that it can be complex to analyze the escape/control sequences coming from the host, and the communication may be much slower than using a transaction-oriented protocol since time is wasted transmitting the screen layout with prompts and attributes which are not needed by the voice processing application.

2. Transaction-oriented protocol.
The alternative to terminal emulation is to design a protocol specifically for the voice processing application. For example, the caller may be offered an option to listen to the current dollar balance of his or her account. The VRU might send the bytes:

    B1234556<CR>

where 1234556 is the account number and <CR> is the ASCII code for a carriage-return. The host computer might respond with the dollar balance:

    456.78<CR>

The amount of data transferred is much less than would be used by a terminal emulation approach. In practice, additional information would probably be required: a sequence number to match a request to a reply, and a checksum to guard against transmission errors (serial communication is not usually error-correcting at the data transport level).

On a UNIX computer a transaction-oriented protocol may be very simple to implement: a shell script of a few lines might be sufficient to do the job.

Synchronous.
IBM mainframe and midrange computers use synchronous protocols. Mainframe computers use terminals in the 3270 series, midrange computers (S/34, S/36, S/38 and AS/400) use terminals in the 5250 series. IBM provides a standard programming interface called *HLLAPI* (High Level Language Application Programming Interface) which allows a PC to emulate one or more 5250 or 3270 terminals. HLLAPI takes function calls from a program executing on a PC and sends them to a 3270 or 5250 communications session (sometimes called an LU or Logical Unit). The HLLAPI programming commands for 3270 and 5250 are quite different in the details.

A PC may be connected to terminal sessions in three different ways:

>Direct (coax, sometimes called "local") connection.
>Remote (SDLC) dial-up connection.
>Token-ring network connection (TIC).

In the case of a direct connection, there is an adapter card in the PC which is connected, via coax cable, to a mainframe controller.

Using SDLC (Synchronous Data Link Communications), the PC adapter card is cabled to a synchronous modem. At the remote IBM installation, a corresponding modem completes the connection.

With a token-ring connection, the PC is connected to a mainframe controller via a Local Area Network. There is no special adapter card (other than a LAN card) required in either the workstation or gateway PC.

With each of the three connection options, there are two alternatives: *Stand-alone* configurations, where all 3270 software and hardware is in the VRU PC, and *Gateway* configurations, where a second PC (the gateway PC) manages the sessions. The gateway and VRU PCs are connected through a LAN.

In a gateway configuration the VRU contains a LAN adapter card, a LAN redirector shell and a HLLAPI module which sends requests to the gateway PC. The gateway PC contains the communications adapter card (in a direct or SDLC configuration) and a software module responsible for managing the sessions. In the case of a token-ring connection, the gateway PC will, of course, not require any adapter cards, it will just require the gateway software.

Chapter 3

Telephony Applications

Introduction

This chapter describes voice processing applications which have been used in different industry segments. Industries are arranged in alphabetical order. The list is by no means comprehensive, but will hopefully stimulate ideas for the new and creative solutions. Each example is presented through a description of a problem, followed by an example of how call processing can be a solution to the problem.

In more complex solutions, a variety of other products are needed, such as local area networks, PBXs (Private Branch Exchange), ACDs (Automatic Call Distributors). In many instances, mainframe and minicomputer connections are required to put a solution together.

Automotive

Repair Status

Have you ever dropped off your car for repair at the local dealership and had a hard time finding out what the status of the repair was? This is a universal problem, because most repair department calls are routed through the main switchboard at a dealership before you can talk to the foreman or repairperson about what's going on with your car. Many times, you're put on hold or get a busy signal. This can be particularly difficult if you must have the status in order to make arrangements to get a friend or colleague to drive you over to the dealership to pick up your car. How about the cost of the repair? How much money should you bring? Do you need to authorize further work to be done?

Car dealerships are starting to solve these problems by applying call processing technology. Here's how it works. When you drop off your car for repair, a repair order number is given to you. You are then instructed to call back on a special telephone number and input the repair order number in order to authorize repairs, check the status of your car, and to find out when it will be finished and how much it will cost. The dealership installs a call processing system that either has its own dedicated outside lines, or has it hooked-up to an extension on their internal telephone system. The call processing system is connected to an onsite repair processing system and to the invoicing system as well.

When your car's problem is diagnosed, the repair person picks up a phone and calls into the system to leave a message for you under your repair order number. This message indicates the scope of the repair and any unforeseen costs. You can then call into the system and type in your number. You will then be prompted to type digits to indicate that

Telephony Applications 49

you authorize the repair. Subsequent calls into the system will provide you with information on the status of the repair and costs. More sophisticated applications may allow for the automatic entry of your credit card number to pay for the service, and an automatic outbound call to your phone number to alert you that the car is ready to be picked up.

Parts Order Status

A special problem is presented both to car dealers and customers when parts and accessories have to be backordered because they are not in stock. Some dealers are linked via computer to a parts locator network, much like the ones used by auto parts outlets. On many occasions, parts are simply ordered with no indication of how long the delivery will take.

A call processing system can help to alleviate this problem by centralizing access to a parts inventory database. A group of dealers can cooperatively support such a system, or one can be designed to work exclusively with one dealership. The call processing system can be hooked up as a remote terminal to the main inventory system via a remote data line or dial-up line. There are a variety of network access and modem cards available that can coexist with voice processing cards in the same computer to allow this. A customer can dial into such a system and enter his or her invoice or order number to initiate a query of inventory system. The call processing system can then respond with messages that tell the caller when the part will be delivered. In some cases, an alternate or substitute item can be quoted by the system, so the caller can choose whether or not to wait for (or cancel) the original order.

Construction

Time Reporting

Construction foremen and subcontractors have a difficult time in tracking and reporting the status of jobs and employee activities at job sites. It is typical for some kind of a paper log to be kept by the foreman, so that an individual's working hours can be logged manually and later transcribed by a central payroll employee for paychecks to be prepared. This process is time consuming, and often not accurate due to human error in the transcription process.

A call processing system can be used to solve this problem by automatically collecting "time card" data over the phone. This can be especially helpful for remote construction sites that have to be managed by a central contractor's office. Telephones can be temporarily installed at construction sites which act as a time punch terminal. A worker is given an access codes which is keyed to their employee number. This code is entered when an employee reports to work, and also when they leave for the day. In order to curb fraud, automatic time stamping can be done by the system, and employee codes can be handed out by the foreman at the beginning and end of the day. In this fashion, the central system can automatically compile time reports and payroll data and even automatically calculate overtime and job status as well. A more sophisticated system could also prompt the foreman to input information on job materials and completion in order to track job progress and profitability. A benefit of this call processing solution is that it not only streamlines a complicated process, but that it also increases accuracy and saves the time previously needed for transcribing records.

Corporate

Benefits

Ask any Human Resources professional what problems are presented by benefits inquiries and he or she will describe what seems to be a never-ending list of headaches. Some 20 to 25% of all calls into a human resources department are inquiries about 401K distributions, retirement benefits, vacation time, and savings plans. The amount of time spent over the phone to relay this information is a waste for most of these professionals, because other pressing matters that are less clerical require their "real time" attention.

Most benefits information is stored in a computer system or group of systems for medium to large sized companies. A call processing system can interface with these computers in order to relay this data to employees. A caller can access the system over regular telephone lines and then be prompted to enter his or her Social Security number and personal identification code. A menu of prompts can guide the caller to choose from a variety of benefits issues, including the accrued vacation days, cash value calculations on retirement plans, amount of stock options and value, etc. More sophisticated systems will automatically create a special report for the caller, who can have the data mailed confidentially. One benefit of a system like this is that sensitive and private information can be accessed by authorized persons without having to discuss the same information with another employee. There are companies who have extended this type of system to include start and stop date information, in order to better control disbursement of "last paychecks."

Communication

Corporate communication systems are often overloaded, slow, and extremely frustrating for callers and employees alike. Callers are often put on hold several times and transferred to a number of extensions before they reach the intended party. This not only clogs the telephone lines and causes more busy signals, but also creates frustration for callers who may hangup in disgust.

Call processing systems that solve these corporate communication problems are prolific and diverse. The most common is an automated attendant coupled with a voice messaging capability. When a caller dials a company with such a system, he or she is presented with options to either access the intended party through an automated personnel directory, or is prompted to enter a digit which indicates what department is required. These systems free the receptionist or switchboard operator to deal with visitors or callers that require live assistance.

The automated attendant can be programmed to ask the caller to speak his or her name in order to prioritize, or screen the call so it can be handled better. Some more sophisticated systems will prompt the caller for their customer number or even collect it automatically by using ANI. Customer numbers can then be forwarded to the company's computer system, so that a service screen can be transmitted to an agent's computer terminal in preparation of receiving the caller's transaction or request.

The most common solution for corporate communications is the voice mail system. These systems allow callers to leave detailed messages that can be picked up by the recipient from anywhere. These systems are especially helpful after business hours, when no switchboard operator is available to alert employees about messages. Employees can leave detailed greetings which indicate when he or she will be back from a meeting or available. Additionally, these greetings can indicate travel schedules and alternate phone numbers to dial. Employees who travel can pickup and receive messages from any telephone, and even respond to messages sent by colleagues who use the same system. New systems are being installed now that allow users to not only pickup and send voice messages, but also do the same with fax messages — all in the same phone call.

Corrections

Collect Call Service

Special problems arise in prisoner communications. Perhaps the most difficult is the use of coins to make payphone calls to relatives and friends who live in the outside world. Carrying money to make these calls is not encouraged, and can be downright dangerous, yet telephone calls are sometimes an inmate's only connection to friends and family. Some correctional institutions have designed elaborate debit card systems, and other (non-cash) means to solve this problem. Collect calls can be particularly expensive, so alternate means of billing and collection are being sought.

There are a number of solutions that call processing can offer to solve these problems. One such solution is the automation, tracking, billing, and collection of inmate-placed collect calls with private operator services systems. These systems use call processing to prompt the inmate for his or her voice, which is then forwarded to the called party without the use of a live operator.

With these new call processing systems, the called party listens for the inmate's recorded name prompt and then either types a digit to indicate acceptance of the call, or speaks "Yes" or "No" to accept or reject the call. A voice recognition subsystem inside the call processing system can determine whether or not the called party said yes or no, and then allow the call to be completed if it was "yes". In this manner, the overall cost of the call is lowered due to the absence of operator assistance, and the use of coin to initiate the call is eliminated as well. The call processing system can produce logs that are automatically sent to a billing system which either debits an established account, or prepares an invoice that is sent to the called party for collection.

Parole Reporting

With overcrowded jails, and an increase in parolee case loads for parole officers, it is increasingly difficult to keep track of cases and meet with parolees on a regular basis. A number of new electronic monitoring devises have been tested over the past several years which help to control "house arrest' subjects, but the problem of scheduling, reporting, and general contact are still apparent.

Call processing technology can solve many of these problems. A parole reporting system can be programmed to either receive calls from parolees, or to regularly call them at a certain time or place in order to solicit input which verifies activity or adherence to parole guidelines. With such a system, the subject is provided with a special access code and personal identification number. The system will prompt the parolee to record messages which indicate how he or she is coming along on a job or project, where he or she is at the time of the call, or any other information which may be appropriate. More sophisticated applications of call processing technology could use ANI to verify the originating location of the call, or even provide for full-scale voice messaging capabilities. Other applications could include the automated scheduling of appointments, or special instructions or requests from the parole officer. The system could provide a verbal (over the phone) status report to the parole officer of his or her cases, and even compile this information to be faxed to a central location.

Education

Truancy

The task of tracking and reporting on truant students is difficult, especially for large school populations, and parents who have irregular working hours. Often, students will either destroy or alter notes from a school principal or teacher in order to hide the fact that they have been "skipping" school. The work involved in calling parents on the phone to report this information can be insurmountable, especially if there are many legitimate absences due to sickness.

Call processing technology can ease the burden on truancy reporting for school districts by automatically making outbound calls to parents at either work or at home after hours. A system like this would place telephone calls from a list that was either manually input by the principal's office, or automatically from an exception database that is updated after morning roll call. Parents would answer the phone and be prompted to input a special code, and then informed that their child was not in school. If the absence is authorized by the parent, they could enter digits which indicate the reason for the absence, which would then be logged for the student's absentee records.

Course Registration

College course registration is a time consuming and often frustrating task for both students and administrators alike. The typical scenario is for an entire student population to line up in the school gym or auditorium in front of temporarily installed computer terminals, so that course selection and availability can be ascertained on a case-by-case basis. Students are given a booklet with course numbers on it along with a computer card which is marked with the students selections. If there is a problem with a course being "closed out" or if credits are miscalculated, manual changes and entries are then resubmitted to the system after discussing the problems with counselors or professors. If all goes well, the student will have made one or two passes at standing in line (usually several hours in the process), and a printout of the courses is then provided to the student.

Call processing systems are helping to solve some of the biggest problems for registrants and administrators. A centralized system can be installed that has many telephone lines attached. Students are supplied with a course booklet complete with course code numbers and instructions on how to access and use the system. Each student uses his or her Social Security number and personal identification code to access the system over the phone. The student is then prompted to enter the course numbers and class time selections. The system will repeat the choices back to the student in order to verify the selections.

Telephony Applications

During the phone call, the student is either informed that everything is accepted, or is asked to make alternate choices if there are any conflicts. A hard copy of the selected courses and class times is then provided to the student either by mail or direct pickup. The benefits of this type of system is that the number of temporary workers who are stationed at computer terminals is virtually eliminated, the students don't have to wait in line, and the maintenance of the computer equipment is simplified. In addition, transcription errors are kept to a minimum, because most of the paperwork is eliminated.

Report Cards

Delays in getting report cards, falsified information, and the lack of interim reporting have always been a problem for parents. Report cards sometimes take several weeks to a month to be issued, which could otherwise be time spent in corrective action or tutoring.

Teachers and administrators can take advantage of call processing by inputting a student's progress so that a parent may access this information over the phone at any time. A system of this nature would prompt the parent to enter a special code in order to provide authorized access to the system. The call processing system could then provide a verbal report on the grades a student earned in each class. More sophisticated applications would allow the teachers to record a voice message to the parents which detail concerns, congratulations, or a request for a conference.

Electronic Media

Interactive TV

There's plenty of competition between networks and cable companies who constantly wrangle for audience viewership. Much of the feedback that networks relay on is non-realtime information collected by rating services, who place "black boxes" in selected viewer's homes. These black boxes record what shows people are watching and then transmit the data to a central location for processing. The trouble is that these services do not necessarily provide a representative sampling of what viewers are thinking at the moment they are watching a particular program. Another problem that network promoters face is the increasing demand for more entertaining programming, and the desire of viewers to express their feelings and feedback on particular programs.

An automated call processing system can be dynamically programmed to receive calls from viewers in order to "vote" on certain issues posed by a program. In some cases, program outcomes have been designed to change according to the wishes of a majority of callers. This spurs great interest in "audience participation" and tends to boost ratings for particular shows. A call processing system can be designed to take thousands of calls each hour, so that large samples of viewer disposition can be compiled quickly in order to make changes in programming. A system may, for example prompt a caller to input digits on his or her overall satisfaction with the programming, or even ask the caller to indicate what outcome of the program he or she wishes to happen. A number of 900 lines have been used to solicit viewer input on contests, lost children, and crime cases as well. The number and frequency of these services has increased significantly over the past few years.

Cable TV Pay-Per-View

PPV (Pay-Per-View) services allow cable company subscribers to order certain premium-rate movies or events that are broadcast via cable to their addressable converter. These special events or shows are then billed on the subscriber's monthly cable bill. One of the biggest challenges in processing these orders is that most viewers wait until the last minute to order the show or event. The problem that occurs is that many orders cannot be taken due to busy signals at the cable company, and the lack of a sufficient number of operators to take the orders in time of the event.

Call processing systems have become increasingly popular in solving this problem. The caller is connected to a computer that automatically sends the request to the cable company local computer which immediately "unscrambles" the special event signal to the customer's addressable converter. At the same time, the call processing system sends the caller's customer information to the cable billing system in order to invoice the caller automatically. The subscriber is either prompted to enter his or her account number and show selection, or a more automated system will collect the caller's number using ANI service so that their converter box is pre-identified.

Radio Contests

Radio stations receive hundreds, and sometimes thousands of calls a day for contest information, viewpoints, and special requests. It is almost impossible to answer all of the calls that come in each day, so many callers become disgruntled, and may even stop listening to the station.

By installing a special call processing system a radio station can allow listeners to access information on contests and other subjects automatically. Call-in talk shows can have callers screened automatically for subject matter, before the calls are presented to the producer of a show. An automated system can also "unclog" the telephone lines by providing regular information that does not require human interaction. One example includes special contests, where listeners can dial into a call processing system that will prompt them to enter the answers to quiz questions, or a system that will ask them to record the "funniest joke" of the week, or a song parody. DJ's and other workers can then review this information at their leisure from any phone, so that winners can be announced later in the day or the next day.

Entertainment

Ticket Sales

Inbound telemarketing centers that specialize in ticket sales take a large number of calls each day. Staffing requirements for agents are very complex, because at some times of the day, workers are not busy at all, and other times of the day, there are not enough agents to handle all of the calls that come in. This is troublesome, because many tickets are purchased over the phone from callers who see an advertisement, and call to buy on impulse. If there is no one to handle the call, the potential buyer may "cool" and then hang up and not call back again.

Call processing technology can take out the "peaks and valleys" of staffing in these inbound call centers, and provide a way to automatically take orders over the phone. Callers can be presented with a menu of choices for tickets and shows that are available, and then be prompted to enter a payment method. The payment method could either be COD delivery or credit card, for example. When the caller is prompted to input their credit card number, the credit card is verified automatically through a clearing house computer. The caller is then prompted to record his or her name and delivery address for the tickets. The verbally recorded information can then be transcribed by an agent during a less busy period of the day. In some cases, the agent may call the buyer in order to verify the order later.

Concert Schedules

Concert and event schedules are typically sponsored by local theater consortiums, convention centers, radio stations, and civic organizations. In the past, these services have been offered by loading prerecorded information onto a bank of answering machines, so that the caller has to listen to the entire tape in order to find the desired information. This information is sometimes out-of-date and does not go into sufficient detail for each event. Often, the event calendar details information about concerts or shows that are offered by different establishments, yet the recording is in the same person's voice throughout.

An interactive call processing system can solve many of these problems. First, a centralized system can have many telephone lines installed, which all access the same "tape." In the case of a call processing system, the "tape" is a computer disk that can provide dynamic and non-sequential access to the desired information. Secondly, the sponsoring establishment of each concert or event can be provided with a special security code in order to call in and record updated information on a timely basis. This provides for not only current information, but also a more interesting dialogue, because the personality and voice of each sponsoring establishment can be different.

Horoscopes

Many of us have called a line to hear horoscope information which is "canned" and the same for each caller. The problem for the providers of such a service is to keep the content of the information interesting enough to keep callers coming back. Most of the horoscope lines are accessed by dialing a pay-per-call (976 or 900 number).

A special call processing computer can be designed to calculate a caller's true horoscope by asking the caller to input the exact time and date of his or her birth and other personal information. This provides a level of interaction with the caller that is more satisfying than a simple one-way tape recording. Once the information is input by the caller, the computer program matches the data with stored "star charts" in order to customize a horoscope that can be different each day the caller accesses the system.

Crossword Puzzles

Many newspapers publish games including crossword puzzles which are entertaining for readers, and attract a loyal following. Many of the game players actually call the newspaper in search of clues for solving the crossword puzzle, which causes a telephone traffic problem.

Some papers have installed a centralized call processing system which allows callers to input the "down" and "across" numbers of individual puzzles in order to get clues read out to them automatically. These clues are updated daily for subsequent puzzles, and old puzzles can be stored and accessed by the date of publication. The readers find these services to be valuable enough to actually pay for the clues by dialing a pay-per-call number.

Lottery

Lottery players are an anxious bunch, and most of the time, they have to travel to the retail outlet where they bought the ticket in order to find out which number won the lottery. In some cases, a player will wait until the 11pm television news. Many lottery players buy tickets from two or more states, which compounds the problem of finding out who won.

Call processing systems can solve this problem by providing a centralized telephone number to dial in order to listen to the results of a variety of lottery games. Some of these systems not only provide information on winning numbers, but also provide tips on playing, offer "magic" number combinations, and jackpot amounts. These systems are either sponsored by state agencies, retail outlets, or information providers via 976 or 900 lines.

Financial

Banking Checking And Loan Balances

Bank customers are faced with a variety of problems in handling their loans, checking accounts, and savings accounts. In most cases, these problems occur as a result of requests for balances, or transfer of funds from one account to another. In the case of over-the-phone requests, the problems have to do with after hours access. ATMs (Automatic Teller Machines) will provide access to cash, but sometimes don't provide critical information such as loan balances or checking account balances.

Call processing systems can solve these problems, which can significantly increase service levels for bank customers. A "Bank By Phone" system is designed to hookup to a bank's mainframe computer system. This is achieved by using terminal emulation cards which are commonly available. These emulation cards act in the role of a bank teller's mainframe terminal, and are placed inside the same computer as the voice processing cards that makeup the call processing system. A bank customer will call a telephone number and then be prompted much the same way a bank teller would prompt the caller over the phone.

The customer is asked by the system to enter his or her account number and identification code, and then to choose from a variety of services. These services may include the quoting of the available monies in a certain account, or the outstanding balance of a loan. In some cases, a caller can ask the call processing system to respond with the "payoff" amount of a loan. Some bank-by-phone systems will let callers transfer money from savings to checking and vice versa.

Stock Market

Investors who have a diverse portfolio of stocks sometimes have a difficult time getting hold of their broker to find out how well the portfolio is performing. It is typical for the investor to wait until the evening paper is delivered to read about this information. This is sometimes more difficult while traveling.

A call processing system can provide a highly personalized service to each investor by storing his or her portfolio profile in a computer database which is automatically updated throughout the day. An investor can call the system, enter the portfolio number, and then the system will read out the details one at a time. This saves time for investors, and provides 24-hour a day access. Systems of this nature have been available through pay-per-call lines and also are sponsored by some brokerage services.

Cash Management

Many nationwide retail outlets and regional enterprises are forced to wait anywhere from one to two days in order to make cash management decisions based on night deposits. Store managers and regional directors sometimes deposit cash receipts at a local bank and then forward deposit slips to headquarters accounting office. The headquarters accounting department is in need of this information so that decisions can be made on what funds should be transferred in order to solve local or centralized cash flow problems.

A centralized call processing system can provide financial officers with a "jump start" on night deposit information to solve this time-delay problem. A cash management system that uses call processing allows store managers to call in from any area to report the amounts that are being deposited. This data can be calculated by the central system and presented much quicker than waiting for receipts. In addition, the need for individual computer systems at each region is eliminated. Store managers can call from any phone in order to send this information. More sophisticated systems can allow the managers to send and receive special voice messages as well.

Fund Liquidation

Many large corporations are presented with special problems in liquidating funds such as special savings plans, 401K, or other disbursements as a result of regulatory changes, or change in corporate ownership. The means to do this is mostly manual, and fraught with inaccuracies. The typical scenario for a fund liquidation is for the benefits department to send out letters explaining the decisions that have to be made by each employee, with a request to fill out a form that has to be sent back. This form is then transcribed and input into a computer system. Later, checks are dispersed, accounts are "rolled over" and confirmation slips are sent back out to the employees. Depending on the number of employees, and the time required to facilitate the liquidation, many temporary employees are hired to handle the transactions.

A customized call processing system can be installed which prompts callers for their employee ID and security code. The system can then provide verbal instructions and explanations about the employees options. The options could range from the purchase or liquidation of stock options, transfer of funds from one account to another, or outright disbursement of all funds to the employee in the form of a check. This data is then read back to the caller in the form of digitized voice prompts for confirmation. If the employee confirms the transaction, the call processing system transfers this data to the benefits system and follow-on transactions occur. These additional transactions include automatic check writing, and hard copy confirmation slips which are mailed to the employee. A system of this type can take thousands of calls a day, thus eliminating the need for temporary help, and also streamlining a difficult process.

Government

Surveys

Constituents of local governing bodies and councils have encountered many roadblocks in telegraphing their concerns and issues in a timely and collective fashion. Whether the issue is a survey of opinions for immediate action, such as a petition, or a long term item such as zoning ordinances, it takes much effort and time to compile a representative sample of the peoples' ideas. Congresspeople have funded monies which are set aside for special mailings to constituents. These mailings typically solicit recipients for survey information and participation at "meet the people" meetings on a regional basis. It is difficult to survey a constituency on its various issues in order to prepare for press conferences and face-to-face meetings.

Call processing offers a variety of solutions, especially when used as an adjunct to special mailings. A call processing system telephone number can be published in a mailing or newspaper article in order to ask constituents to provide their input on certain issues. The system could prompt callers to enter their name, address, or phone number for follow-up to their concerns, and then forward a list of their opinions to the government body, council, or politician. The system may prompt the caller to enter digits which indicate what issue or category they are interested in airing, and then further prompt them to leave a detailed message on that item. This data can be transcribed and compiled in order to put together a profile or abstract of citizens of a certain town, borough, or county.

Telephony Applications

Polling

Opinion polls are gathered by both news agencies and third parties in order to report on issues regarding government actions and those actions or decisions of politicians. Much of this information is gathered by individual solicitation, and then compiled manually. In order to provide timely reporting on this information, many interviewers are "put on the street" to conduct face-to-face interviews, or asked to call hundreds of people on the phone. This process is an arduous, and painstaking task which sometimes takes so long that the data may no longer be newsworthy or germane when finally compiled.

Call processing systems can be programmed quickly and loaded with prompts to receive hundreds and thousands of calls simultaneously in order to collect polling information. Callers can express their opinions on a variety of topics which are accessible by pressing digits on the phone, or perhaps by speaking certain key words using voice recognition technology. In this manner, a variety of questions can be asked of callers, and the answers can be automatically tabulated for publication. The call processing system can collect and report on this type of information in a much more accurate and speedy fashion than a manual system.

Tax Office

Before the capabilities of call processing were applied, most tax offices, such as the IRS, were faced with the prospect of answering each tax return inquiry by manning phones with hundreds of operators. A large percentage of the calls are requests for tax refund information, and while this information seems to be a reasonable request, it chews up thousands of man-hours on the telephone, and increases the work load for any agency. Unfortunately, calls of this nature have actually slowed down the ability to process returns, thus making the entire process a burden on both the agency and the taxpayer.

Call processing systems are being installed that relieve much of the burden for tax return inquiries. These systems are designed to serve hundreds of callers simultaneously, thus freeing the agents and operators to speak with taxpayers who need live assistance. Systems of this nature are connected to the tax agencies' mainframe computers in order to access tax return status, information about monies owed, or monies to be refunded. As a security measure, the system will prompt callers to enter not only their Social Security number, but also the amount of money that the taxpayer submitted as being refunded on their return. The call processing system will query the mainframe system and speak back the amount to be refunded and whether or not the refund check has been mailed. In some cases, the system will report that the return has been received, but not yet processed.

Health Care

Patient Information

Hospital telecommunication lines are treated with great sensitivity. In some cases, a busy signal or a busy switchboard can mean the difference between life and death. It is for this reason that routine calls about how a patient is doing, or what phone number to call in order to reach a patient, are particularly burdensome to the hospital switchboard. But hospitals are service organizations, and therefore attempt to process these calls as best they can.

A specialized automated attendant or messaging system that uses call processing can be very helpful in solving this problem. When a patient is checked in to a hospital, he or she is typically assigned a room number and a telephone number, just as a hotel would do for a guest. The check-in procedure may include the administration of an automated attendant system, which provides prompting for relatives and friends to type in the spelling of the patient's last name. The call processing system can use call blocking features in order to keep calls from disturbing patients who should not be disturbed for medical reasons, or may allow the call to be automatically placed to the patient's room. A call processing system can also be used to provide information to callers about visiting hours, visiting rules, and directions to the hospital, so that the operator is not kept busy relaying this information.

Appointment Scheduling

Patients and doctors sometimes have a difficult time in agreeing to appointments for routine visits and follow-up visits. Much of this is as a result of the office staff being busy, or a desire on the patient's part to make an appointment after office hours, for example.

A call processing system can aid in the scheduling procedure by providing access to patients around the clock, so that they can enter the day they wish to visit and then select from a list of available hours. In the same way, a patient who must cancel an appointment can call in at any time to reschedule an appointment. The information can be processed by the computer and printed out on a report that can be reviewed by the office staff upon returning to the office at the beginning of business hours on the following day. A more sophisticated system can be programmed to call patients automatically to remind them to reschedule for a follow-up appointment, or to prompt them to indicate how they are doing. A system of this kind can be especially helpful for clinics which handle hundreds of patients each day, and in those cases where individual patient follow-up is virtually impossible.

Hospitality

Wake-Up Service

Hotel wakeup service is sometimes accurate, and sometimes not. On many occasions, a guest will request a wakeup call that is then written down on a piece of paper and then lost by the front desk staff. These mistakes can cause great hardship for guests who have traveled a long distance for an important meeting or appointment which is then missed because the wakeup call never occurred.

There are numerous call processing solutions that are available that take care of the wakeup problem. In most cases, a call processing system is connected to the hotel's PBX, so that guests can pickup the phone in their room and dial a predetermined extension number. The call processing system then asks them to enter digits indicating the time of the wakeup call including AM and PM designations. The system then automatically makes an out-dial to the guests extension at the time indicated and plays a recorded message to announce the guest's wakeup call. As an option, the wakeup call may include a local weather forecast and announcements about special events in town or special meals on the hotel restaurant menu.

Concierge

Every seasoned traveler has had occasion to request the services of the concierge desk in having a repair made to a garment, or to arrange dinner reservations at a local restaurant. The trouble is that many concierge desks are only open during business hours, thus leaving the guest to fend for him or herself in making reservations or getting directions to local establishments. This can be frustrating, because the night clerk or front desk staff are often not equipped to handle many of these requests, especially during busy times.

A popular call processing solution can provide both an audiotex and locator service capability to answer this problem. Guests can call a special hotel extension in order to be connected to an interactive system which prompts the caller to input digits indicating restaurant choices, entertainment, or other services. Many of these services are sponsored by establishments who may have an arrangement with the hotel itself, or perhaps the local Chamber of Commerce. The system may prompt the caller to enter restaurant choices by type of cuisine (Italian, Greek, French, Continental, seafood, for example), or perhaps by how expensive the restaurant is. More sophisticated systems will provide an option for the guest to be actually connected to the establishment in question.

Convention Message Board

Trade shows and conventions offer special problems to convention participants and customers alike. Because most of these individuals are traveling, they may have to make repeated calls to the home office to retrieve messages, or make trips to a centralized message board to pick up messages from colleagues and customers. Many times, these messages are fairly private, and not the kind of communication that the recipient wants posted on a public bulletin board (especially if they contain competitive information, for example).

An ad hoc voice messaging system can be customized to accommodate attendees at conferences, so that colleagues, friends, family, and customers can call at any time to leave recorded messages for them. Most trade show coordinators have registration desks to sign in participants, so the agents can easily input their name on a messaging system and handout temporary ID numbers and security codes to the registrants. At this point the convention-goers can call back to their home offices and let the switchboard operator know what the telephone number is to leave messages for the traveler. From time to time during the day, the trade show participant can call into the system from the temporary phone that is installed in the trade show booth, or take a break and call into the messaging system from a local payphone. This type of solution can be most helpful in following up on leads that were generated at the trade show itself in order to close business with the most interested prospects.

Insurance

Claims Reporting

Insurance companies process many claims for service and repair of automobiles that have been involved in accidents. Many times, the customer needs information on the closest "authorized" insurance adjuster location or auto repair shop. In addition, claims must be reported within a certain period of time, which is difficult on occasion due to the fact that some accidents occur after hours, when claims agents are not taking calls. Due to the fact that many calls are processed at claims centers, some callers are forced to either wait on hold or hang up and call back at a later time. This causes a level of frustration that can be unnerving, especially just following an accident.

A call processing system can augment the efforts of live claims agents by providing automated access to claims processing procedures and helpful information. Firstly, a system of this type can prompt the caller for the vehicle identification number, policy number, and time and date of the accident. Further instructions may prompt the policy holder to input the zip code of the area where the accident occurred in order to determine the closest repair shop. The repair shop location can then be spoken out to the caller by the call processing system. More sophisticated applications may prompt the caller to leave a detailed message about the nature of the accident, and even route the caller automatically to a repair shop or gas station to have the car towed.

Claims Status

As with the auto insurance industry, medical insurance claims create numerous calls each day that take a long time to service. Often, the nature of the call has to do with an inquiry regarding the simple disposition of a claim, and whether or not the claim has been processed and paid. The typical length of such a transaction if done manually is anywhere from three to five minutes, including holding time. This causes a service level problem for callers who get frustrated and hang up. Live operators and claims adjusters are required to explain policy provisions, track down special problems, and deal with exceptions. These adjusters and operators can do a superior job in servicing special calls if they are relieved from the duties of servicing simple informational calls.

Call processing alleviates much of this problem by allowing policy holders to access claims information on their own. In most cases an automated system will qualify the call, and include provisions for routing the caller to a live operator. This type of system will ask the caller to input the date of the claim and sometimes the dates of service in question. When this data is matched up with the policy holder's account number and identification code, an inquiry is made from the call processing system to the insurance company's mainframe computer in order to come back with the claims status. Information is then spoken out to the caller, which may include a request for more data from the policy holder in order to process the claim.

Manufacturing

Test Gear

There are a variety of problems that face manufacturing environments, but perhaps one of the most difficult to overcome is the testing of telecommunications equipment before shipment. Many switch manufacturers have to exercise telephone systems by making repeated calls into them in order to test the various features of the system as a quality measure before delivering the system to a customer. To do this testing manually introduces human error into the process, so that certain inconsistencies can be overlooked, thus allowing faulty equipment to be released.

A special "traffic generator" call processing system can be programmed to make hundreds and thousands of calls into a new telephone system in order to test it. These load testers can put a large amount of use on the new system that simulates the kind of traffic it will have to bear in the worst conditions when the system is installed on a customer site. This kind of testing can include automated reports that indicate the time it took for the telephone system to answer the phone, transfer a call, and

Telephony Applications 69

provide different kinds of progress tones. These reports can easily identify the performance of a system, and pinpoint problems that can be corrected before final shipment.

Military

Mobilization Orders

When a military initiative is undertaken, a massive mobilization effort occurs. It is typical for telegrams, mailed orders, or chain-phone calls to be made in order to mobilize military personnel. It is difficult to contact hundreds, even thousands of people for quick mobilization, especially if repeated phone calls have to be made in order to locate someone.

A special call processing system can be programmed to help out in mobilization orders by providing both for incoming and outbound calls. In the first case, a general announcement can be made by television, newspaper, and radio for personnel to call a certain dedicated telephone number to receive their orders. The system will prompt the callers for their military ID number in order to provide security before telling them when and where to report. An outbound version of this system can be programmed to automatically call predetermined telephone numbers for the same purpose. Unanswered or busy calls can be retried later by the automated system, or it can call alternate phone numbers that are on file, such as friends, neighbors, or relatives, depending on the level of security required for the orders.

Maintenance

Military mechanics are trained to repair and maintain a dizzying array of equipment and vehicles. There are literally millions of spare parts and procedures that have to be ordered and replaced in order to keep gear in tiptop shape. Due to the fact that manufacturer specifications for equipment and procedures are constantly changing, it is difficult to keep track of the proper procedures for maintaining equipment.

A call processing system can be used to provide interactive voice response capability in order to identify the part or procedure in question. Both verbal or faxed instructions can then be forwarded to the caller in order to provide the most up-to-date information available. Military contractors and manufacturers can be managed in order to keep the system current by providing changes in procedure prompts, part numbers, and schematic diagrams, for example. Diagrams and pictures can be sent to the system automatically, or can be scanned in manually by a system administrator.

Flight Simulators

Pilot training needs to be as realistic as possible in order to orient both transport, reconnaissance, and fighter pilots. Prerecorded messages are usually played into the cockpit which simulate radio traffic from other planes, command from the control tower, and vibration sounds. Simulators even record the spoken words and actions of the pilot in order to rate his or her decision making prowess and reaction time. The difficulty is in tracking the exact time of a simulation scenario versus the reaction, or spoken word of the pilot in question.

A flight simulation and recording system can be enhanced by using call processing equipment. In the case of simulation, there are no telephone lines attached, but rather direct connections to computerized audio speakers and recording equipment. A multi-channel system can play a variety of messages at the same time it is picking up sounds and recording them in the cockpit. A system of this type can be synchronized so that an occurrence on each channel can be time stamped with great accuracy. An instructor can sit down with the pilot after the simulation and pinpoint any stage of the simulation by monitoring a replay of the digitized "tape."

Power Utilities

Service Outage

Power companies receive thousands of calls which indicate that a power outage or other problem has occurred in a certain area. Most of the time, a large power outage will generate so many calls, that not all of them can be answered. This can be dangerous if an emergency call comes in that can not be serviced due to the "outage" calls.

A call processing solution provides the capability to handle most of these calls automatically. Once an outage is determined by the utility, the service area affected is identified and matched to the telephone exchange for that area. Callers are prompted to enter their exchange number. If the exchange number matches the affected area, an automatic message is played which indicates that the utility is aware of the outage and is working on the problem. The callers are also given an option to escape to a live operator for assistance if the purpose of their call was not related to the outage. As an option, some call processing systems can be equipped to handle ANI so that a caller's geographic area can be automatically determined in order to play the prerecorded outage message.

Emergency Notification

Nuclear power plants and reservoir authorities have a special responsibility to local residents because of the possible danger of nuclear disaster or flooding. Emergency procedures for such occurrences include short-wave radio broadcasts, telephone calls, sirens, police radio calls, and other (more direct) means. The goal in an emergency notification is to contact as many residents and authorities as possible in a short period of time.

An outbound dialing system uses voice and call processing by placing calls to many numbers simultaneously in order to deliver a uniform message which may include evacuation procedure information. These calls can be placed in order of priority by creating a "phone chain" by automatically calling designated citizens who are trained to understand and believe in the credibility of the call, so that they can subsequently call a predetermined list of neighbors themselves.

Print Media

Newspaper Subscriptions

Outbound telemarketing centers provide for the bulk calling of thousands of subscribers in order to renew, or get new people to subscribe to a paper. These calls are sometimes made manually, one at a time by each telemarketer. Each call can take from 30 to 45 seconds before it is determined whether or not the call can be completed, or whether or not the call is being made to an answering machine. What this means is that a good 25 to 30 percent of an agent's time can be taken up by simply placing calls, before a prospect is even spoken to.

Call processing systems have revolutionized the way these calls are made. Sometimes called *power dialers*, or *predictive dialers*, these special systems make calls in a totally automated fashion from a predetermined list that is stored in a computer. These calls are placed and then switched over to an agent only when it is determined that the called party has answered the phone. This frees the agents to spend their time speaking with subscribers rather that dialing the phone numbers and waiting for someone to answer much of the time. More sophisticated systems will "pace" the rate at which calls are placed in order to keep up or slow down according to how long the agents are taking to finish each transaction.

Yellow Pages

Telephone book yellow pages have long been the most prolific and constant source for information about products and services that are available to local residents. The problem for users of these books is that the publication is only updated annually, so the information that is available is fairly generic and not very time sensitive. On many occasions, basic information can not be detailed easily on the page itself without incurring great cost for the service or product provider. In addition, it may be difficult to find "yellow page" information outside of one's immediate geographic area.

Call processing systems have been installed to solve the timeliness problem of yellow pages advertising. These systems are called "Talking Yellow Pages" and provide a special telephone access number which appears somewhere within the advertisement. When a reader calls the telephone number, they are greeted by a custom message which can be recorded by the business in question. The messages can include information about current specials, sales, events, or directions to a store for example. In some cases, a caller can be automatically routed to the business in questions, or can leave a message for the business if it is after business hours. These messages can be picked up by the business so that the reader can be contacted the next day.

Classifieds

Have you ever wanted to know just a little bit more about an item listed in the classified section of the newspaper before making a call to the person who listed the item? Many people are not quite convinced by the short listings and may not inquire any further.

Many newspapers are solving this problem by enlisting the help of call processing equipment. These "Talking Classifieds" systems allow the sellers of items to provide a verbal description of the item in question in order to generate more interest in the readers. Some newer systems are providing the capability to do "blind transfers" of readers to the person who listed the item in the classified at a predetermined telephone number. The caller is not given the phone number, because it is not listed in the paper. In this way, the person who posted the listing can keep his or her home telephone number private, and still converse with the interested party in order to qualify the inquiry.

Product Distribution

Inventory Inquiry

Wholesalers and retailers alike have the problem of satisfying a large number of patron requests for certain products that may be out of stock. It is typical for the availability of an item to be ascertained either manually, or by the help of a clerk who is at a computer terminal. Even with the help of a computer terminal, clerks do not always have an accurate picture of inventory if other orders are being processed simultaneously.

An inventory inquiry and ordering system can be used to automate much of this process. When a customer accesses the system, he or she is asked to enter an account number which is used to automatically invoice the caller for any items that are ordered. Of course, this assumes that the caller has an established account. The caller may inquire about the number of items available, and even be prompted for alternatives if required. The call processing system will automatically tally the results of the inquiry, and then send a report either to a printer for manual "picking" of the stock, or to another computer which actually debits the stock from an accounting standpoint. A system like this both streamlines the ordering process, and provides after hours service to callers.

Locator Service

Many consumers find it difficult to find local dealers or service centers for products that have been previously purchased, or for products that they are thinking about buying. Sometimes, the consumer will call the 800 operator in order to get a toll-free number and then make several subsequent calls to find the closest location to his or her home.

A locator system can provide not only the automatic routing to the closest dealer but also provide alternate choices for the caller as well. A call processing system of this type provides an computerized zip code or phone number sort in order to match the caller with an outlet or dealer address and phone number that is previously stored. The caller is asked to input his or her zip code and/or telephone number in order to do this matching. In some cases, ANI (Automatic Number Identification) is used in order to provide a quicker connection to the local dealer.

Recall

Product recalls can be very expensive propositions for a manufacturer. This is especially true in the case of unfortunate public relations or fear factors caused by tainted food scares, for example. In the case of mechanical recalls, consumers want to know what to do to fix the problem, where to go for help, or perhaps how to get a refund. Short of a massive advertising and bulk mail campaign, this information is difficult to convey quickly and inexpensively.

An audiotex system can be used to provide specific instructions to hundreds or thousands of callers each day. Callers can be prompted to enter the serial number or lot number of the item in question, and then receive specific instructions on what to do. In the case of a partial recall, a consumer may hear a message that says: "According to the number that you entered, 3334445, your unit is not identified as one of the problem units. Thank you for taking the time to check..."

Retail Services

Catalogue Sales

There are a variety of ways retail businesses can allow customers to order product. The methods span from "catalogue window" walkup purchases, to mail order, and inbound telemarketing service. In most

Telephony Applications

cases, either written or verbally conveyed orders must be transcribed and entered into a computerized order entry system of some kind. If an order is mailed in from a customer, there is a delay of anywhere from several days to a week to process the order. In addition to mail delays, the number of errors in transcription can add days and even weeks before problems are sorted out and the customer receives his or her order. This is particularly annoying if the item to be purchased is in stock and the customer has a credit line with the retailer.

Many innovative call processing solutions have been applied over the past several years which eliminate some of the problems in remote order entry and have also sped the entire process significantly.

The first means to offer better service with call processing is to provide telephone access to an inbound telemarketing service center with a voice response unit. A voice response unit can be installed in order to allow callers to get more information on certain products. A message will include instructions on how consumers can call a certain number to receive more information on a product. Each catalogue page has clearly written product codes or information access codes next to each item. When a consumer calls into the system, they are asked to either type in the catalog number and page number of the item they are interested in, or to simply press the product code number. A customized message can then tell the caller more about the product, and perhaps whether or not the item is sold out in different sizes or colors.

The second, and more complex approach is to prompt the caller to enter the size, color, quantity, and so on of each item desired and then to also enter credit card or account number information as well. If the consumer has an established account with the retailer, the rest of the transaction can be automated since a customer record can be matched to the order which includes a shipping address. This means is used with some cable TV product channels, allowing callers with established accounts to order products in a simple transaction that takes less than one minute in some cases.

Tape Rental

Video tape rentals have become very popular, and a number of problems arise as a result of feature film shortages. Most tape rental outlets stock numerous copies of favorite films so that there are enough to go around for all of the customers. Customers sometimes call the store to inquire about the availability of a certain film before making the trip to the store. It is frequent that a customer will arrive to pick up a film only to find that someone just walked out with the last copy.

Since many of the tape rental outfits are equipped with computerized systems today, the application of a call processing system could provide additional levels of customer service. A call processing system could be linked to the tape inventory system in order to allow customers to find out whether or not a certain film is available.

In some cases, a customer with an account may be able to reserve a film by calling into the same system and entering the tape number and a customer number. A grace period could be provided in order for the customer to have enough time to reach the store to pick up the tape before it is rented to someone else. In addition to reserving tapes, the system could provide customers with the option of extending the period they wish to rent the tape, as well. It is also possible to provide an "audio movie review" by recording a short synopsis of the available films so that callers could decide on which films to rent before coming to the store.

Social Services

Neighborhood Watch

Many local citizens groups have formed neighborhood watch programs in order to protect their family and property from criminals. One of the most difficult aspects of coordinating a neighborhood watch and then reporting on trouble in the neighborhood is that of communicating to a large number of homes in a short period of time. Say, for example, that a burglary has occurred in a neighborhood of 300 homes. A sophisticated door-to-door alert or phone chain would have to be established in order to let everyone know to be on guard.

A call processing system can be programmed to take both incoming calls and even make outgoing calls to many homes simultaneously in case of trouble in the neighborhood. In the case of an inbound calling scenario, a system could be programmed to provide a recorded message on the incident in question when residents call into the system and type in their phone number or other identifying number. Everyone could be trained that they should call the system when alerted with a fire whistle or some other loud alarm. In addition, the system could be made to make phone calls to many homes simultaneously in order to alert those who had not called in on their own. The system could repeat the dialing of numbers to telephones that were not answered or were busy. It is even possible for such a system to provide an update on which homes did not either call in or get called by the system, so that someone could contact those people in another way.

Job Lines

Most job hunters must wait until the weekly or daily newspaper comes out in order to scan the pages for job listings in the want adds. Others wait in line for hours to be interviewed at temp agencies and local unemployment agencies. It is not only frustrating, but sometimes a desperate experience for those in search of employment.

A community job line is one application of call processing technology that can not only speed up the search for employment, but also provide a much easier means of accessing information for the unemployed. A special system can be programmed to accept the recorded input of employers, who could call in to the system and type digits that categorize the type of job, job description, salary, and other items that the system would then automate into prompts for job-seeking callers. A recorded message could even be recorded by employers which provides a careful explanation of the qualifications required for the job.

Job-seeking callers could then call the same system and input digits at the prompts to select the type of job they're interested in, and then receive verbal descriptions that were recorded by employers. This can be most helpful in allowing quick updates to postings that either have just been filled or ones that have just become available in a more efficient manner then waiting for the newspaper to be published.

Unemployment Verification

Unemployment agencies are required to keep careful records of a client's job search efforts, and also the progress of temporary job assignments in order to continue providing benefits to a client. In most cases, forms have to be filled in by the client, or regular phone calls and visits have to be made to the agency in order for the client to continue receiving benefits. The process is slow and frustrating to both the clients and agency workers, who have to keep track of thousands or cases each month.

A call processing system can go a long way to providing relief in the verification of activities. For example, each client can be instructed to a dial call processing system at the beginning and end of each week. The callers could enter their Social Security number for identification, and then type in the telephone numbers of the places of business that they interviewed for work. The client could then provide a recorded message which indicates if they are still looking, if they got part-time work, etc. This information could then be listened to by the case workers and transcribed or archived if necessary. This type of system would allow for 24-hour access over the phone, thus extending the "working hours" of the agency for its clients.

Emergency 911 Service

Some municipalities provide a centralized emergency number for police, fire, or health related incidents to be reported by citizens. In most cases, emergency calls are transferred to a PSAP (Public Safety Access Position) operator, who then talks to the caller about the nature of the emergency and their location. Many times, it is difficult if not impossible to dispatch emergency vehicles and services to the scene of an incident, because the caller is either too upset, or no longer able to relay location information to the PSAP operator.

Call processing systems can be used in E911 telephone systems in order to detect the ANI (Automatic Number Identification) of a caller. The way it works is that the telephone company signals into the telephone system a string of multi-frequency digits that indicate the telephone number of the calling party. The call processing system then extracts these digits for the telephone line when the emergency call comes in, and then sends the digits to a name and address database computer which is connected to terminals at the PSAP positions. In this manner, the PSAP operator is able to pinpoint the location of the caller at the same time the caller is greeted by the operator. The operator is then able to dispatch emergency vehicles by typing commands on the terminal while he or she is assisting the caller over the phone.

Sport

Olympics

People all over the world are interested in the outcome of certain events at the Olympics, but unfortunately, it is difficult to provide up-to-the-minute reporting on events in the many languages that are spoken by Olympics fans.

A centralized call processing system can easily act as an interpreter for event results and schedule changes. A series of prompts can be recorded in separate playback "bins" of a call processing system. Each of these playback bins will hold the same prompts indicating events, country names, and scores, except that they are recorded in a different language. A system administrator will regularly call into the system and type in the results of each event either by remote terminal or telephone. The call processing system software will then be able to link the status of each event to the correct language prompts that are separately stored in the system. Callers can either be prompted to enter the digits which represent their language, or separate telephone numbers can be published which will automatically branch callers to the correct playback bin.

Golf Tee Scheduling

Golf clubs have always been challenged in scheduling enough support staff to accommodate golf parties, and to schedule tee-offs in such a way that most club members are satisfied in playing just about when ever they want to — and sometimes at the last minute. The quick scheduling sometimes causes mistakes to occur, which are compounded when one person out of a party does not show up in time, or does not show up at all. More problems are introduced by golfers in search of a party who would like to play, but need to connect with other "singles".

Call processing systems are beginning to provide flexible scheduling for golfers, and a way of matching up singles to parties, as well. Let's say that a member calls into a system in order to punch in information about the day, time, and number of persons in the party. The call processing system may prompt the individual for whether or not he or she would like the system to attempt a match for the forth player. A fourth player may call in and say that he is flexible about when he wishes to play, but would just like to connect to a party that already has a scheduled tee-off, but is missing one person.

Callers can easily be identified by typing in a membership number, in addition, they could be prompted to record their name as well as their telephone number. A more sophisticated system would call the parties when a match is found and speak the confirmed tee-off time and perhaps also the prerecorded name prompts of all of the people in the golf party. Players could also be asked to type digits which confirm the schedule.

Skiing Reports

Every ski enthusiast knows how difficult it is to get comprehensive reports on ski conditions from time to time. Many skiers prefer to get a report from many different resorts in a certain area before choosing where to stay during a ski trip.

Call processing systems can provide an automated means to providing comprehensive weather, travel, a promotional information. A centralized system can be programmed to accept status reports from numerous ski resorts, who call in several times a day to provide information. Skiers could be prompted to enter the slope or resorts of their choice in order to hear current information on conditions. The call processing system could also be programmed to accept input from "frequent skiers" who access the system with a special code in order to hear information in a predetermined order of preference.

Telephone Companies

Cellular

People on the go want to be in touch, and cellular phone users are just one example. Many cellular subscribers don't give out their car phone number, and choose to just make outgoing calls at their convenience. The trouble with this is that subscribers sometimes miss very important messages that could have been acted on right away if the subscriber had only known of the urgency of a certain call.

A special cellular call processing system can provide not only the message-taking capability of a regular voice mail system, but also the privacy that some cellular subscribers demand. Some subscribers are

Telephony Applications 81

able to program their cellular phones to always forward calls to another number. This forwarded number can be the access code to a "cellular mailbox" that is provided by the cellular telephone service provider. The forwarded calls are answered by the call processing system, which then automatically collects digits from the cellular system which indicates the number that was dialed. The messaging system then plays out a custom greeting that is prerecorded by the subscriber. This custom greeting asks the caller to leave a message which can be picked up by the subscriber from his or her cellular phone.

Paging

Pagers are simple but helpful devices that let subscribers know that it's either time to call their secretary for messages, or that a certain event has occurred. The problem that many "beeper" subscribers face is that they don't know how important a certain "beep" is compared to another.

Paging and call processing systems can work side-by-side in order to provide subscribers with not only notification of a message, but also a very detailed recorded message from the caller. Many voice messaging systems now offer the option for callers to record their spoken message, and then to direct the messaging system to also page or "beep" the recipient to let them know of the urgency of the message. Still other systems allow the subscriber to turn the beeping function on and off. This can be helpful in screening calls and prioritizing them when business hours are over.

Directory Assistance

It has been calculated that for every second of time that is saved in each call for a bank of operators at any Regional Bell Operating Company (RBOC), that $1,000,000 is saved each year. Naturally, this is a high incentive for the telephone companies to save a lot of money each year by streamlining the Operator Services function. Roughly 3,000 operators are staffed at each RBOC to handle subscribers' calls for directory assistance, toll assistance, and coin credit. In addition to saving money, the RBOC wish to generated revenue by providing premium services when possible. This is difficult to do with the thousands of calls that are received each hour.

Call processing systems are being used today which significantly upgrade the capability of regionalized Operator Service Centers. Virtually every Operator Service center is now equipped with some form of automation. In most cases, a live operator answers a call for directory assistance and locates a listing by typing commands on a terminal which is connected to a centralized SIDB and LIDB (Subscriber Information Database and Line Information Database). The SIDB and LIDB listings are cross-referenced to a telephone number which is then automatically sent to a voice response unit which plays the digits out to the caller, who is transferred to the unit automatically. The voice response unit also allows the caller to have the number repeated, or in some cases will route the caller back to an operator for further assistance. More sophisticated systems will use ANI in order to bill the caller a certain premium if they want to be connected to the number they asked for automatically. This is referred to as *call completion.*

Payphone Message Forwarding

Many travelers are faced with the problem of getting a message to someone while they are in transit, or about to board a plane or train. When these payphone users call a party and they encounter a busy signal or get no answer, telephone company revenue is lost on a non-completed call, and the caller is not able to communicate as was the original desire.

Special call processing systems are now providing a service called message forwarding. When a busy signal is reached, or there is no answer, the caller is prompted to record a message to the called party. This message is then stored in the call processing system so that it can be spoken to the called party in subsequent call attempts that are made automatically by the system. This type of service is provided either at the cost of a three-minute call to the intended number, or at some premium which is posted for review on the payphone placard.

Customer Service

There are a variety of services provided to telephone company subscribers at regionalized customer service centers. They include new orders for service, trouble reporting, custom calling feature upgrades, billing problem resolution, and general assistance. The biggest challenge for a telephone company is to be able to answer calls quickly and to process requests for service as courteously as possible. Customer service efficiency is one of the most important measures in the state PUC (Public Utility Commission) decision-making process of rate increase grants. Many customers are put on hold for tens of minutes on occasion which causes great frustration, especially if they are calling on a simple matter, reporting trouble or asking about the status of a repair.

Call processing systems can provide much relief for both the telephone company and subscribers. A system can be programmed to accept caller input in the form of their telephone number and the nature of their call. Much of the information a caller is looking for can be prepared by computer while an operator or agent is finishing up a previous call. For example, the system may ask the caller to input his or her telephone number and then indicate how a certain repair is progressing before forwarding the call to a service agent. The subscriber information can be automatically sent to an agent's terminal so that the subscriber is greeted properly when a live person is ready to receive the call. This cuts down on holding time, and also provides a level of comfort to the caller that something is being done to help them while they are on hold.

Transportation

Airlines: Frequent Flyer

It has been estimated that only 25% of the calls coming into a reservation center are revenue-producing calls. Many customer calls are actually requests for general information, or even requests to get an update on frequent flyer mileage credits. Frequent flyers want to have quick access to this information to plan trips and to also verify that certain flights were credited to their account. Frequent flyer service centers are typically only open during business hours, thus forcing after-hours callers to contact the 24-hour reservation center for this information.

A call processing system can automate almost all these calls. The system could be programmed to ask for the caller's frequent flyer number and security code. After this information is entered, the caller can be prompted to enter digits to verify flight credits, number of accrued points, and perhaps even listen to special mileage offers or trips that are being offered. An additional level of service can be provided by allowing the caller to record his or her new address or phone number information as well.

Crew Scheduling

Imagine the complexity and logistics involved with all of the domestic and international flights that take off and land by the thousands each day, then consider the task of scheduling the crews for each of these flights. Such is the everyday coordination effort required for personnel assignments including pilots, copilots, navigators, and attendants. Criteria used to choose crews include not only their availability, but their tenure at the airline, and the number of contiguous hours already logged during a pay period. More tenured airline employees get "first dibs" on the more desirable (least exhaustive) schedules, and much wrangling over the schedules causes delays in schedules and general headaches for the employees and management.

Call processing systems can be attached to airline computers to automatically provide access to both crews and management for a variety of tasks. As system of this nature can be programmed to ask each crew member for his or her employee identification number and password. This would be followed by a series of prompts which request the desired flight scheduling, vacation time, and stopover intervals. The system can then interact with the scheduling system to match the crew member's request with other employee requests in order to calculate the best fit for each set of flights. Crew members would then be verbally quoted by the system whether or not their request is logged and authorized, and if they need to call back for verification at a later date.

Flight Status

As is the case with frequent flyer requests, the calls that come into reservation centers requesting flight status number in the thousands each day. These calls are placed by travelers, limousine companies, travel agents, and family members who wish to know when a certain flight has actually landed or will be departing. Although this information is carefully tracked for scheduling and safety reasons, most airlines have no efficient means for letting callers know about this information. This creates a level of frustration for customers and airline employees alike.

Telephony Applications 85

A flight status computer can be connected to a call processing system in order to automate the delivery of this information to callers. Every major airline has a sophisticated means of reporting every step of a plane's flight to a centralized flight status computer. This is the same computer that illuminates terminals around an airport which indicate to travelers what gate a plane is arriving or departing from. A call processing system then can send a request to the flight status computer system for current information which can be spoken to the caller.

Trucking: Dispatch

Transportation companies instruct their drivers to periodically park their rigs in order to call a centralized dispatcher for instructions about trailer pickup, drop-off, and general messages. Dispatchers are very busy, so drivers sometimes make repeated calls that either go unanswered or force them to stay on hold for long periods of time. This is a costly proposition, especially when the nature of a pickup or delivery is time critical, which is almost always the case. Often, a driver will call dispatch to report that a load has been successfully delivered, only to find that the line is busy. The driver wastes precious time that could be spent traveling to pickup the next trailer for delivery. In the case of a perishable load, or a competitive situation, minutes can count.

Special dispatch and messaging systems are now starting to take advantage of call processing technology. Drivers are provided with identification numbers and job numbers for each load. A series of codes are provided by the call processing system which are input by the driver to indicate the status of a certain delivery. These codes include information such as the time of pickup, traveled distance, estimated time of arrival, and so on. This information is compiled with other data to determine the most optimal coordination of return loads, alternate trips, and contingency plans in case of delay. In addition, special voice mailboxes are made available, so that both the dispatcher and driver can leave detailed messages for one another. One advantage of the call processing system for dispatching is that critical information can be captured by the system from the driver when the dispatcher is busy, so that he or she is better prepared to handle the call when the system connects the driver to the dispatcher's extension.

Trucking: Driver Messaging

Long distance drivers are often times "tied to the wheel" for hours at a time, thus unable to call home to friends and family to talk or let them know how things are. Family members often have a difficult time getting information from a dispatcher about the exact status of a job or driver, which adds a level of frustration for family members.

A special call processing system in the form of "Driver Voice Mail" can alleviate this communication problem. How it works is for both the driver and his or her family members to be assigned a personal mailbox number. When a driver can get to a phone, he or she calls a central number and accesses the appropriate mailbox to check for messages. These messages can be responded to at any time of day or night. The responses can be stored for later retrieval by the recipient, or the system can be made to deliver the message automatically. This is helpful for late-night calls. For example, a driver may have a chance to leave a recorded message for a spouse at 2am which is then forwarded by the system in the morning when the driver is on a limited access highway. The message may say something like: "Hi, it's really late, and I didn't want to wake you, so I figured I'd have the system send this message to you in the morning. By then I'll be just outside of Chicago and on my way to Boise. I'll give you a call at home around dinner time. Talk to you then..."

Shipping: Global Tracking

Shipping lines track the progress of cargo ships for thousands of ports worldwide. The containers and cargo of these ships are typically tracked from a central location and stored on a large computer system. Ship captains maintain radio contact with the Line's central dispatch in order to report on weather conditions, delays, repairs, and other important information. The number of calls that a shipping line receives each day concerning the whereabouts of containers are numerous. The problem that shipping lines face is not only the hundreds of calls that are received, but also the amount of time it takes to research and report the tracking information to each caller.

A call processing system can be connected directly to the line's central computer system, which has information stored in it on container numbers, cargo ship location, weather delays, and shipping manifestos.

Telephony Applications   87

A caller can be prompted by the system to enter his or her shipping manifest number, or the number of the cargo ship or containers in question. The call processing system can prepare a query to the main computer so that the answer can then be spoken out to the caller. The caller may be asked: "Please enter the manifest number. Please enter the container number. One moment please... Your container will be arriving in the port of Philadelphia on November 21, 1991. Please call in several days to verify. This arrival date may change as a result of weather conditions."

Railway: Train Scheduling

Most calls which come into bus or railway reservation centers are for information on schedules. Most of this information is provided on printed train schedules, but many callers don't have this information with them, or would like to verify a schedule before leaving for the train station. Problems occur when some callers wish to make a reservation or are in need of help and so many other calls are tying up the lines with simple information requests.

Regional and nationwide train lines are already beginning to implement specialized call processing systems to solve this problem. Callers are asked to listen to a menu of "corridor" services and then to choose which line they are interested in. The system will then play train numbers and departure and arrival times to the caller on the train in question. More sophisticated systems will then take a customer's call and transfer it to an agent so that tickets can be purchased or seats can be reserved. This allows the agents to quickly serve customers, who know what they want when connected to the agent.

Postal: Package Tracking

Any package that's important enough to send overnight may be important enough to follow up on the next day. Postal customers of both public and private postal services flood carriers with requests for package information each day. Most of the time, callers are advised that package confirmation can not be provided until after the "promised" delivery time has expired. Callers usually call back minutes after the scheduled delivery time in order to provide the postal agent with a tracking number. Lost packages receive priority consideration, and a special tracer number is provided to the customer at that time.

Many postal companies are now installing sophisticated computerized tracking and inventory systems for major clients. These systems allow customers to generate their own packing slips which are logged automatically by the postal company. These automatic logons allow for connection to the postal company tracking computer, so that requests for delivery verification can be entered at will by customers on their local terminal. Call processing systems can be programmed to emulate these same onsite customer terminals, so that customers without terminals can access the same information by entering the tracking number themselves by using their telephone. More sophisticated systems may be programmed to provide a lost package tracer report, which can be automatically faxed to a customer who calls into the system with a tracer number.

Chapter 4

Telecommunications

The Phone Line

Most domestic telephones are connected to the telephone company's nearest exchange using a cable containing two conducting wires.

The telephone company exchange is called a *Central Office* or simply *CO*. The CO is similar to a business phone system but on a much larger scale. Phone systems and other devices which can connect calls are called *switches*. Your phone is therefore connected directly to the *CO switch*. Business phone systems which function like CO switches are called *Private Branch Exchanges*, or *PBXs* (sometimes *PABX*, with an extra "A" for "Automatic").

The connection to the phone company using two wires is called a *two-wire* connection. To distinguish it from a digital connection, it may also be described as *analog*, since sound is represented by varying current rather than by streams of bits. A regular business or domestic phone line without special features may also be called a *POTS* line, for *Plain Old Telephone Service*. A line to the phone company lets you connect to a number anywhere in the world through the *Public Switched Telephone Network*, or *PSTN*.

Some business phone systems (PBXs) use more than two wires to connect a phone on a desk (the *station set*) to the PBX. The extra wires are used to send signals between the station set and PBX, which can be used to implement message-waiting lights, LED displays, conferencing and other features. A "standard" voice card will not support the additional wires, and uses a connection to the PBX which is like a domestic wall socket, requiring a *two-wire analog station card* in the PBX. Some PBXs use a two-wire connection but are still not compatible with domestic telephones since high-frequency signals are used to control buttons and lights. The CO provides a small DC voltage across the two wires, called (for obvious reasons) *battery*.

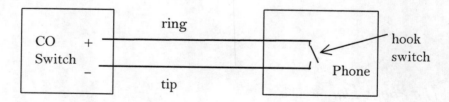

Connection of CO Switch to a Telephone. The Central Office switch applies voltage ("battery voltage") to the two ends of an analog phone wire, which loops through the subscriber's telephone. If the telephone is off-hook (handset is picked up, hook switch closed), there is a complete circuit and current flows through the wire. Variations in the current carry sound. Telephone cable for a single line carries two conducting wires and is generally of the type known as "twisted pair".

The ends of the wires are known as *tip* (connected to battery -) and *ring* (connected to battery +). For most purposes, it doesn't matter which way round tip and ring go, but it is wise to get it right anyway.

Starting An Outgoing Call: Getting Dial Tone

When the handset is taken out of its cradle, the phone is said to be *off-hook*. The action of taking the phone off-hook closes the *hook switch*, so that there is a complete circuit to the CO along the phone wire. The phone wire is known as the *local loop*. This causes current to flow, known as *loop current*. The CO switch will usually react to this by making a sound (a combination of 350Hz and 440Hz tones), known as *dial tone* which indicates that you may dial. The process of taking a line off-hook to ask for dial tone is called *seizing* the line.

Most analog lines are *loop start*, which request dial tone in this way — the alternative is *ground start*, where service is requested by grounding one of the two conductors in the two-wire loop.

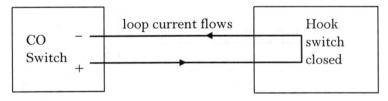

Phone goes off-hook to get dial tone ("seizing"). When the telephone handset is on-hook, the circuit is broken and no loop current is flowing. To make a call, you pick up the phone (go off-hook), which closes the hook switch and completes the circuit. Loop current starts flowing. The CO switch detects this and plays a dial tone to indicate that you may begin dialing.

Announcing An Incoming Call: Ringing

When the handset is in its cradle, the phone is said to be *on-hook*. When the phone is on-hook, the main circuit is broken, but there is still a circuit made through a capacitor, shown as ═══ in the diagram. The capacitor has a very high resistance, so there is effectively no current flowing. When a call arrives, the CO applies an A/C voltage of typically 90V at 20 Hz to the circuit (*ring voltage*), and the phone rings.

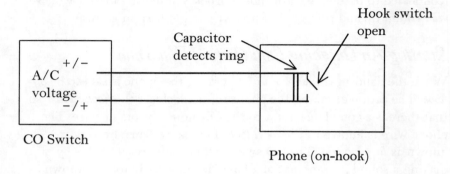

Incoming call: CO switch rings line, handset is on-hook. To announce an incoming call, the CO will apply an alternating (A/C) voltage to the telephone. The telephone is on-hook, so the circuit is broken except for a capacitor (shown as a double line) which is able to detect the A/C and cause the bell to ring.

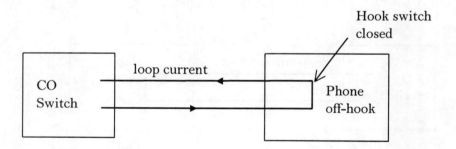

Telephone answers call: handset lifted. The handset is lifted to answer the call, this is called "going off-hook". The action of lifting the handset closes the hook switch, completing the circuit. The CO stops applying A/C voltage and resumes normal battery (D/C) voltage. Current, called loop current, therefore flows along the line. Variations in this current carry the sound of the conversation.

Dialing A Number

There are two fundamentally different ways of dialing numbers: *tone dialing* and *pulse dialing*. Pulse dialing is sometimes called *rotary* dialing because that is the method used by old-style rotary phones.

Telecommunications

Tone dialing uses sounds to represent digits (0 through 9, # and * all are called "digits"). Each digit is assigned a unique pair of frequencies, hence the name *Dual Tone Multi Frequency* (*DTMF*) digits, or *Touch Tone* digits. There are four *DTMF* digits (in the *fourth column* — because the usual tone pad has three) which are not usually found on telephones in the US but are used in some European countries; these are named a, b, c and d.

There are actually two standards for tone digits: DTMF and *MF* (for *Multi Frequency*). MF is used internally by the phone company, but is also used by some phone companies for *ANI* (*Automatic Number Identification*) and occasionally for other services. ANI is the business equivalent of *Caller Identification*, where the telephone number of the calling party is transmitted to the telephone receiving the call. MF is very similar to DTMF except that different pairs of frequencies are used for each digit. Some speech cards have the ability to detect MF as well as DTMF, others, especially older models, do not.

Pulse dialing uses the loop current itself to send digits. When the dial of a rotary telephone rotates, it briefly turns the switch AB on and off, thus turning the loop current on and off, resulting in "pulses" of loop current. Count one for each pulse to get the digit being dialed. You know when a digit is finished and the next one starts by the longer pause between pulses.

Pulse dialing has two major disadvantages. The first is that pulse is much slower than tone dialing, and the second is that most switches will not transmit pulses over a connection. If you make a call from a rotary phone, and dial a pulse digit in the middle of the conversation, clicks will be heard at the far end, but no interruptions will be made in the loop current. This means that the only way to detect pulse digits from a remote telephone is to try to analyze the sound patterns and to "guess" when a digit has arrived. Imagine distinguishing a pulse "1" digit from the click caused by static on the line, for example — not an easy problem. Components capable of recognizing pulse digits from the sound they make are called *pulse recognizers*. They may convert these digits to DTMF, in which case they may be referred to as *pulse to tone converters*.

Completing The Call

When dialing is complete, the person or equipment which dialed the number can listen to the line to determine when and if the call is completed, i.e. if the called party answers the phone. Along the way, a number of *call progress* signals may be generated to indicate how things are going in the process. Call progress signals are mostly sounds (tones) generated by a switch, some signals are made by dropping loop current briefly.

Ringing tones (called *ring-back* in the business), indicate that ring voltage is being applied to the line corresponding to the number dialed. Ring-back is generated by the CO switch which is attached to the number that you called, not by the called phone (there may be no equipment attached to the number at all).

If the dialed number is off-hook when the connection is attempted, a *busy* signal will be generated instead. If the phone company's network is busy and the local CO (the CO you are attached to) fails to make a connection to the distant CO (the CO connected to the dialed number), a *fast busy* may be generated. You don't hear fast busy tones very often — they sound similar to busy, but the pause between the beeps is shorter.

If you dial a bad number (an area code that does not exist, or a disconnected number), you will get an *operator intercept* signal (three rising tones) followed by a recording: "doo-doo-doo We are sorry, ..."

Less obvious but still important are brief drops in loop current which are sometimes generated when making long-distance calls; these can be used by the phone company as an acknowledgment that the distant CO has been reached, and are often used to indicate that the called number went off-hook. It is all too easy to confuse these brief loop current drops with the drop in loop current that signals a disconnect (end of a call due to the called party hanging up). Finally, if all goes well, the called party will answer the phone and say "hello."

Call Progress Analysis

Automated recognition of call progress signals is called *call progress analysis*, and this is a difficult area. Most voice boards do have the ability to detect some call progress signals. (An important exception is the older Dialogic D/40 series). Call progress analysis is a complex procedure, and high accuracy is hard to achieve if different environments are to be supported, such as when dialing local, long distance and international calls. To make matters worse, the signals, such as the tone frequencies used for ring and busy, vary considerably in different environments. For example, a business phone system (PBX) will likely produce different tones than the phone company, and the voice board must be configured to recognize these sounds if call progress analysis is to be used.

A typical call progress algorithm reports the results of call progress analysis as one of the following:

Ring no answer	Ringing tones were detected, but after a pre-set number of rings there was still no answer.
Busy	A busy tone was detected.
Fast busy	A fast busy tone, indicating that the network was unable to reach the desired CO.
Operator Intercept	Three rising tones followed by an informational message, something like "We are sorry but that number is no longer in service", or "That number has been changed to...".
No ring-back	After waiting for a set length of time, ringing tones were not detected.
Connect	The call was answered.

The Flash-Hook

When a call is in progress, a service from a CO, PBX or other switch may be requested by making a *flash-hook*. A flash-hook puts the phone on-hook briefly—long enough for the switch to detect it, but not long enough to disconnect the call. You will probably be familiar with a flash-hook from the Call Waiting feature many people have in their home phone service.

On a business phone system, a flash-hook will generally give you a second dial tone, allowing you to make a three-way conference (you stay on the line, flash-hook a second time to complete the conference), or a transfer (you hang up when the second number answers).

Most local phone companies in the US offer *Centrex* features, which allow transfers and three-way conferencing using the same flash-hook and dial sequence as a typical small business phone system.

Terminating A Call: Disconnect

To terminate a call, one end goes on-hook, i.e. hangs up the phone. In the US, if the caller at the other end hangs up while you stay on the line, then sooner or later (there may be a delay of twenty seconds or more, less for a local call), the phone company will notify your phone of the disconnect by dropping the loop current for about one second. PBXs, and the public telephone networks in some countries, generally give a tone signal, perhaps a dial tone or a different tone for signaling a disconnect by the distant party. This tone must be detected if an automatic disconnect notification is required. Some phone systems don't transmit a disconnect at all, so you may have no reliable way of detecting that the distant party hung up.

T-1 Digital Trunks

A T-1 digital trunk carries 24 telephone connections on two *twisted pair* (two-wire) cables. The set of 24 connections is called a *T-1 span*. Each of the 24 connections is referred to as a *time-slot*. Each time-slot carries sound, digitized at 64Kbps, and two *signaling bits*, referred to as the *A and B bits*, which play a role similar to loop current signaling on analog lines. One time-slot is sometimes referred to as a *DS-0*, for Digital

Signal level 0, signal; the T-1 span with 24 channels is then called
DS-1.

While use of the A and B bits is not the same in all T-1 equipment, a
common convention, called *E&M* signaling, keeps the A and B bits
equal, and uses the A bit to indicate whether or not a connection is
active. Thus, A bit high (set to 1) corresponds to loop current flowing,
A bit low (set to 0) corresponds to no loop current.

Digits can be dialed on a time slot using the same methods as analog
lines: DTMF, MF and pulse. Pulse digits are sent by turning the A and
B bits on and off, just as rotary pulse dialing turns loop current on and
off on an analog line. With E&M signaling, a pulse would be sent by
briefly changing the off-hook state (AB = 11) to on-hook (AB = 00) and
back again.

E-1 Digital Trunks

Outside the US, most countries use E-1 rather than T-1 digital trunks.
Notable exceptions are Japan and Hong Kong. E-1 is similar to T-1
except there are 30 instead of 24 voice channels per circuit, and four
signaling bits named A, B, C and D are used rather than two.
Unfortunately, there are a large number of different signaling protocols
used on E-1 circuits, this makes hardware and software design for E-1
much more difficult than for T-1.

ISDN Digital Trunks

There are two main types of ISDN trunk: *BRI (Basic Rate Interface)*, and
PRI (Primary Rate Interface). ISDN carries data in 64 kbps channels.
There are two types of channel: *B Channel (Bearer Channel)* and *D
Channel (Data Channel)*. Bearer channels carry digitized voice in almost
exactly the same way as a T-1 channel. Data channels are used for
signaling information, such as the dialed number, disconnect
notification, and so on. BRI trunks run at 192 kbps, carrying three 64
kbps channels, two B and one D. BRI trunks are sometimes referred to
as 2B+D trunks for this reason. There are two main types of PRI
trunks: one based on a T-1 type of connection, the other based on an E-1 type of connection. T-1 type PRI usually carries 23 B channels and
one D channel, E-1 type PRI usually carries 30 B channels and 2 D

channels (non-ISDN E-1 dedicates also carries a total of 32 channels, with two channels dedicated to signaling). It is possible for two or more PRI ISDN trunks to share a single D channel, this feature is called *NFAS*. Using NFAS makes an additional B channel available on all except one trunk.

DNIS And DID Services

Many phone companies offer a service which allows you to terminate several different phone numbers on the same set of lines. When an incoming call arrives, it searches (*hunts*) for an available line and generates a ring signal. Your equipment responds with a *wink* signal, which causes the phone company switch to send you some or all of the digits that the caller dialed, usually as DTMF digits. This is called *Dialed Number Identification Service*, or *DNIS* (pronounced "dee-niss"). A T-1 wink is a brief off-hook period, which will usually be signaled using the A bit on a T-1 channel.

DNIS is used by answering services and service bureaus, which may be answering calls from many more numbers than they have operators or IVR lines.

An analog wink is done by reversing battery voltage. Analog DNIS is called *DID*, for *Direct Inward Dial*, or *DDI* for *Direct Dialing In*. A typical use of DID is to provide direct numbers to employees in a large company. The company phone system obtains the last few (usually three or four) digits of the dialed number via DID and routes the call to the appropriate extension. A service bureau can use this feature to provide services for many more clients than the bureau has phone lines: each client is given one or more phone numbers, all of these client numbers are routed to the bureau's set of DID or DNIS trunks. In phone company jargon, a number is said to *terminate on* a trunk, more than one number may terminate on the same trunk if DID or DNIS is used.

Automatic Number Identification (ANI) And Caller ID

ANI (pronounced "Annie") delivers the phone number of the calling person (or machine). It functions in a similar way to DNIS: digits are transmitted at the start of the call, with ANI the digits are the number of the originating telephone line. Both **ANI** and DNIS may be provided by a trunk. The ANI can be used by a **VRU** to block a call, automate client billing, route the call to a responsible sales agent or and/or to do a database lookup to retrieve the caller's account information. There are several standards for the way phone companies deliver ANI digits. Local phone companies, like New York Telephone, deliver ANI digits on domestic analog phone lines between the first and second ring. This service is known as *caller ID*. They sell it as a service for their residential subscribers to ward off harassing calls, but caller ID also has business applications. Caller ID information will usually include the caller's phone number and may have additional fields such as the subscriber's name as text string. Caller ID is now available throughout the US and in many European countries.

Protocols for sending ANI and DNIS vary widely. A typical protocol for T-1 is wink + MF digits. When an incoming call arrives, your equipment sends a *wink*. The CO will respond with a string of MF digits, which is often like this:

* I (7 or 10) #

The digit string is as follows:

*	The MF KP tone (written as *)
I	An informational digit
(7 or 10)	The caller's number, usually 7 or 10 digits
#	The MF ST tone (written as #)

MF tones define the usual ten digits 0 - 9, plus two special tones known as KP and ST, which are sometimes written as * and # respectively.

In summary, the sequence to answer a typical ANI call would be:

1. Wait for ring
2. Send wink
3. Enable MF detection
4. Get ANI digits
5. Disable MF detection
6. Go off-hook

The application can then complete the call by going off-hook and proceeding as for any other in-coming call. Some providers given ANI and DNIS digits using DTMF rather than MF.

PBXs: Private Telephone Systems

Most businesses have their own private telephone switch, called a *PBX* (*Private Branch Exchange*) or *PABX* (with an extra "A" for "Automatic"). Smaller businesses sometimes have simple switches called *key systems*. You can easily recognize a key system because each telephone has a row of buttons, one for each central office line. To pick up a line, you press the appropriate button. Key systems require manually-operated buttons to process calls, and are therefore not suited for computer integration.

A true PBX contains a computer and often has many capabilities which can be programmed by the user. The main unit is called a *Central Processing Unit* or *CPU*. Two sets of phone lines will be connected to the CPU. One set of lines is connected directly to the phone company. These lines are called *trunks* or *CO lines*. The other set of lines are connected locally, usually to users' desks within the company and to fax machines, modems and other devices. These lines are called *PBX extensions*.

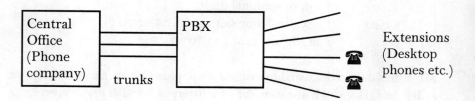

A PBX connects trunks on one side to extensions on the other.

There are typically two types of extension: one for *feature phones*, and one for modems, fax machines and other devices which require standard analog phone lines. Feature phones are the desktop telephones with blinking lights, special buttons and other features not found in a standard telephone such as the one you have at home. Each type of PBX has its own type of feature phone, and one model of PBX will usually be unable to use a feature phone from another model. Often different PBXs from the same manufacturer have incompatible feature phone specifications.

If you want to add voice mail or other computer telephony features to your PBX, you will generally need "modem" type extensions, rather than feature phone extensions, in order to hook up the computer. Most telephony cards from Dialogic and other vendors support only standard phone lines and are unable to process the proprietary signals on feature phone lines. There are exceptions: the Dialogic D/42 series of boards supports feature extensions from a few of the most important PBX types, including the Northern Telecom SL-1 and others.

Many PBXs support a programmable capability called *hunt groups*. A hunt group is a set of extensions which can be searched in a pre-defined order for a free line. For example, a group of extensions might be defined as the "sales" group. An operator might transfer a caller to this group, and the PBX would hunt for (search for) the first free extension and send the call to that salesperson. This is a convenient feature for adding a computerized auto-attendant feature. A set of extensions is connected to the computer and defined as a hunt group. The PBX is programmed to send incoming calls to the auto-attendant group. When an incoming call is detected on a trunk, it is sent to the first free auto-attendant extension, and the computer answers the call. The computer plays a menu, and can transfer the caller to the requested extension by using a flash-hook sequence.

A major problem for computer integration with PBXs is disconnect supervision. In many cases, when a caller hangs up, no signal is sent through to the extension. Or, the only signal that is sent may be a tone, such as a dial tone. On the public network (at least in the US) disconnect is always signaled by a one-second drop in loop current; equipment that relies on this signal will not work correctly when connected to a PBX extension.

The Telephone Network

The public telephone network is a vast and highly complex web of different technologies: it is well beyond the scope of this book to describe the network in detail. We'll be satisfied with a brief sketch which covers the basics. Generally, the voice processing applications developer doesn't need to know very much about the network. All he or she needs to know are the characteristics of the telephone line (trunk) connecting the voice processing equipment to the service provider: the PBX or telephone company.

When you pick up your telephone at home and call a neighbor down the street, chances are that both your telephones are connected to the same switch at the central office.

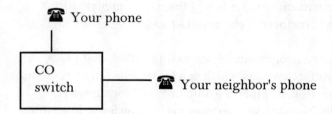

Two phones in the same neighborhood are probably directly connected to the same CO switch.

The connection to the switch (the *local loop*) is likely to be a cable carrying two twisted copper wires (so-called *twisted-pair*). The switch applies battery voltage so that current to carry the sound will flow when the telephone is off-hook. The situation is very similar to two extensions on a business telephone system.

In their turn, CO switches are connected via switching "nodes" which are connected via high-speed, high-density digital trunks. When you make a long-distance call, say between San Francisco and New York, you will be connected via one or more of these switching nodes.

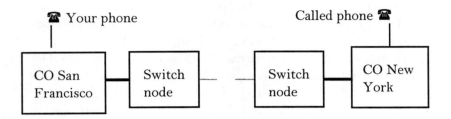

A long-distance call will be connected via at least one "switching node". Each node is a digital switch which operates on high speed digital trunks; these nodes do not connect directly to subscriber trunks.

In the US, there is a five-level hierarchy of switches in the network. The CO switch which connects directly to subscriber lines is called the *end office*, to distinguish it from other types of "office". Each end office will be connected to one *toll center*, which provides the next level in the switching hierarchy. Further levels are referred to as *primary centers*, *sectional centers*, and *regional centers*. All levels except the highest are connected in a "star" formation, which means that a phone is connected to only one CO, a CO to only one toll center, and so on: a node is connected to no more than one node at the next level up. The highest level is composed of ten *regional centers* which are completely connected: each regional center has a trunk to each of the other nine, making a total of 45 links.

To complete a call, a path through the network will be established at as low a level as possible. If both lines are attached to the same CO end office, then a direct connection will be made. If lines are attached to different end offices but both end offices are directly connected to a common toll center, then the call will be routed through that toll center, and so on.

In the US, the breakup of the AT&T monopoly over the telephone network has divided the ownership of the network over several companies. The seven *Baby Bells* (*Regional Bell Operating Companies* or *RBOCs*) now have (more or less) monopolies over local telephone services. For example, telephone service for a home in San Francisco from Pacific Bell, which is part of Pacific Telesis, the RBOC covering California and some neighboring states. "Local telephone service" includes calls made within a *LATA* (*Local Access Transport Area*), one of the 161 geographical areas defined across the US. AT&T may not offer service within a LATA (intra-LATA), though other long-distance carriers such as MCI and Sprint may do so in some regions. On the other hand, the RBOCs may not provide calls across more than one LATA (inter-LATA service): these are carried by the long-distance companies AT&T, MCI, Sprint and others. Thus, a call from San Francisco to Los Angeles is carried by a long-distance company even though both these cities have local services from Pacific Bell. The advent of new technologies which will build on cable television and other related networks is blurring the line between traditional telecommunications and other types of service and will doubtless result in dramatic changes to many of these network features. Regulatory changes are also in the works which are likely to break the monopoly of the Baby Bells over local phone service.

Telephone Numbers

On my business card, it says that my office number is:

+1 (415) 332-5656.

This number exemplifies the three-level system which is used in most countries of the world:

Country code,
Area code,
Local number.

Microsoft has defined a standard representation of telephone numbers as part of its Windows TAPI telephony programming standard. This "canonical" representation is as follows:

+*CountryCode* **Space** [(*Area Code*) **Space**] *SubscriberNumber*

The area code is optional, indicated by the square brackets [..] which are not part of the representation. The symbol **Space** represents a single ASCII space character. For ISDN numbers which include "subaddress" and "name" fields (special for ISDN), the above number is followed optionally by:

| *Subaddress* ^ *Name*

Punctuation characters, such as "-", are permitted in the subscriber number as long as they are not special dial string characters (listed in a later section). In this canonical representation, the following is my office number:

+1 (415) 332-5656

as quoted earlier from my business card.

Country Code.

The country code, often written following a "+" sign, is a number with from one to three digits identifying the country of origin. Since there are fewer than three hundred countries in the world today, it seems like a safe bet that no more than three digits will be needed in the foreseeable future. The country code in my fax number is +1, which is shared by the US and Canada.

Some general rules for two digit codes are:

3x, 4x	Europe
5x	South America
6x, 8x	Australasia, Far East

Some examples of country codes are:

South Africa	27
Netherlands	31
France	33
Ireland	353
UK	44
Denmark	45
Guatemala	502
Argentina	54
Brazil	55
Australia	61
Singapore	65
Thailand	66
Russia	7
Japan	81
Hong Kong	82

Area, Or City Code.

This is a numerical code which identifies the region. Generally, callers from this same region will not need to dial this code, all other callers will. For example, callers in the San Francisco Bay 415 region only need to dial 989 0441 to reach my fax, but a few miles away in Oakland (which has the 510 area code), the 415 must also be dialed.

In the United States, most area codes are assigned to distinct geographical regions and follow these rules.

The area code is three digits long.
The area code has 2 through 9 as the first digit.
The area code has 0 or 1 as the second digit.
The area code does not have 00 or 11 as the last two digits.

The rule that the second digit must be 0 or 1 used to be binding but is now being relaxed as the explosion in demand for phone numbers has required an expansion of the number of area codes. So far, this has always been accomplished by "splitting" area codes so that subscribers in one part of the geographic region are changed to the new area code (with both the old and new code both working in a transitional period).

Telecommunications 107

Some phone companies are considering assigning new area codes to a given region so that all subscribers can keep existing numbers. The main drawback with this scheme is that subscribers within that region will then often have to dial an area code to make a local call.

A less explicit rule is that "similar" numbers are assigned to regions which are widely separated geographically. This is intended to help people remember which area code applies to a particular number and reduce incorrectly dialed numbers. For example, the following similar codes are assigned as follows:

201	New Jersey (Newark, Paterson)
202	Washington DC
203	Connecticut
501	Arkansas
510	Oakland, San Francisco Bay Area (East).

The numbers assigned to a contiguous region are quite "different", for example:

213	Los Angeles (Central)
310	Los Angeles (Santa Monica to Long Beach)
714	Los Angeles (Orange County, Riverside etc.)
805	Southern California
818	Los Angeles (Burbank)

It used to be a rule that the seven-digit local number would never have 0 or 1 as a second digit, which would allow a switch to distinguish a seven- from a ten-digit number by looking at the second digit. This results in faster call completion, since the switch can begin routing a local call immediately upon receiving the seventh digit without having to wait to see if more digits are coming. Now, however, as with area codes, the US phone system is running out of three-digit prefixes for local numbers, and some area codes now include local prefixes with 0 and/or 1 in the second digit position. (My advice: ask the phone company for a different number if you are assigned one of these. Otherwise you'll likely get more mis-dialed numbers where people are trying to dial long distance and forget the initial "1").

Note that a local call may require an area code, as in calling from San Francisco, CA (415) to Oakland, CA (510), and a long-distance call may not require an area code, as in calling Sacramento, CA (916) from Redding, CA (916).

In some countries, area codes vary in length. For example, in England, area codes may range from three to five digits.

A few smaller countries, such as Singapore and Panama, do not use area codes.

Some area codes may indicate special services rather than regions of the country. In the US, the following are in current use:

700.
Reserved for special services which may be defined by the long-distance companies. For example, AT&T provides a service called "EasyReach" which can provide you with a fixed phone number even when you change addresses.

800.
Toll-free numbers. The call is billed to the called party rather than the caller. 800 numbers can be routed to local phone lines or to dedicated lines (known as WATS lines) purchased from the long-distance companies.

900.
Premium-rate or pay-per-call numbers. The caller pays an extra charge for the call which appears on his or her telephone bill, a portion of this charge is paid to the *information provider* who provides a service through the call such as providing weather forecasts, sports scores etc.

Other countries use reserved area codes for similar purposes.

Local Number.

This is the smallest number of digits which can be dialed over the public network to reach a subscribers phone. In the US, this is always a seven digit number. The first three digits, referred to as the *NXX* code, specify the CO switch to which the line is attached. The last four digits identify the line itself: just like the extension number on a business phone system. A single switch will typically have several different NXXs assigned. For example, the 332 code of my fax number tells me (and the telephone network) that my fax line is attached to the Pacific Bell end office at the intersection of Bush St and Pine St in the Financial District of downtown San Francisco. Our neighbors in the next floor of our office building have telephone numbers 415-332 ????, 332 is also an NXX code belonging to the Bush/Pine switch. As a rule, if you move to a new location served by the same switch, you can take your phone number with you. If we were to move out of the downtown Financial District, we would have no choice but to change our phone number because we would be attached to a different switch.

Local numbers in other countries may also vary. In England, local numbers vary from seven digits in London and other metropolitan areas to as few as three digits in the remote countryside.

There is an NXX value which has been reserved in the US, like 800 and 900 area codes, for a special service. The 976 prefix to a local number is a premium rate service, analogous to a 900 number, which is provided through the local phone company (RBOC) rather than through a long distance carrier. The NXX prefix 555 is never assigned to subscriber lines, which explains why American films and television shows often use 555 for telephone numbers used in the plot.

NXX codes are generally restricted from value x00 and x11. Emergency service is obtained almost everywhere in the US by dialing 911. Local directory information service is often obtained through the three-digit code 411, and almost always through the local number 555-1212.

Dial Strings

A *dial string* is the sequence of characters sent to a device which can dial a telephone number, such as a modem or Dialogic card. The dial string may include the digits of the telephone number and also special characters which are interpreted as commands for the dialer. As an example of a special character, most readers are probably familiar with the comma which can be inserted in the dial string for a Hayes-compatible modem which means "pause".

The following list of typical dial string special characters is based on the dial string specification from Microsoft's TAPI (Windows Telephony programming standard, TAPI is described in more detail in a later chapter), and the Dialogic programming interface.

Character	Meaning
0 - 9	Digits to be dialed (pulse or DTMF).
A - D a - d # *	Additional DTMF tones (there are no pulse equivalents for these tones).
!	Flash-hook (TAPI standard).
&	Flash-hook (Dialogic standard).
P p	Use pulse dialing (applies only to current dial string and to digits following the P).
T t	Use tone dialing (applies only to current dial string and to digits following the T, tone dialing will usually be the default).
,	Pause. The length of the pause will be device-specific: half a second is typical.
W w	Wait for dial tone (TAPI).
L	Wait for dial tone (Dialogic).
I	Wait for internal (PBX) dial tone (Dialogic).
X	Wait for external dial tone (Dialogic).
@	(ASCII 40 hex: occasionally replaced by a different character on some keyboards or displays) Wait for quiet answer. Means to wait for at least one ring tone followed by a pause long enough to break the ring pattern,

Character	Meaning
	indicating that the phone has been answered (TAPI).
$	(ASCII 24 hex) Wait for billing tone. A typical billing tone is the "bong" tone used to prompt for credit card digits (TAPI).
|	(ASCII 7c hex) Subaddress follows. A *subaddress* is a part of an ISDN destination phone number (TAPI).
^	(ASCII 5e hex) Name follows. A name is an ISDN destination phone number field which is passed through to the recipient of the call (TAPI).

Access Codes

If I'm at home and I want to send a fax to my office from my laptop, I'll use the dial string 332 5657. If I'm at a trade show in Dallas and I want to reach my office fax, I'll use the dial string 1 415 332 5657. The "1" at the beginning is not the country code +1 shown on my business card, it's the access code which is universally used in the US to indicate that an area code and number is following. In many countries, a "0" (zero) is used for this purpose. If I'm in Manchester, England and I want to dial central London, I dial 0171 followed by the local number. In many countries, this "0" will be shown on business cards, magazine advertisements etc: a London number might be shown as (0171) 123-4567. This can be confusing when you want to reach the London number from another country. When dialing from the US, I need the country code, area code and number: the "0" is a prefix which only applies when calling from within England. From San Francisco, I'd dial 011 44 171 123 4567. The "011" prefix indicates that I'm about to dial an international number, 44 is the country code for England, and 71 is the area code. If I dial 011 44 0171 ... I may get a dialing error (although it does sometimes work both with and without the "0").

In some parts of the US, where NXX prefixes are still restricted to values other than x0x and x1x, it still may be possible to dial long distance numbers without the 1 prefix. However, using the 1 will almost always work.

There several prefixes which are used in the US public network.

1 area code number
Direct dial long distance. These types of calls are sometimes referred to as "1+" (pronounced as "one plus") calls.

0 area code number
Operator assisted long distance. When the number has been dialed, a *bong* tone will be heard. The caller can then dial a phone credit card number which will be used to bill the call, or can stay on the line and wait for live operator assistance. These types of calls are sometimes referred to as "0+" ("zero plus") calls.

011 country code area code number
Direct dial international. Many people don't know this, but in many parts of the US you can dramatically speed up the completion of an international call by dialing a pound tone (#) at the end of the number. This tells the switch that you've done dialing. If you don't do that, the switch will wait for a few seconds to see if more digits are coming before it starts routing the call.

01 country code area code number
Operator assisted international. Like 0 for operator assisted long-distance.

10xxx 1 area code number
The xxx is an *equal access code* which specifies which long-distance carrier should carry the call. Following the breakup of AT&T, each telephone line is now assigned a "default" long-distance company which carries calls if no equal access code is dialed. If, for example, at home, I use MCI as my default carrier, and I dial 1-201 ... to reach New York, MCI will carry the call. If, however, I dial 10288 1 201 ... the call will be carried by AT&T. The number 288 is AT&T's equal access code (it spells ATT on a keypad which shows the letters). MCI's code is 222,

Telecommunications

and so on. Each long-distance carrier has a three-digit equal access code and can be reached from any phone on the public network. Similarly, dialing "10xxx 0 area code number" requests an operator assisted call from a specific long-distance carrier. This feature allows a form of *least-cost routing* (*LCR*) where a caller can choose the carrier which offers the lowest rate to a given area at a given time. Some PBXs can be programmed to determine automatically which carrier to use based on a rate and time of day table.

If your equipment is attached as an extension on a PBX, it may need an additional prefix to get an "outside line" (direct connection to dial tone from a CO line). The prefix which is dialed to the PBX asking it to find a free outside line and make the connection is often a 9, which should be followed by a pause to wait for a new dial tone.

Chapter 5

Telephone Signaling

Introduction

For many simple applications, a detailed understanding of telephone signaling isn't needed. However, to really get control of network interface devices, you can't avoid getting down to the details of how telephone equipment communicates.

"Signaling" refers to information carried by a connection in addition to the voice or audio of the conversation.

Common examples of signaling include:

> Request to start an in-bound phone call (*ring*).
> Request to start an out-bound phone call (*seize*).
> Request to terminate a phone call (*disconnect, hang-up*).
> Send digits to switch (*dial*).
> Request service from switch (*flash-hook*).

Signals may be carried by the audio channel itself. Such signaling is called *in-band*. Examples include touch-tone digits and call progress tones.

Signals may also carried by a channel separate from the audio. On a T-1 trunk, signaling information is carried by the A and B bits, which take the values 0 or 1, generally corresponding to the presence or absence of loop current. Actually, it's debatable whether the A and B bits are truly separate from the audio channel, since the method usually used by the Dialogic PEB bus and by T-1 to carry signaling bits is known as "robbed bit" signaling, where least significant bits are occasionally borrowed from the digital audio samples to carry the A and B values. ISDN is a kind of modified T-1 where one time-slot (the "D-channel") is dedicated to signaling information, this is a true *out-of-band* signaling scheme where signaling is completely separate from the audio.

A signal is sent from one device to another device. To distinguish the two devices, we will refer to them as the CO (Central Office) or CPE (Customer Premise Equipment). A CO device acts like a typical phone company switch. A CPE device acts like a typical phone at a subscriber site.

A D/41ESC board, for example, is a typical CPE device. An MSI board with station cards, on the other hand, is a typical CO device since it acts like a phone company switch and can be directly connected to D/41ESC boards or to telephones. T-1 equipment is symmetrical electrically, like a null-modem RS-232 connection, but one end will be the "client" (CPE) and one end will be the "server" (CO).

Telephone Signaling 117

The following paragraphs outline the most common signaling procedures. The discussion of digital signals assumes the most common signaling convention where A and B bits are set to the same value, and represent the presence or absence of loop current, i.e. if A bit is 1, a connection is active, if A bit is 0, no connection is active. Bit values are given from the perspective of the equipment making the signal. For example, the transmit A bit at the CO is the receive A bit at the CPE. Be aware that, while the great majority of equipment does follow this (E&M) convention, yours may not.

Seize

A *seize* is where a CPE grabs a line from a CO and says "I'd like to place a call". This is what you do when you pick up a phone at home prior to dialing a call. There is a small chance you pick up the phone just as the phone is about to announce an in-coming call. In this case, you may hear the caller at the other end saying "Hello? Hello?" instead of getting the expected dial tone. This is called *glare*. All trunks which support both in- and out-bound calling have the potential for glare.

Seize: CPE requests dial tone from CO.

Analog seize

1. CPE: Goes off-hook.
2. Loop current starts to flow because circuit is now complete.
3. CO: Responds with dial tone.

T-1 seize

1. CPE: Changes transmitted A bit from 0 to 1.
2. CO: Responds by changing transmitted A bit (that is, the bit received by the CPE end) from 0 to 1 and plays dial tone. Some T-1 services will respond with dial tone but will not set the A bit to 1 until an out-bound call is dialed and the call is answered.

Disconnect

A *disconnect* or *hangup* is an indication that one party to the call wishes to terminate the call. Both the CPE or CO may issue a hangup signal.

The usual disconnect signal for an analog line is to interrupt the loop current for one second or more. This is done by putting one end on-hook, breaking the circuit between the CPE and CO.

Disconnect: terminates call, also called hang-up.

a) Analog disconnect (CPE disconnect).

1. CPE goes on-hook.
2. Loop current stops flowing.
3. CO recognizes hang-up, sends disconnect signal through public phone network to signal the other party.

b) Analog disconnect (CO disconnect).

1. CO drops battery voltage for period of one second.
2. Loop currents stops flowing for one second.
3. CPE recognizes the loop drop and goes on-hook ready for the next call.

c) T-1 disconnect

1. Party that wishes to disconnect changes transmitted A bit from 1 to 0.
2. Opposite party recognizes this change and responds by also changing their A bit from 1 to 0.

Many business phone systems do not pass a disconnect signal through to an analog extension when the caller hangs up. Instead, a tone may be generated, or there may be no signal at all. The same applies to central office switches in countries outside the US. In these environments, a VRU may need the ability to monitor for one or more tones throughout a call in order to detect a disconnect, or may need to wait for a time-out (no digit dialed in response to a menu) to determine that there is no longer a caller on the line.

Ring

A *ring* is a notification from the CO to CPE that there is an incoming call. On an analog line, this is done by applying ring voltage (AC) rather than the usual, DC battery voltage. On a T-1 or E-1 digital line, ring is signaled by changing the value of a signaling bit. On a typical T-1 line, ring is signaled by the CO changing the A bit sent to the CPE from 0 to 1.

On an analog DID circuit, the situation is a little different. On a DID trunk, the CPE is applying battery voltage instead of the CO. The CO, which is initially on-hook, signals the arrival of an incoming call by going off-hook, which starts loop current. The loop current start is detected by the DID equipment: this corresponds to a ring signal. The CPE DID equipment will respond with a *wink*, a brief reversal of the polarity of the battery being applied to the line, and the CO should then send the DID digits. The CPE will usually acknowledge with a second wink, and the CO will then complete the call.

Ring: CO notifies CPE of incoming call.

a) Analog loop-start (regular phone line) ring.

1. CPE: Is on-hook.
2. CO: Applies AC voltage (ring voltage) to line.
3. CPE: Detects ring voltage and goes off-hook.
4. Loop current starts flowing, audio can be transmitted and conversation starts.

b) Analog DID incoming call sequence.

1. CPE is applying battery voltage to line, is off-hook. CO is on-hook, so no current is flowing.
2. CO goes off-hook, loop current starts to flow.
3. CPE reverses battery voltage (+/- to -/+), this is an analog wink.
4. CO responds by transmitting DID digits.
5. CPE reverses battery voltage again to acknowledge receipt of digits.
6. CO switch completes connection, conversation starts.

c) T-1 ring

1. Before ringing, CPE sends transmit A=0 to CO, CO sends transmit A=0 to CPE.
2. CO changes A=0 to A=1.
3. CPE responds by changing A=0 to A=1.
4. CO completes connection.

d) T-1 DID (DNIS) incoming call sequence.

1. Before ringing, CPE sends transmit A=0 to CO, CO sends transmit A=0 to CPE.
2. CO changes A=0 to A=1.
3. CPE responds by sending digital wink, setting A=1 for about 0.5 seconds then setting A back to 0.
4. CO responds by sending DNIS digits, usually as DTMF or MF.
5. CPE acknowledges digits by sending a second digital wink.
6. CO completes connection.

Flash-Hook

A *flash-hook* is a request by CPE for an additional service from a CO while a call is in progress. A typical example is a call transfer. When a flash-hook is made, the caller is put on hold, and new dial tone is given by the CO. A new number may now be dialed. When this number answers, the CPE hangs up. This results in the original caller being transferred to the newly dialed number. The flash-hook is an on-hook period of about half a second: long enough to be distinguished from a glitch in the line, but not long enough to become a disconnect.

Flash-hook: CPE requests additional service from CO.

a) Analog flash-hook

1. Conversation is in progress, CO and CPE are both off-hook and loop current is in progress.
2. CPE goes on-hook for brief period (about 0.5 seconds, not long enough to make a disconnect/hang-up signal).
3. CO responds by putting caller on hold and providing new dial tone to CPE.

b) T-1 flash-hook.

1. Conversation is in progress, CO and CPE are both sending A=1 and loop current is in progress.
2. CPE sets A=0 for brief period (about 0.5 seconds, not long enough to make a disconnect/hang-up signal).
3. CO responds by putting caller on hold and providing new dial tone to CPE.

Wink

A *wink*, as explained earlier, is a signal from a CPE to a CO to acknowledge receipt of an incoming call on a DID or DNIS trunk. On an analog DID line, it is sent by briefly reversing loop polarity (+/- on the battery voltage). On a digital trunk, it is sent by briefly changing the transmitted A bit from 0 to 1 and back to 0.

Disconnect Supervision

Disconnect supervision, also called *hang-up detection*, refers to monitoring a telephone connection to detect the action of the calling party terminating the connection by hanging up the telephone. This information may be transmitted to the called party in one of the following ways:

1. Analog connection.

   a) By a short interruption (usually about one second in length) in loop current. This is the most reliable method in common use, and is always provided on telephone circuits connected to the public network in the USA, though not in some other countries. Many PBXs do not provide this loop current supervision.

   b) By a tone which is transmitted from the switch to which the called party is connected. This may be a dial tone, fast-busy, "re-order" tone or other tone. This is the supervision provided by most PBXs.

2. Digital connection.

> By a change in the signaling bit(s) transmitted from the switch to which the called party's equipment is connected. In a typical T-1 connection in the USA, the A bit will be changed from 1 to 0.

Analog Dialogic boards such as the D/2x and D/4x have loop current disconnect supervision "built in".

PEB-based boards such as the D/81A and D/12x are not connected directly to a trunk, they receive a signaling bit via the PEB: when this bit changes from 1 to 0 this is interpreted as a disconnect signal and a "loop current drop" event is generated by the driver. The signaling bit is put onto the PEB by the PEB interface board, which will be an LSI/xx, DTI/xx etc. In the case of the DTI/xx series, the DTI board itself has the ability to detect signaling bit transitions received in the digital signal from the trunk, and may generate an independent event such as an "A bit OFF" transition.

SC bus-based voice-only boards such as the D/240SC do not handle signaling at all. Unlike the PEB bus, the SC bus does not carry robbed-bit signaling. Therefore, signaling detection and generation is always handled by the interface card or by an interface device on a combined card such as a D/160SC-LS.

The Dialogic boards offer the following features for disconnect supervision:

> 1. Loop current drop. Any card with a loop start analog interface device can detect the one second drop in loop current which signals a disconnect.
>
> 2. Signaling bit transition. Boards with a T-1 or E-1 digital interface report when a signaling bit changes. On a PEB-based system, a voice board may report a "loop drop" event when the robbed bit passed via the PEB drops to zero. On an analog system, the LSI/xx PEB interface card translates the presence or absence of loop current to the robbed bit on the PEB (on or

off). The voice channel reports a drop in the PEB robbed bit as a "loop current off" event (the voice channel cannot see any difference between an LSI/xx interface card reporting a change in loop current or a DTI/xx card copying the A bit from the T-1 span).

3. Tone detection. All DSP-based D/xx boards can use Global Tone Detection to monitor for a tone which may be sent by the switch as a disconnect signal. A "Tone ON" event will be generated when the tone is recognized in the received audio.

4. Continuous no-silence detection. All D/xx boards, including non-DSP boards such as the D/xxA and D/xxB series, will monitor the received audio for continuous sound while performing a multi-tasking operation (recording, playing, getting digits etc.). If the switch produces a continuous tone (such as dial tone) or a rapidly repeating tone (such as a re-order tone), then this may provide an option if loop current supervision is not provided by the switch. It is a less reliable method because it is only active during a multi-tasking function, and because it may produce a "false detection" due to other sound sources. The Dialogic driver reports continuous no-silence as a terminating event for a multi-tasking function. This hang-up detection method is obsolete since in most cases tone detection can be used and is much more reliable.

Chapter 6

Call Progress

Introduction

Call progress addresses the important and difficult question: what is the state of the telephone connection?

Call progress analysis, sometimes known as *call supervision*, is most commonly performed immediately following the dialing process which communicates the desired routing of a call to a switch. The results of the attempt are indicated to the calling person or equipment originating the call in the form of *call progress tones*. Common examples include ringing and busy tones.

Call progress analysis presents a difficult challenge to automated equipment such as voice processing hardware because of the antiquated technology used in signaling. The most commonly used telephone connection, the analog two-wire interface, provides no separate signaling channel — all call progress signaling must be done using *in-band* signals. In other words, the only means available of signaling the progress of a call is to use the sound (audio) on the line, or the loop current. More modern digital connections such as ISDN and the European E-1 standard provide for *out-of-band* signaling, where signals completely separate from the audio channel can be used.

While some switches do provide loop current signals to indicate the completion of a call, in most cases the voice processing hardware will need to analyze the received audio on the line to match the sound against expected patterns of tones. This process is complex, and requires a great deal of tuning and flexibility if it is to be completely robust, especially when many different environments may be encountered. For example, a voice mail unit may be required to transfer a call within a PBX, dial pager units through local and long-distance calls, or forward voice mail messages over the public network to other national or international offices. It is a major challenge to provide reliable call progress features for all these different environments. Each PBX will have its own characteristic tones for ring and busy, these may even vary between different models from the same manufacturer.

Call Progress Monitoring

The term call progress analysis or supervision is usually applied to the period immediately following an out-dial or transfer. A related but more general feature is *call progress monitoring*, which refers to the entire length of the call. The simplest, and generally most reliable, aspect of call progress monitoring is hang-up, or disconnect, supervision.

In analog environments, hang-up supervision is usually implemented by dropping loop current for a brief period. Digital environments will

Call Progress

generally signal a disconnect by changing the value of a signaling bit. On standard T-1 circuits, for example, the A bit transmitted to a device will generally be changed from A=1 (corresponding to loop current active) to A=0 (corresponding to loop current absent).

The bad news is that some switches, including unfortunately many business phone systems, do not provide loop current disconnect supervision. To detect a disconnect in such an environment, it may be necessary to monitor the line continuously for the tone generated by the switch when the call has been terminated and the line is left on hook — generally a new dial tone, or a re-order tone (beeps, possibly followed by a message like "Please hang up and try dialing your number again").

Loop Current Supervision

Loop current supervision is the simplest method of signaling a call result: the loop current is dropped briefly to indicate a connect. Within the US, this method is not widely used except when completing a long-distance call, when the long-distance carrier's switch may provide a loop current drop. Out-dial applications should be careful not to confuse this drop in loop current with a caller hang-up, and should also be careful to use this method only when it is known that loop current supervision is provided.

It may often be preferable to disable the loop current supervision feature of your voice board's call progress algorithm unless you are sure it is going to work correctly.

Cadence

Cadence is used to refer to the pattern of tones and silence intervals generated by a given audio signal. Examples of tones with cadence patterns are busy tones and ringing tones. Dial tone is a continuous tone which has no cadence (unless you live in Italy, in which case your dial tone sounds like a US busy tone). Ideally, the call progress analysis algorithm should be able to take advantage of the known characteristics of the tones, including:

1. Frequency or frequencies making up the tone
2. Duration of tones
3. Duration of silences between tones

A typical cadence pattern is the US ringing tone, which is one second of tone followed by three seconds of silence. Some other countries, such as the UK, use a double ring, which has two short tones within about a second, followed by a little over two seconds of silence. A busy signal in the US is a half second of tone followed by a half second of silence. Another type of busy, the *fast busy* which indicates that the network cannot reach the called central office, is similar to busy but at twice the rate. A third type of signal is *SIT*, or *Special Information Tones*, which are three rising tones followed by an informational message which explains a problem: for example, that the requested number was disconnected, or that the area code dialed was invalid.

Some typical cadence patterns are sketched in the following diagrams.

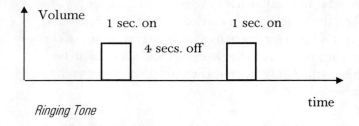

Ringing Tone

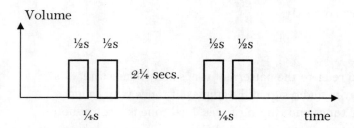

Double Ring Tone

Call Progress

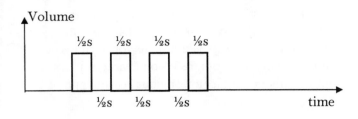

Busy Tone

The D/xx Call Progress Algorithms

There are two types of call progress algorithm which are incorporated into Dialogic voice boards. The newer, DSP-based boards include an algorithm called Perfect Call, which includes detailed analysis of the frequencies and cadences of call progress tones. By default, this algorithm is disabled. In addition, there is an older algorithm called Basic Call which does not examine the frequencies (except of operator intercept tones), it works by matching the cadence pattern of sound and silence. Generally, Perfect Call is more accurate and reliable, so newer applications should use this algorithms.

Either algorithm provides one of the following overall results as the output from the analysis:

Connect	Call completed.
Busy	Busy tones detected.
Ring no answer	Ringing tones detected but call not answered after a set number of rings.
SIT	Special Information Tone, such as operator intercept, detected.
No ring	Ringing tones not detected after set waiting period.

Additional information may also be available, depending on how the result was determined. For example, the duration of sound and silence periods making up a cadence pattern, or the length of the sound period assumed to be the caller's "Hello". If Perfect Call is used, there will be an additional indication of whether a human voice was detected (*Positive Voice Detection, PVD*), or an answering machine (*Positive Answering Machine Detection, PAMD*). PVD relies on detecting a characteristic spectrum of frequencies as found in the spoken human voice, and is quite reliable. Answering machine detection relies on characteristics which Dialogic does not document, and is (for obvious reasons) not 100% reliable.

Basic Call progress relies on three separate algorithms working in parallel:

1. Loop current detection.
2. Frequency detection.
3. Cadence detection.

Loop current detection watches for the brief drop in loop current which is used by some switches to indicate a completed call (connect).

The frequency detection element has rather limited functionality, that of detecting one instance of a single tone of a minimum duration in a given frequency range. This is primarily used to detect a chosen intercept (SIT) tone.

Cadence detection is the most important element, and in the great majority of cases is the part of the algorithm which produces a result determination.

The cadence detection part of the algorithm starts by classifying the incoming audio into periods of silence or non-silence (Dialogic's inelegant term for "sound"). The resulting pattern of sound and silence periods is then matched against expected sequences for ring, double ring or busy tones. If a pattern is recognized, the algorithm continues to monitor the audio until the pattern is broken, which is assumed to indicate that the call was answered.

Finally, the algorithm can be configured to monitor the first period of non-silence following a connect, assumed to be the caller's "Hello". This feature has two advantages: the application can wait for the caller to finish speaking before playing a greeting rather than interrupting

Call Progress

immediately or waiting for a pre-set period, and the length of the greeting can help distinguish between residences, which tend to have brief greetings ("Hello, this is Mike"), businesses, which have longer greetings ("Good afternoon, this is Parity Software, how may I direct your call?"), and answering machines/voice mail, which often have longer greetings ("Hello, this is Elaine, sorry I can't take your call at the moment, please leave a message at the tone").

The outline of the cadence detection part of the algorithm is as follows:

1. Wait for a set period after dialing before beginning analysis.

2. Try to match the initial silence, non-silence, silence sequence to a cadence pattern. If a pattern can be established, go to step 4, otherwise go to step 3.

3. If no sound was detected, return No Ringback result, otherwise return Connect result.

4. Was the established cadence pattern broken before a set number of repetitions? If so, return Connect result, otherwise go to step 5.

5. If pattern matched the expected busy sequence, return Busy result, otherwise return Ring No Answer result.

Since the cadence detection part of the algorithm only distinguishes sound from silence without examining the frequency components, there are several situations which can give misleading results. For example, if the called party picks up after one ring and says "Hello" where the duration of the greeting is about one second, the "Hello" may look to the algorithm like a second ring, and a connect will not be signaled before a couple of seconds of silence or a second utterance from the caller ("Is anybody there?").

Full details of the call progress algorithm is found in the Dialogic Application Note AN002, "Customizing D/41 Call Analysis". A thorough study of this document is strongly recommended if your application is going to require tailored call progress analysis. This note is included in some Dialogic Voice Software Reference manuals in a section entitled "Call Progress Analysis", or (obscurely) "D/41 Theory of Operation".

Global Tone Detection

DSP-based boards, including all the SC bus voice boards and the older D/xxD, D/81D, D/121A and D/121B, feature down-loadable firmware known as SpringWare. New features can be added through software upgrades rather than by purchasing additional hardware. Among the SpringWare capabilities introduced in 1993 were *Global Tone Detection (GTD) and Global Tone Generation (GTD)*. By analyzing frequency components as well as cadence patterns, GTD is able to detect tones such as ringing and busy with greater reliability than cadence analysis alone. Global Tone Generation allows specified tones to be played on a channel.

GTD recognizes four types of tones:

Single tone with no cadence.

Single tone with cadence.

Dual tone with no cadence.

Dual tone with cadence.

A single tone is a tone with one frequency, a dual tone is a tone with two frequencies. A cadence pattern is a specified period of sound followed by a specified period of silence.

Cadence patterns are specified based on the following model:

Call Progress

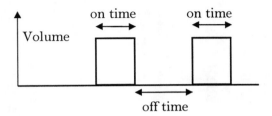

There is only a single period of tone and a single period of silence. Acceptable ranges for the detected on and off times may be specified.

Note that each Dialogic board has a minimum spread of frequencies which can be resolved into a dual tone. If a dual tone is composed of two frequencies which are closer together than this, then the tone will be detected as a single, not a dual, tone. This causes particular problems with US dial tone, which is a dual tone with frequencies 350 Hz and 440 Hz. Some Dialogic boards ("low horse power") can resolve a dual tone with a difference of 125 Hz between the tones, some ("high horse power") can resolve up to 62.5 Hz. This means that dial tone must be defined as a dual tone on high horse power boards, but as a single tone on low horse power boards. It is not possible to make one definition which works on all boards.

You might think that it is not possible to define double-ring tones, such as UK ring-back, because of the single on/off cadence pattern which is used. However, by specifying large ranges for the on- and off-time permitted deviations, it is possible to define a cadence pattern which will match.

There is no way to specify acceptable *twist ratios*. The twist is the ratio in amplitude between the two components of a dual tone.

Chapter 7

Fax Integration

Introduction

Fax is a growing element in many voice processing applications. Just as voice can be stored, retrieved and forwarded to new destinations through a PC-based voice processing systems, so can fax documents be stored, retrieved, forwarded and generated directly from word processing, database, graphics and other file formats.

Major categories of fax applications include fax mail, fax broadcast and fax on demand.

Fax Mail.

Analogous to, and perhaps a feature of, voice mail. Fax mail allows a caller to fax a message rather than speaking a message. Fax messages may be retrieved by the mailbox owner from a fax machine or desktop PC which is able to access the stored file and display it as an image on the computer screen. Some voice mail systems allow fax messages to be incorporated into mailboxes.

Fax Broadcast.

Broadcast systems send a copy of one document to several phone numbers. For example, a company with many offices in the US and in Europe might use a fax broadcast to save on telephone toll charges. If an office in London wanted to send a fax to all regional offices in the US it would be less expensive to send one fax to the New York office and then have New York broadcast the document to all other US offices. Other applications of fax broadcasting are in distributing information such as press releases, questionnaires or promotions. Some e-mail services such as MCI mail offer fax broadcast options.

Fax On Demand.

A typical use for fax on demand is to provide product information to potential customers. A caller dials a voice processing unit and selects one or more documents of interest using touch tone menus or voice recognition. If the caller is calling from a fax machine, transmission can being immediately (this is called *one-call* or *same-call* faxing). If the caller is using a telephone rather than a fax machine, a fax number can be entered in response to a menu prompt and the fax on demand system will make a later call to that number to deliver the document.

A Brief History Of Fax

Most people probably think of fax as being a recent invention. Certainly, it is only in recent years that fax has become so ubiquitous. In the United States, fax has almost made the letter obsolete for routine business communications. However, the technology of transmitting images over communication lines is more than a century old.

Fax Integration

The beginning of facsimile technology is traceable to 1843, when the first successful fax device was patented. Commercial fax service began in France, in 1865. Over the 1930s and 1940s, fax evolved into the form we recognize today.

As fax became more widely used, it soon became obvious that standards would be needed to enable different fax machines to communicate with each other. In 1966, the Group 1 standard made possible the more generalized use of fax. At that time it took about six minutes to transmit a single page of text, and resolution was poor. In 1978, the international telecommunications standards body CCITT published its Group 2 recommendation, which was adopted by all fax machine manufacturers. Then, in 1980, the Group 3 digital fax standard was implemented, leading to today's familiar fax devices. Group 3 machines are faster, requiring about 30 seconds to transmit a page of text, and offer better resolution. As a digital standard, Group 3 lent itself naturally to computers, and the explosive growth of the standard PC architecture has resulted in a wide range of PC-compatible fax modems.

At first, the fax modem was viewed as a way for PCs to emulate fax machines. Now PC users can not only send and receive and print faxes, but also create them, using graphic and text files. A major advantage is in the reproduction quality at the receiving end. Documents scanned by an ordinary fax machine will generally be noticeably inferior to computer-generated faxes. Time and paper is saved, and the cost of a fax modem is typically well below the cost of a fax machine. Toll charges can be saved by scheduling fax transmissions outside business hours when rates are lower. Integration with voice processing has opened up new applications such as fax on demand, fax re-broadcasting and fax mail. Current estimates give fax/voice applications only perhaps a 5% share or less of the voice processing market. However, fax is expected to grow in importance as *integrated messaging* applications, which combine voice, fax and E-mail, reach maturity.

Eventually, industry observers believe there may be a merging of fax and X.400, the store-and-forward messaging protocol defining a framework for the distribution of data between networks. Group 3 information would then be packaged as part of an Email message and sent over the worldwide X.400 hierarchy of circuits.

The future of fax is evolving toward higher transmission speeds and added features, such as color. Group 4 is an ISDN-based standard written to implement some of these features into future fax devices. However, Group 3 machines today do most of the things envisioned for Group 4, and are compatible with the more than 25 million machines across the world, whereas Group 4 is not.

As requirements increase and faxes become more fully digitized, a form of multimedia fax will appear providing high resolution, color, voice and motion, working with Email and offering encryption for sensitive data.

Fax Transmission Standards

It was not until October 1966 that the Electronic Industries Association proclaimed the first fax standard: EIA Standard RS328, Message Facsimile Equipment for Operation on Switched Voice Facilities Using Data Communication Equipment. This "Group 1" standard as it later became known, made possible the more generalized business use of fax. Although Group 1 provided compatibility between fax units outside North America, those within still could not communicate with other manufacturers' units or with Group 1 machines. Transmission was analog, typically it took between four to six minutes to transmit a page, and resolution was very poor.

US manufacturers continued making improvements in resolution and speed, touting the "three-minute fax." However, the major manufacturers, Xerox and Graphic Sciences, still used different modulation schemes (FM and AM) so, again, there was no standard. Then, in 1978, the CCITT came out with its Group 2 recommendation, which was adopted by all companies. Fax had now achieved worldwide compatibility and this, in turn, led to a more generalized use of fax machines by business and government, leading to a progressive reduction in prices.

When the Group 3 standard made its appearance in 1980, fax started well on its way to becoming the everyday tool it is now. This digital fax standard opened the door to reliable high-speed transmission over ordinary telephone lines. Coupled to the drop in the price of modems, the Group 3 standard made possible today's affordable desk top unit.

Group 3 is a flexible standard, which has stimulated competition among manufacturers by allowing them to offer different features on their machines and still conform. The improvement in resolution has also been a factor. The standard resolution of 203 lines per inch horizontally and 98 lines per inch vertically produces very acceptable copy for most purposes. The optional fine vertical resolution of 196 lines per inch improves the readability of smaller text or complex graphic material.

Group 3 fax machines are faster. After an initial 15second handshake that is not repeated, they can send an average page of text in 30 seconds or less. Memory storage features can reduce broadcast time even more. Contemporary machines offer simplicity of operation, truly universal compatibility, and work over regular analog telephone lines, adapting themselves to the quality of the phone line by varying transmission speed downward from the standard 9,600 bps if required.

Fax Call Protocol

With small variations, a modern fax machine's call is made up of five definite stages as defined by the CCITT T.30 specification: *Phase A*, the call establishment; *Phase B*, the pre-message procedure; *Phase C1* and *Phase C2*, consisting of in-message procedure and message transmission; *Phase D*, the post-message procedure; and *Phase E*, call release.

Phase A takes place when the transmitting and receiving units connect over the telephone line, recognizing one another as fax machines. It is here that the "handshaking" procedure begins between the transmitting and receiving units. This is accomplished through a piercing, 1,100 Hz tone (the *calling tone* or *CNG*), sent by the caller machine. Everybody who has sent a fax is well acquainted with this whistling sound.

The caller machine replies to the CNG by sending its *Called Station Identification* (*CED*) and the answerer replies with its own, a shrill 2,100Hz tone. Once this has been accomplished, both machines move on to the next step.

In Phase B, the pre-message procedure, the answering machine identifies itself, describing its capabilities in a burst of digital information packed in frames conforming to the High-Level DataLink Control (HDLC) standard. Always present is a Digital Identification Signal (DIS) frame describing the machine's standard CCITT Group 3 features and perhaps two others: a *Non-Standard Facilities* (*NSF*) frame which informs the caller machine about its vendor-specific features, and a *Called Subscriber Identification* (*CSI*) frame. The CSI typically consists of the telephone number of the fax machine. At this point, optional Group 2 and Group 1 signals are also emitted, just in case the machine at the other end is an older Group 1 or Group 2 device unable to recognize digital signals.

This is very rare, however, and chances are that both devices will be Group 3 machines, in which case they will proceed from the identification section of Phase B to the command section. Here the caller responds to the answerer's initial burst of information, with information about itself. Previous to transmitting, the caller will send a *Digital Command Signal* (*DCS*) informing the answerer how to receive the fax by giving information on modem speed, image width, image encoding, and page length. It may also send a *Transmitter Subscriber Information* (*TSI*) frame with its phone number and a *Non-Standard facilities Setup* (*NSS*) command in response to an NSF frame.

Then the sender activates its modem which, depending on line quality and capabilities of the machines may use either V.27ter (PSK) modulation to send at a rate of 2,400 or 4,800 bps, or V.29 (QAM) modulation to transmit at 7,200 or 9,600 bps. A series of signals known as the *training sequence* is sent to let the receiver adjust to line conditions, followed by a *Training Check Frame* (*TCF*). If the receiver receives the TCF successfully, it uses a V.21 modem signal to send a *Confirmation to Receive* (*CFR*) frame; otherwise it sends a *Failure-to-Train* signal (*FTT*) and the sender replies with a new DCS frame requesting a lower transmission rate.

Phase C is the fax transmission portion of the operation. This step consists of two parts—C1 and C2—which take place simultaneously. Phase C1 deals with synchronization, line monitoring, and problem detection. Phase C2 is devoted to data transmission. Since a receiver

Fax Integration 141

may range from a slow thermal printing unit needing a minimum amount of time to advance the paper, to a computer capable of receiving the data stream as fast as it can be transmitted, the data is paced according to the receiver's processing capabilities. An *Error Correction Mode (ECM)* procedure encapsulates data within HDLC frames, providing the receiver with the capability to check for—and request the retransmission of—garbled data. If there are errors in the transmitted data, ECM adds time; however, although ECM adds an overhead, it uses no fill, so this overhead does not lengthen transmission time. *Fill* refers to transmitting sequences of zero bits to allow a receiver time to perform an operation such as feeding the next page into the printing unit.

Once a page has been transmitted, Phase D begins. Both the sender and receiver revert to using V.21 modulation as during Phase B. If the sender has further pages to transmit, it sends a frame called the *Multi-Page Signal (MPS)* and the receiver answers with a *Message Confirmation Frame (MCF)* and Phase C begins all over again for the following page. After the last page is sent, the sender transmits either an *End Of Message (EOM)* frame to indicate there is nothing further to send, or an *End Of Procedure (EOP)* frame to show it is ready to end the call, and waits for confirmation from the receiver.

Once the call is done, Phase E, the call release part, begins. The side that transmitted last sends a *Disconnect (DCN)* frame and hangs up without awaiting a response.

Compression Schemes

When digital techniques were developed for fax technology, it became practical to use encoding techniques to remove redundancy from the page being scanned, and restore the data at the receiving end. This helped reach the sought-after results of error reduction and shorter transmit times.

The CCITT specifies that Group 3 machines incorporate *modified Huffman run-length encoding* of scan lines. In a scan line a white picture element, or pixel, is likely to be followed by a long string of the same, before reaching a black pixel. In typed or printed material, these white strings (or runs) generally continue across the entire page. *Run-length* encoding represents a run by storing the length of the run rather than each bit in the run.

A white or black pixel is represented by 1 bit. The early fax machines would transmit 1728 bits of information for every white line across a page. Now, instead of sending all 1728 bits to represent a white line across a page, modified Huffman (MH) produces a 9bit code word representing that white run, thus compressing this information 192 times. The result is that only 92 binary codes are needed for white runs of 0 to 1728 pixels, and another 92 for the black ones. The shorter binary codes are assigned to the longest, and most commonly occurring runs, in this way saving even more transmit time. The receiving fax units then decode this information, reproducing the original run.

A typical Group 3 fax image can have a total of 1,973,376 pixels—1728 from side to side and 1,143 from top to bottom. Were this image to be sent uncompressed at 9,600 bps, under ideal circumstances, it would take almost three and a half minutes to transmit. This is not practical and it can get very costly, in terms of telephone time and personnel, to take over ten minutes to fax a three-page document.

Today, Group 3 fax machines use two compression technologies. One of these, the modified Huffman code, is an encoding scheme that takes advantage of the similarities between pixels on the same line but not between pixels in succeeding lines. It uses a combination of run-length and static (not adaptive Huffman encoding). In this process, the sender searches for transitions from black to white (or vice versa) and reports the number of pixels since the previous transition. To prevent blotching in the event of an error, the line must always start with a "white" run length (which can be zero if the first pixel is black). The code for a run of black pixels is not the same as for an equal number of white ones because they are not as statistically likely: runs of black pixels tend to be shorter than runs of white ones in most printed documents. A mostly black document takes longer to fax than a mostly white document, as you might intuitively expect.

If the run is longer than 63 pixels, a special "makeup code" is prefixed to the code word. Each makeup code adds a multiple of 64 pixels to the run length. One makeup code plus one normal run-length code is

Fax Integration 143

enough for a run of up to 2623 pixels long (wider than a normal A4 page fax transmission).

Modified Huffman compresses only one scan line at a time; it looks at each scan line as if it were the only one. Each line is viewed by it as a separate, unique event, without referencing it in any way to previous scan lines. Because of this single-line operating characteristic, modified Huffman is referred to as a "one-dimensional compression" encoding technique.

The second encoding scheme, *MR*, or *modified READ (Relative Element Address Differentiation)*, uses the previous line as a reference, since most information on a page has a high degree of vertical correlation; in other words an image, whether it is a letter or an illustration, has a continuity up and down, as well as from side to side, which can be used as a reference. This allows modified READ to work with only the change between one line and the next. This results in a rate of about 35 percent higher compression than is possible with modified Huffman.

Therefore, the difference between modified Huffman and modified READ is, essentially, that the latter uses a "knowledge" of previous lines to reference its vertical compression. Because modified READ works vertically as well as horizontally, it is called a "two-dimensional compression encoding technique."

Two-dimensional encoding, even with all the necessary safeguards against errors, can reduce transmission time by as much as 50 per cent. Actual sending time depends on the number of coded bits per line and modem speed. It now becomes possible to compress the image data Modified Huffman down to a more compressed form, *Modified Modified READ (MMR)*, providing as much as twice the compression of one-dimensional encoding. (Guess what the next generation of READ is going to be called).

Printing The Fax

When the receiver modem decodes the received analog fax signal it regenerates the digital signal sent by the fax transmitter, the MH/MR/MMR block then expands this fax data to black-white pixel information for printing.

Overall, there are two ways to convert a fax into hard copy: through a thermal printer, or a regular printer (the latter covering everything from pin to the preferred laser printer formats).

Each inch of a thermal printer's print head is equipped with 203 wires touching the temperature-sensitive recording paper. Heat is generated in a small high-resistance spot on each wire when high current for black marking is passed through it. To mark a black spot, the wire heats from nonmarking temperature (white) to marking temperature (black) and back, in milliseconds.

About the only advantage of thermal paper is that, since it comes in rolls, a thermal printout has the advantage of adapting itself to the length of the copy being transmitted: if the person on the other end is sending a spreadsheet, the printout will be the same size as the original. With sheet-fed printers, such as a laser printer, the copy is truncated over as many separate sheets as necessary to download the information.

In general, thermal printouts offer less definition, thermal paper is inconvenient to handle due to its tendency to curl, and if the copy is to be preserved, it must be photocopied because thermal paper printouts tend to fade over time. Then, most important of all, there is the matter of cost: thermal paper printouts are between five to six times more expensive than those made on plain paper.

Plain paper fax machines using laser printer technology are becoming increasingly popular, particularly in the US.

Computer-Based Fax

Computer fax boards enable users to receive faxes, print them, store them on a hard disk, as well as to create them using text and graphics files, and then transmit them. Depending on system configuration, these boards can act as fax servers for networked users.

One of the major advantages of computer-based fax is in the quality of reproduction at the receiving end. Text and graphics that are run through a fax scanner are inevitably degraded in sharpness. It does not matter whether the reproduction medium at the other end is thermal paper or a laser printer—it is the scanning process at the transmitting end that degrades the image. Text and graphics produced on a PC, however, are almost invariably superior. Also, it is possible to preview received faxes on the computer display before printing or discarding them, an increasingly useful feature as fax comes to be used more frequently for purposes of every kind.

Another major advantage is in the capability to broadcast large numbers of faxes. Often, it may be necessary to do a mass broadcasting of faxes to clients, suppliers, etc. A list is created using the phone book utility, and the computer is instructed when to send out the faxes to the list. This last feature is useful if the user prefers to transmit at a time when phone rates are at their lowest. Besides this automation feature, other benefits lie in the ability to track fax expenses as closely as any other business expense and the customization of CBF for different purposes, enabling the user to find novel applications for faxes as selling tools, marketing purposes and, in the case of fax on demand applications, to make information available to staff and customers at remote sites 24 hours a day.

A final benefit to having machine-readable representation of the fax image is that *Optical Character Recognition (OCR)* can be applied to the image, converting it to ASCII text or even a word processing file with detailed formatting information. The main drawback of OCR is that accuracies are, at best, in the 95% to 99% range at present, which means that human intervention is required for final proofreading of most such material.

The Future Of Fax

In the swift evolution of fax technology, the Group 4 standard is viewed by some as the next significant step. Proposed in 1987 by the CCITT to allow the transmission of fax over non-analog circuits using X.25 packet switching networks as well as others, the standard has been set. Group 4 is expected to work with ISDN, switched data networks, or dedicated digital circuits. It is a layered standard, laid out in terms of the Open Systems Interconnection Reference Model (OSI), and is expected to provide high-speed transmission, outstanding error control, and features such as gray scale and color imaging.

Group 4, however, is a standard written before a real machine was made and only a group of experts well-versed in its arcane ins and outs know how to implement it. Currently, there is no way to use it over ordinary voice grade phone lines, and it will not inter-operate with Group 3, 2, or 1 machines. At present, there is no available modulation standard or scheme for Group 4. Absent, also, is a protocol for initial handshake. Applications for Group 4 seem rather limited, particularly in view of Group 3 technology's continued evolution. Group 3 is now expected to provide many of the features once promised only by Group 4, such as tighter compression, gray scale, color, and faster transmission. Work has been concluded by the CCITT on a new, faster modem standard, V.17, which gives Group 3 the capability to operate in the 12,000 to 14,400 bps range. Many industry experts are dubious about Group 4's future, arguing that the continued evolution of the T.30 protocol and Group 3's capability to deliver speed, error control, and higher resolution—the very features that made Group 4 attractive—has rendered Group 4 outmoded even before its implementation in the real world.

Primary objections to Group 4 are its inability to communicate with Group 3 machines and work over ordinary telephone lines. Before any consideration can be given to any radical alteration in fax standards, the first thing that must be dealt with is the existence of somewhere between 20 to 25 million Group 3 machines all over the world today. This is a considerable installed base, particularly when one considers these machines are not upgradable nor can they be changed in any significant way. What is more, they run satisfactorily, and users show no intention to discard them for several years to come.

Given that there is such an enormous installed base of unalterable machines, what is going to happen? They cannot be upgraded by adding new keyboard features or new LCD displays, they will not accept new software, there is nowhere one can send in a few dollars and get a new control panel for them. Even if the protocol committees were to come out with faster modem definitions, higher resolution pages, etc., a fact of life is that these machines are going to be around well into the 21st century.

Fax Integration

The only way users might be led to discard this installed base in any significant way would be if there were some truly fundamental economic reason to do so. This happened to the majority of Group 2 fax machines which were disposed of because transmission time for Group 3 was a third or less than for Group 2. There was a significant telephone cost difference in going from a three-minute to a one-minute fax. There is no consensus, however, that this will happen with Group 3.

Once the fact that there exists a large infrastructure of installed machines that is not going to change is accepted, and that any software innovation and fax system planning has to take place taking that reality into consideration, it becomes easier to plan what uses to put the technology to, as well as what sorts of services to offer with it. Thus, services like fax on demand and others have come into being around that reality.

The Evolution Of Group 3

Group 3 technology is still evolving. Agreements by the standard-making bodies—from various countries and administrations—and, ultimately the world's manufacturers, are making additions to the Group 3 protocol set. Group 3 fax machines now may incorporate faster, 14.4 kbps modems. It is estimated that now perhaps two to four percent of the installed base consists of fax machines with the higher speed modems, and this is increasing. Also coming are new resolution capabilities. It is expected that in 1993 or 1994 300dotperinch resolution will be approved by the standard-making bodies.

Other new capabilities include routing codes, binary file transfer capabilities, and superfine resolution, just to mention a few. Although market inertia is considerable, some of these new features may be of particular importance to some users, who will invest in machines offering them, slowly obsoleting existing machines.

It would appear from all this that computer-based fax by being, in effect, part of a computer and therefore more amenable to enhancements and software upgrades would find itself in a privileged position. This is the case, up to a point. The fact remains that an estimate (93) of all LAN fax servers, and even all the PC fax boards, sold barely reaches the half-million mark. It is projected that by the end of the current year there will be about a million of various versions of computer-based fax installed. Although this is by no means a small amount, and market acceptance is growing literally by leaps and bounds, the fact remains that anyone wanting to use the capabilities of these computer/fax hybrids must still deal with an overwhelming installed base of external stand-alone units that cannot change. So, while software for laptops or desktops can add features to fax, unless these features can find their way to the rest of the industry, into the installed base, they will evolve very slowly.

Probably the biggest growth area for fax today is in the area of its digital trunks. There are several proposals out now for consideration by the various standard-making bodies, which provide for the transmission of fax over non-analog circuits. Group 4 was the first one of these. There are Japanese fax machines sending Group 4 fax over ISDN B (Bearer) circuits at 64 kbps. However, in practice, it is only in Japan that any significant population of these fax machines is found, and even there it is very small, in the tens of thousands. The reason is that Group 4 requires digital circuits that are not commonly available. Also, these fax machines have 400dot print scanners and sell for about $10,000 or $12,000 (93), a price most businesses find difficult to justify, particularly when the ongoing additions and expansions taking place within the Group 3 protocol continue making Group 4 seem less desirable.

While Group 4 machines are intrinsically faster, the speed difference between a 14.4 kbps and 64 kbps message may not be all that critical. While it is true that instead of taking 30 seconds to send a page a Group 4 fax can do it in ten, it is still necessary to wait 20 seconds for call setup procedures; Group 4 is a very complex protocol and machines must exchange some 20 messages before they begin transmitting and receiving. An industry observer compared Group 4 to the Concorde: "A three-hour flight to Europe does not make that much difference when you spend four hours waiting in the airport." And, of course, there is that $12,000 price tag and the fact Group 4 machines do not operate over ordinary telephone lines and have no compatibility with the great infrastructure of installed Group 3 machines.

There are other digital standards, such as Group 3 bis and Group 3/D, which keep the standard Group 3 protocols and resolutions, provide a fast handshaking, and make all digital, full-duplex transmission possible. These standards are also ISDN-compatible. The X.38 and X.39 standards offer another way of digitizing Group 3 fax, working with an X.25 transport layer.

Fax And X.400

X.400 is a store-and-forward electronic messaging protocol defining a framework for distributing data from one network to several others. It allows end users and application processes to send and receive messages which it transfers in a store-and-forward manner. An X.400 message consists of a message envelope and message content.

The message envelope carries addressing, routing, and control certification information. The message content part involves methods of encoding simple ASCII messages, as well as more complex data. It may contain fax, graphics, text, voice, or binary data structures. X.400 is being increasingly viewed as a delivery platform for a variety of services, including Email, electronic data interchange (EDI), and others.

Industry observers agree X.400 fax will play a major role in the next generation of facsimile machines. They look at a combination of fax and digital fax with the X.400 Email standard. In such a setup, a Group 3 message would be packaged as part of an X.400 Email message and sent over the X.400 hierarchy of circuits. There are presently over 100 international carriers that can exchange this type of traffic. X.400-based fax store-and-forward networks are already in operation. The user would transmit a fax locally to a store-and-forward switch that would capture, encapsulate, and send it over high-speed circuits later to be recreated as a fax and delivered.

The concept of sending fax raster images as part of an X.400 Email message makes sense, because X.400 already exists as a world standard. It is now in its third division and no further standards have to be implemented to put fax raster into X.400 traffic. X.400 fax traffic would parallel Email traffic generated by corporations. It is not unlikely that in the near future, traveling side-by-side, there would be an Email message followed by fax, and a combined fax and text message, then by a voicemail mail message, etc., all of these intermixed in one circuit.

Compressed TV and animation are not too far off, either. In short, X.400 will become the ideal pipeline for multimedia communications.

The problem with X.400 today is that it is principally used as a backbone to connect disparate Email systems. There are few good user interface packages available, addressing is extremely complicated, as are the options. The tendency has been to install LAN mail systems that use X.400 if needed, and X.400 has not gained the role it deserves. As an industry observer put it, "It is just the Esperanto used to get all these Email systems communicating with each other."

Another problem with X.400 today is that it is primarily a store-and-forward system. One of the things people like most about fax is its immediacy. After receiving a transmission confirmation, the user walks away from a fax machine with reasonable certainty that a piece of paper is in the person's machine at the other end, and that if he picks up the telephone and calls to discuss the fax, it will be there. (Actually, I'm not so convinced of this as I used to be. Some modern fax machines have multi-megabyte memories which may be used to store incoming faxes when the paper runs out. If a power failure or operator error occurs before printing, the fax is gone forever despite my fax machine's log of a successful call).

This is not always the case with X.400. Although the user may get a local transmission confirmation, it may be another 20 minutes before the message works itself across the various hierarchies of the X.400 network and arrives at the recipient's mailbox. It may find itself at the end of a queue waiting for channels to clear, or pause behind a long file transmission. The existing X.400 standard does not have the immediacy that fax users have grown accustomed to expect. This is not an insurmountable problem in many applications such as the broadcast of faxes overseas, however, and X.400 will survive and evolve as a real commercial, enduser standard. Eventually, there will be a large EDI population of users. It is a superb backbone for Email systems, and recognition of its capabilities to handle fax and raster traffic is slowly taking place.

Very little color facsimile equipment is available at present. Among the reasons for this are the complexity and cost associated with color

Fax Integration 151

scanning and printing, as well as a long transmission time. Another obstacle to color fax is a lack of standardization for areas such as color components, transmission sequence, bi-level vs. continuous tone, and encoding algorithms.

Widespread use of color fax will probably take place within the X.400 framework, because of this architecture's evolution into a multimedia Email vehicle. X.400 is, for all intents and purposes, evolving into the international standard for multimedia communications.

As requirements increase, and faxes become digitized, a form of multimedia fax will make an appearance, providing high resolution, color, working with Email, and offering the security of encryption. All the components are available now, all that is needed is for consumer demand to reach critical mass.

Fax Boards

Many fax boards are now available for PCs. Most come with dedicated software which enables the keyboard user to send, review and print faxes. In order to integrate a fax board into a voice processing application, there are two principal considerations:

> 1. The programming interface. There needs to be a well-documented set of functions which can be invoked by your software to control the reception and transmission of fax files.

> 2. The telephone interface. The typical fax board will have one or more RJ-11 jacks for a standard two-wire phone line. Optional connections may include voice bus interfaces to AEB, PEB or MVIB buses.

Among the vendors who make fax boards suitable for voice processing integration include:

> PureData (www.puredata.com) who have taken over manufacture of SatisFAXtion boards from Intel. These boards are full-length, AT-compatible cards supporting one phone line with an RJ-11 jack. Up to four boards can be installed in one PC. LAN server software is also available for routing

commands to a fax board on a separate workstation. The programming interface follows the *CAS (Communicating Applications Specification)* specification developed jointly by Intel and DCA. Connection to the RJ-11 jack can be achieved through a dedicated phone line (precluding one-call transmissions), through analog cards such as the Dialogic AMX, or by "hard-wiring" a Y-connector so that the same phone line terminates both on a speech card port and on the fax card port. Intel abandoned the CAS standard. Intel announced that as of December 31st, 1994, Intel was no longer be providing even the minimal support for independent software vendors using CAS and their SatisFAXtion series which had been provided through their BBS. Instant Information (phone 503-692 9711, fax 501-691 1948) now provides support for CAS developers using the SatisFAXtion boards, including selling drivers supporting multiple boards (where Intel had previously been offering them at no charge through their BBS). Brooktrout Technology (see later) also offers CAS-compatible drivers as an option for some of their fax board products.

Gammalink, Inc. of Sunnyvale, CA www.gammalink.com, 408-744-1430. Now a part of Dialogic Corp. Gammalink makes a range of fax board products called Gammafax, which have several integration options and an extensive C library for controlling the boards. Gammalink currently offers direct phone line interfaces and AEB bus integration only. However, Dialogic and Gammalink have announced the intention to merge, so it would not be surprising to see PEB and SCSA product offerings from Gammalink in the future.

Brooktrout Technology, Inc. of Needham, MA, 617-449-4100, also make a range of boards with a proprietary C programming interface. Brooktrout offers both AEB and PEB-compatible products.

Dialogic offers its FAX/xx family of products. This includes the PEB-based FAX/120 and the VFX/40xx series for voice/fax applications with SC bus connectivity. The FAX/120 is a full-length PEB-based board which is in effect a companion

Fax Integration

card to a D/121A or D/121B which adds fax capabilities to each of the 12 voice processing channels of the D/12x.

Features to consider when choosing a fax card include:
Which telephone trunk and voice bus interfaces are supported.
Which operating systems are supported.
What type of API is provided:

- C library (requires re-building of product with each new release)
- Device driver (allows upgrades without re-build)
- Operating systems supported (DOS, UNIX, OS/2 etc.)

Which transmission speeds and compression options are supported.

Software capabilities such as:
- ASCII text to fax conversion off-line or on-the-fly
- Graphics file to fax conversion off-line or on-the-fly
- Range of graphics file formats supported
- Ability to mix text and graphics on a page
- Automatic generation of a cover page
- Automatic generation of forms
(placing specific text information on a page which may include rules and graphics).

The CAS Programming Interface

The CAS programming interface, as supported for example by the PureData SatisFAXtion board, offers high-level queuing and event management facilities to the application program, at the expense of a rather large device driver (typically over 60Kb per channel in a stand-alone PC configuration, though more recent versions do offer an option to load some of this driver into EMS. This may be an advantage for some relatively simple systems if the features offered match the requirements of the application; in other cases, the application itself may need to implement more sophisticated queuing mechanisms, and the CAS manager may represent unnecessary overhead.

The CAS standard has been dropped by its inventors, Intel and DCA. For more details of programming interface options for SatisFAXtion boards, visit PureData's web site at www.puredata.com.

An important advantage of the CAS interface is the ability to combine ASCII text files with graphics files in PCX and DCX format in creating fax documents. The CAS manager will also automatically construct transmission cover pages, including date, time, sender and message text fields. Another useful feature is the ability to perform a high-speed, error-correcting file transfer to another device supporting CAS.

The CAS programming model deals with events. An event is a single phone call involving the fax board and a remote device such as a fax machine or another fax board. An event is one of the following types:

Send — The computer makes a call and initiates a transmission of one or more files to a remote device (fax machine or other fax board).

Receive — A remote device makes a call and initiates a transmission of one or more files to the computer. A program can obtain information about receive events which have taken place by querying the Receive queue.

Polled Send — The computer waits for a remote device to call and then starts sending information to it.

Polled Receive — The computer makes a call to a remote device and then receives a transmission from it.

Each event is assigned an *event handle* by the CAS manager for the given board. The event handle will be a number in the range 1 .. 32767 which uniquely identifies the event. No two events for the same board will have the same handle. Events for different boards in the same PC may be assigned the same handle.

Each event has an associated *Event Control File* which contains information about the event, such as the date, time, phone number, file

Fax Integration 155

name(s) etc. For events initiated by the computer (Send and Polled Receive), Event Control Files are created by the functions making the request. For events initiated by the remote device (Receive), the Event Control File is created by the CAS manager.

A File Transfer Record is included in the Event Control File for each file transfer operation associated with the event.

An Event Control File created by a Receive event will contain one File Transfer Record for each file received, the CAS manager is responsible for this process.
There are three queues of events maintained as linked lists of Event Control Files by the CAS manager for each installed board:

Task Queue   The Task Queue contains an Event Control File for each event which has been scheduled, but which has not yet been executed.

Receive Queue   The Receive Queue contains an Event Control File for each completed Receive or Polled Receive event.

Log Queue   The Log Queue contains an Event Control File for each event which has been successfully or unsuccessfully completed.

Queues are maintained by the CAS manager sorted in chronological (date and time) order.

An event retains the same event handle even when it is moved from one queue to another. For example, a Send event will be moved from the Task Queue to the Log Queue when it completes, but will still be identified by the same event handle.

A completed Polled Receive or Receive event will appear in both the Receive Queue and in the Log Queue. Two copies of the Event Control File exist in this case — if the event is deleted from one queue, it will still appear in the other.

Other events will appear in one queue only:

Completed Receive and Polled Receive events: Receive and Log Queues.

Pending Send and Polled Receive events: Task Queue.

Completed Send events: Log Queue.

If an event is in the process of being executed, it is known as the Current Event, the Event Control File for the Current Event is not in any queue.

A summary of the CAS functions is as follows:

00 Get Installed State
01 Submit Task
02 Abort Current Event
05 Find First
06 Find Next
07 Open File
08 Delete File
09 Delete All
0A Get Event Date
0B Set Event Date
0C Get Event Time
0D Set Event Time
0E Get Manager Info
0F Auto Receive State
10 Current Event Status
11 Get Queue Status
12 Get Hardware Status
13 Run Diagnostics
14 Move Received File
15 Submit Single File
16 Uninstall
17 Set Cover Page Status

The Dialogic VFX And FAX/xx

The Dialogic VFX and FAX/xx board family adds fax functionality to a voice processing system. The FAX/120 adds fax to channels on a D/12x speech card. With the LSI/120, this allows a configuration with twelve channels of voice and fax in only three full-length expansion cards. In a digital environment with a T-1 span connected to a DTI/101, 24 channels of voice and fax are possible with five cards. The FAX/40 (now discontinued) adds fax capabilities to the four voice processing channels on a D/41E card. The VFX/40, VFX/40E and VFX/40ESC are D/41E or D/41ESC cards with FAX/40 or FAX/40E daughter-cards pre-mounted by Dialogic.

The programming interface for the VFX/xx and FAX/xx is "lean and mean" — no higher-level functions such as queuing or, more significantly, file format conversion other than ASCII, are currently provided by the device driver, which requires only about 36 Kb DOS memory to control multiple boards.

Fax data for the FAX/xx must be in one of three formats: ASCII text, TIFF/F, or "raw" T.4 one-dimensional fax data. *TIFF (Tagged Image File Format)* is widely-used bit-mapped (raster) graphics file standard. Note that there are many different types of TIFF file — most will not be acceptable to the fax driver as input. The "/F" in TIFF/F refers to the class of TIFF file — Fax class, a very recent addition to the TIFF family of file formats, the specification for which is maintained by Aldus Corp. Other classes include TIFF/G for gray-scale images, for example.

If the files you wish to transmit are in a format other than ASCII or TIFF/F, such as or word processing files, you will need conversion software to create files ready for transmission. Dialogic maintains a list of third-party software vendors who provide file format conversion software supporting TIFF/F as output.

You can combine text and TIFF/F files into a single fax transmission by using *striping*. This refers to the technique of sending files one after the other making horizontal bands across the page. More than one file may appear on a single page, but the boundary is always a horizontal line, you cannot overlay two files, tile vertically or use other more complex combinations.

Fax and voice operations may be mixed freely on a D/12x channel, with the usual restriction that two multi-tasking functions (functions which take a significant time to complete) cannot be active simultaneously on the same channel.

The FAX/120 requires at least revision level A of the D/121 board — note that older D/121 and D/120 boards will not work with the FAX/120.

Chapter 8

Voice Recognition

Introduction

Voice recognition, also known as speech-to-text, provides an alternative to digit dialing for the caller to input instructions to a voice processing system. One or more words in the caller's spoken voice is analyzed by the voice recognition subsystem and a match attempted to the "known" vocabulary. The designation *Automatic Speech Recognition* or *ASR* has become a popular alternative to "voice recognition".

Voice recognition is important for the following reasons:

1. Voice recognition can be used where the caller does not have a touch-tone phone. In the US, it is estimated that more than 30% of calls are made from rotary or pulse dial phones (remember that pulse digits are not transmitted through a switch, except as clicking sounds, which creates a separate recognition problem). In other countries the percentages are higher. In other countries, the percentage may be much higher.

2. Many callers find it more natural to speak commands rather than dialing them, your system may be perceived as more "user-friendly" and enjoy a higher rate of completed transactions if it employs voice recognition.

3. Voice recognition permits "hands-free" operation, which allows the handicapped, callers on car phones and others who are unable to use their hands to take advantage of your service.

4. Vocabularies other than digits can be provided: astrological signs, months, colors, letters of the alphabet and so on. This provides a degree of flexibility not available to the touch-tone pad.

The reduced bandwidth (range of frequencies transmitted) of a telephone line, the poor quality of many telephone microphones, line problems such as echo, static and background noise combine to create a much more difficult environment for telephony-based voice recognition compared with desktop and other environments where a high-fidelity microphone or recording can be directly input to the computer. This explains many of the limitations of over-the-telephone voice recognition compared to other environments.

Voice recognition technology can be classified in three main areas, as follows.

Speaker-Independent

This is the type most commonly used in telephony applications. The recognizer is trained to accept a wide range of different voices with varying regional accents and line conditions (local, long-distance, international, background noise etc.). Speaker-independent systems have relatively small vocabularies, twelve to fifty separate words are the typical limits with the current widely-available technology. The words in the vocabulary may need to be clearly distinguishable: for example, there might be problems including both "Austin" and "Boston" in a vocabulary of major US cities, or both "F" and "S" in a vocabulary of letters. Speaker-independent systems can often be *trained* by the user to create new vocabularies, this will require collecting a wide range of different speakers to create samples of each word.

Speaker-Dependent

Not to be confused with *speaker verification*, to be described in the next section, this technology can achieve much greater vocabulary size at the expense of having to be trained to recognize one given speaker or a small group of speakers. The system may also demand narrow tolerances in line conditions (background noise, bandwidth and so on). This technology is more commonly used in desktop situations than telephony applications. A new speaker-dependent application promises to be a major market for voice processing systems in the next few years: hands-free cellular phone dialing. Optimized for a particular subscriber's voice, this technology allows the caller to dial a number without pressing the buttons on the phone, an advantage especially for car phones and the disabled.

Speaker Verification

Verification systems are designed to provide positive identification of a single speaker, usually for secure access to some resource. The emphasis is on eliminating "false positives", where another speaker is able to fool the recognizer into reporting a match. If the authorized user has a cold or other temporary speech impediment, he or she may be unable to give a recognizable speech sample.

Touch-Tone Availability

The importance of voice recognition is naturally greater in those countries where touch-tone is not widely available. The following list indicates the approximate percentage of telephones which do not have touch-tone in some major markets.

Country	% phones without DTMF
Argentina	95%
Austria	75%
Australia	75%
Belgium	50%
Brazil	10%
Canada	35%
China	95%
Columbia	95%
Denmark	15%
France	55%
Germany (West)	85%
Germany (East)	95%
India	95%
Japan	70%
Italy	90%
Luxembourg	60%
Portugal	90%
Netherlands	65%
New Zealand	75%
Mexico	95%
Norway	30%
Portugal	80%
Spain	90%
Sweden	35%
Switzerland	75%
United Kingdom	70%
US	30%
Venezuela	95%

As this table shows, the US, Canada and the Scandinavian countries have excellent touch-tone penetration, other markets have much lower penetrations. Of course, these numbers are approximate, and are likely to change as countries modernize their telephone networks. For example, in England, almost all exchanges are now digital and are capable of supporting touch-tone, it is to be expected that the fraction of telephones with DTMF dialing will increase rapidly in the next few years.

Grunt Detection

There is a primitive type of voice recognition which can be done on some basic voice processing boards: *grunt detection*. This relies on the ability of the voice board to report the presence or absence of audio on the line. The approach is to ask the caller to speak when he or she hears the desired option, something like: "if you'd like option A, say Yes now...if you'd like option B, say Yes now...", and so on. This obviously is not a very reliable method since it does not discriminate "Yes" from static noise, airplanes flying overhead, "No", "Help" and so on. However, it is much less expensive than adding voice recognition.

Multiple Matches

Most voice recognition technologies allow for multiple matches to a given audio sample. This feature reports two or more possible matches to the sample with a *score* or *confidence level* which allows the application's program to select the best match and evaluate the relative exactness of the alternative match. In many cases, of course, it will be the match with the highest confidence level that is selected. However, if this match is not a valid response, and the second most likely match is a valid response, the program could choose to accept the second match, possibly prompting for verification with a Yes/No question.

With single-word responses, matches to invalid responses are usually eliminated by specifying "subvocabularies". For example, if a menu only allows responses "one", "two" or "three", then the programmer can instruct the recognizer not to attempt a match with any other digit. This may not be possible when obtaining a string of digits such as a zip code, in which case the option of rejecting the strongest match should be considered if the response is not valid.

The User Interface

If you have an application which gets input via touch tones, you should be aware that there is much more involved in adding voice recognition than simply making spoken digits an alternative to dialed digits. The menu prompts may need different wording, especially if the user cannot interrupt the menu to speak a selection before the prompt is completed. Error handling is also a critical issue. Alan Murphy of Voice Control Systems likes to tell the story of some early studies performed by his company where they experimented with voice recognition user interfaces. The example he cites is that changing "Enter the number, one digit at a time" to "Speak the number, one digit at a time", where changing just the one word produced better response.

If the confidence level of a match is not high enough for unconditional acceptance, there are several ways to handle the situation.

> "Sorry, we didn't understand your reply. Please speak your responses slowly and clearly when you hear the beep."

This could be followed by a repetition of the menu. Alternatively, you could take one or more of the attempted matches, and try for a Yes/No verification, since Yes/No is easily understood and more reliably recognized. Something like:

> "Did you say 'Five', please say Yes or No [beep]."
>   (User says "No")

> "Did you say 'Nine', please say Yes or No [beep]."

The application should keep a count of the number of re-tries to avoid getting into a frustrating loop.

Error situations which may be detected by the voice recognizer and which may require different responses from the application include:

> Caller begins speaking too soon.

> Caller begins speaking too late.

> Caller speaks too loudly.

> Caller speaks too softly.

Caller speaks too quickly.

Caller does not speak at all (application times out).

Caller speaks fewer than the required number of responses (eg., speaks four instead of five digits for a US zip code).

Caller speaks "fillers", such as "Yes, five" or "I want five" instead of "five".

Spoken utterance is close to more than one word, such as "S" and "F", "Austin" and "Boston", "five" and "nine" etc.

Caller speaks "garbage", i.e. an utterance which does not match any alternative closely.

Remember that voice recognition is often a slower method of communicating information compared with entering touch tones, especially if the application must request verifications or re-entry of input or if long strings of digits must be spoken. Good user interface design will be critical to the usability and acceptance of your system.

Scott Instruments has developed a high-level library called GoodListener incorporating solutions to many user interface issues, such as how to collect a string of digits using a discrete rather than continuous digit recognizer. The GoodListener Cookbook, available from Scott, is a gold mine of useful ideas in terms of implementing a high quality voice recognition user interface, and is not afraid to digress into many areas of human interface engineering. Highly recommended. VPC also provides guidelines for user interface design in their documentation. (Dialogic has traditionally been sparser on the descriptive and application aspects of its product documentation, preferring to concentrate on the minimal engineering details required to make things work).

Vocabulary Features

The *vocabulary* of a speech recognizer is the set of utterances (words) which can be recognized. The following are some points to consider when choosing a vocabulary technology.

The Words In The Vocabulary.

A pretty obvious point: if you want to recognize digits, the selected vocabulary must include the digits.

Subvocabularies Provided.

There are two different types of subvocabularies: trained, and specified on the fly. If you have a digit vocabulary, but only wish to recognize "one", "two" or "three", you can ask the recognizer to restrict its matches to these options. However, a subvocabulary trained on samples "one", "two" or "three" and specifically trained to reject other digits and other random utterances will achieve better results. A particularly important example is a "Yes"/"No" vocabulary, which should be as reliable and effective as possible. Specific "Yes"/"No" subvocabularies may be provided as options as well as larger vocabularies which may include "Yes" and "No" in addition to other words ("Help", "Cancel", "Stop" etc).

Range Of Accents And Dialects.

There is a trade-off between accuracy and specific targeting of particular accents or dialects. For example, a US English vocabulary might produce acceptable recognition percentages and accuracy for any native born American English speaker. However, a vocabulary targeted on (say) New York, i.e. created from speech samples collected from native New Yorkers, will be more accurate if the target group of users of the application is based only in New York. You should ask the vocabulary vendor about their strategy in terms of selecting regional accents and dialects in the samples used to train the vocabulary. The tolerance of accents and dialects not used in the vocabulary build will also vary depending on the recognizer technology.

Tolerance Of Line Conditions.

Different recognizer technologies display different tolerances of poor line conditions: echo, static, background noise, poor telephone microphones and so on.

Continuous Or Connected Properties.

Several vendors describe some vocabularies as supporting "continuous" or "connected" speech. However, there is significant variation in the ability of these different products to accept natural or rapid speech. With some, the caller is required to leave at least a short pause between words, with others, the words can flow into each other as in natural conversation. Technically, there are three types of recognition: discrete, connected and continuous. Discrete recognizers can recognize a single spoken utterance. Connected recognizers are capable of recognizing multiple utterances in relatively quick succession, but are not expected to deal with *co-articulation*. When speaking a pair of words such as "seven nine" or "test tube" people will generally omit the consonant which starts the second word, and will in fact say something like "seven'ine" or "test'ube". This phenomenon is called co-articulation, and was invented by people to make sentences easier to say and to make life difficult for developers of voice recognition systems.

Voice Cut-Through.
This refers to the capability where a caller may interrupt a menu prompt by speaking before the prompt plays to completion. This may come in two variants, a simple "barge" where the user says anything, the prompt stops playing, a beep is played and the user then speaks a selection, or a "true" cut-through where the spoken selection itself can interrupt the prompt. True cut-through requires more signal processing horsepower than barge cut-through.

Word Spotting.
Rumor has it that the US intelligence networks have voice recognition technology that allows them to pick out certain key words from telephone conversations ("nuclear", "assassinate") so that attention can be focused on the few perhaps more relevant conversations among the many that may be monitored. This technique is called *word spotting*. More conventional voice processing applications might prefer to pick out the word "No" from "No, thank you", or the number "five" from "send me five tickets please."

Sample Size Used To Create The Vocabulary.
There will be a significant difference in quality between a vocabulary created from a sample of one hundred different speakers and one based on samples from ten thousand different speakers.

Ability To Change Vocabularies In An Application.
An application may require several different vocabularies. Depending on the memory capacity of the voice recognition board and other factors,

there may be limits on the number of active vocabularies in the system at any one time. Loading a new vocabulary may require reading in a disk file, which may be a slow process.

Ability To Create Your Own Vocabularies.

The recognizer vendor may have software tools available which allow you to create your own vocabularies. The vendors may also have services to create vocabularies from samples provided by customers: this service may be able to create better quality vocabularies than the tools provided to customers.

Adaptive And Learning Vocabularies.

It may be possible for an application to collect voice samples from the live usage of the system: these samples can later be added to the vocabulary, allowing the system to adapt to its users in a continual process of improvement.

Phonetic Description.

New techniques, just now becoming available to the mass market, allow recognition with typed descriptions of words to be recognized rather than by collecting samples. It's almost like reversing text-to-speech: by typing a description of the vocabulary, the software is able to generate a template of the sounds to be expected and produce a recognizer.

Sharing Recognizers

In many systems there will be fewer voice recognizers than phone lines. There are several reasons for this. Voice recognizers are expensive resources, and are generally only in use for a short duration compared with the length of a call. Sharing one recognizer between two or more lines may only result in a small degradation in service. Also, a recognizer may only be required as a backup in the case of a rotary call, in which case only a small fraction of calls may use a recognizer.

If the ratio of recognizers to lines is not 1:1, an algorithm must be included in the application which can allocate recognizers to lines, and handle a situation where no recognizer is available.

In outline, a suitable algorithm is as follows.

Voice Recognition

1. The application reaches a point where a spoken response will be required to the next prompt which will be analyzed by a voice recognizer.

2. A table of recognizers able to service this phone line is queried to see if a free recognizer is available. If a free recognizer is found: allocate to the current phone line; proceed to step 4. Otherwise, proceed to step 3.

3. Play an explanatory message to the caller — "please wait while we access your information", or a commercial, or other message, and wait for a recognizer to become available. If a recognizer does become available within a set time limit, go to step 4. If no recognizer becomes free, abandon the call or go to an alternative part of the application.

4. Play the prompt inviting the caller to speak a response.

5. Obtain and analyze the response.

6. Mark the recognizer as free.

Looking at this method, you might think that there is wasted recognizer capacity since the (idle) recognizer is tied up in step 4 where the prompt is being played. This is necessary, however, to avoid a situation where a prompt is played but no recognizer is free, resulting in the caller speaking a response which cannot be analyzed.

Chapter 9

Pulse Recognition

Introduction

Automated voice processing systems have become heavily dependent on touch-tone detection for allowing the caller convenient interaction and input. However, the majority of telephone users world-wide do not have touch-tone phones. The United States, Canada and the Scandinavian countries are well-equipped with DTMF phones; the rest of the world uses rotary or push-button pulse dial phones. From 70% to 95% of the population in major markets such as Germany and the United Kingdom are without DTMF.

Voice recognition offers one possible solution. Another option is pulse digit recognition, which is based on direct detection of digits dialed by a caller with a rotary or pulse telephone.

Pulse digits are dialed by interruptions in loop current on an analog line. A short pause between interruptions tells the receiver to keep counting the current digit, a longer pause tells the receiver that the digit has been completed.

Pulse digits are sent by interrupting loop current briefly. This example shows 23 being dialed. Each interruption of the current counts as one pulse. A longer pause between interruptions indicates the end of a digit. When the rotary dial is at rest, or when all digits have been sent by a push-button pulse phone, loop current is flowing in order to allow sound to be transmitted as variations in the current.

The digit zero is sent as ten pulses.

The major problem facing pulse recognition technology is that switches do not transmit changes in loop current across a connection, the recognizer must identify and count the audible clicks which are carried by the audio channel. This turns out to be very hard to do. Imagine, for example, having to distinguish the single click representing the digit 1 from static on the line, or recognizing pulse digits while a menu prompt is being played. High accuracy is hard to achieve, and may require significant fine-tuning of a pulse digit recognizer for a particular environment.

The crudest form of pulse recognition is simply a "click counter" which does not take into account the details of the audio envelope. More sophisticated products use DSP-based signal analysis to distinguish a pulse click from other sounds on the line.

Pulse Recognition Vendors

There are four major vendors of pulse recognition equipment at the present time:

> Dialogic Corp., Parsippany NJ (+1) 201-993 3000.
> Dialogic has pulse detection firmware available for several board models.
>
> Aerotel, Holon, Israel.
> Phone (+972) 3-559 3222, fax (+972) 3-559 6111.
>
> Pika Technologies, Kanata, Ontario, Canada
> Phone (+1) 613-591 1555, fax (+1) 613-591 1488.
>
> Teleliason, St. Laurent, Quebec, Canada
> Phone (+1) 514-333 5333, fax (+1) 514-337 6575.

Pulse Recognition Equipment

Pulse recognition equipment is generally installed "in-line" between the phone company trunk. The recognized digits can be conveyed to the voice response system (VRU) in three ways: by conversion to DTMF tones, in which case the equipment is called a *pulse to tone converter*; by using a separate communications line (probably RS-232) which is separately connected to the VRU; or by using a PC add-in card.

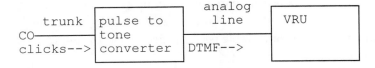

A pulse to tone converter produces DTMF digits corresponding to the recognized pulse digits.

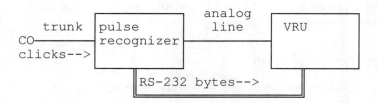

A pulse recognizer may alternatively use a serial port to convey the recognized digits to the VRU.

Features to look for in such equipment might be put into the following categories.

Accuracy.

The detection accuracy rate of the recognizer for the target region, and the types of errors are made when the recognition is not correct.

Tolerance.

By "tolerance" is meant the range of different pulse telephone types that may be used in a single application. The range of pulse make/break times which are accepted, for example. Different countries have different specifications for the exact range of timings which is acceptable. One application may need to interact with rotary callers from several countries, which would require a broader tolerance.

Training.

Some recognizers require *zero training*, where the caller is requested to dial a zero (ten pulses), allowing the recognizer to adapt to the characteristics of that particular caller's telephone. Some recognizers are designed to function without training. Zero training may be an advantage if it increases accuracy and reduces talk-off, or a disadvantage if another product is able to achieve comparable performance without training.

Trunk Types Supported.

Since the recognizer is connected directly to the trunk, it must have an interface compatible with the trunk type: analog, DID or an appropriate digital interface.

Pulse Recognition

Disconnect Supervision.

Since the recognizer is connected to the trunk, the recognizer is responsible for detecting a caller hangup and transmitting a signal to the VRU. The disconnect detection options provided must be sufficient for the PBX or public network in the region to be serviced.

Flash-Hook Transmission.

Just as a disconnect signal must be passed through the unit from the trunk to the VRU, a flash-hook from the VRU must be passed through to the trunk if this is needed to effect a call transfer or other service from the PBX or other switch type.

Dial Cut-Through.

The recognizer may or may not be able to detect digits while a prompt is being played by the VRU.

Indication Of Pulse Detection.

When using a pulse to tone converter, it may be desirable to send a special sequence of one or more DTMF tones to the application to indicate that the caller has a pulse phone. Otherwise the application will not be able to distinguish touch-tone from rotary callers, and will not have the opportunity to change the user interface accordingly.

Talk-Off Suppression.

Just as the frequencies in a human voice can trigger a false positive from a DTMF detector, so can a pulse recognizer be fooled into detecting digits when a human voice is on the line. As you might expect, the false detection of the digit 1 is the most likely. Several strategies are available for reducing pulse talk-off, check with the vendors for the options offered by their respective products. A recognizer with an RS-232 link may offer programming commands which can be sent from the host VRU to select such options on the fly.

The User Interface

As with voice recognition, the particular characteristics of pulse detection require special consideration when designing the user interface. An application which supports DTMF, voice recognition and pulse recognition should have three separate flow-charts to handle each type of call.

The first point to realize is that the accuracy and reliability of pulse recognition is limited. 95% or better accuracy is good. This means that the application must be tolerant of errors and must verify important input with the caller to make sure that an error has not gone unnoticed.

There are several strategies which can be used to improve the reliability of an application using pulse recognition. These techniques may not be necessary, depending on the recognizer technology used.

Use short prompts, collect digits following a "beep".

Asking the caller to dial a digit at the end of a prompt, for example, "Please make your selection following the tone," reduces the problems of talk-off and play-off. By reducing the span of time that recognition is used, it reduces other types of spurious digit reporting such as static on the line.

Avoid the use of the 1 digit.

Using 1 is popular when touch-tone input is expected, however 1 is the most likely pulse digit for a "false positive" detection. When requesting phone numbers, credit card number etc. this is obviously not an option, but menu choices might be adjusted to use other digits. It may be possible to program the recognizer to ignore the digit 1.

Avoid the use of consecutive digits.

The most common pulse recognition errors are "off-by-one" mistakes where the reported digit is one higher or one lower than the dialed digit. This tends to be a characteristic of a particular telephone, and may be repeated throughout a particular call. One way to minimize the damage is to space the digits used for menu selections, for example "for selection A, dial 3, for selection B, dial 6, for selection C, dial 9". You could then accept a reported 2, 3 or 4 for A, 5, 6 or 7 for B and 8, 9 or 0 for C. A more critical menu requiring a Yes/No response could ask for 3 for Yes, 9 for No and use a similar scheme where a reported 5, 6 or 7 was rejected as invalid input.

Chapter 10

Text To Speech

Introduction

Text-to-speech (TTS) is the process of generating spoken phrases from input presented in the form of a text string or text file, typically in the form of ASCII characters. A primitive form of TTS is described in the chapter on Speaking Phrases. Mature TTS systems are able to generate understandable speech from almost any text file. The artificial voice will still be recognizably "robotic", the state of the art at the time of writing does not mimic human speech well enough to fool a listener that a real person is speaking or made a recording.

TTS allows a voice processing application to convey information to the caller which was not recorded in speech files when the system was created.

TTS also allows information to be stored much more efficiently. One hour of speech consumes ten megabytes or more of storage in the form of audio files. The same information would typically require 50 Kb or less in the form of ASCII text, just 0.5 % of the space required for the audio files.

Major application areas for TTS within voice processing include:

News Services.

New information is distributed as large amounts of text data. It can be made available instantly through the telephone.

E-Mail And Fax-To-Speech.

Integrated messaging, i.e. combining voice mail, e-mail and fax into a unified system, can be achieved when TTS enables text information to be retrieved via the telephone. E-mail messages can be read, and faxes can be converted to text using Optical Character Recognition (OCR) software and spoken also. (Berkeley Speech Technologies, Inc. claims a patent on fax-to-speech conversion).

Database Retrieval.

Large databases may make pre-recording of all items impractical. For example, an automated dispatcher system for delivery trucks might store the names of tens of thousands of streets in a number of cities. It may be more practical to use TTS to speak the street names and addresses rather than recording tens of thousands of different street names as separate files.

On-Line Searches.

Searches of large databases such as medical information and legal cases, traditionally performed with a computer and modem, can be performed

Text To Speech

using the ubiquitous telephone provided that the information can be spoken back to the caller once the request has been processed. Once again, the sheer volume and volatility of the information makes pre-recording impractical.

A significant issue with TTS is its cost. The per-channel cost of most TTS solutions is currently around four times or more the cost of basic voice processing services such a voice record and play-back, DTMF detection and generation, answering and disconnecting calls, and so on.

TTS Vendors

The major vendors of TTS technology for voice processing are:

> Berkeley Speech Technologies, Berkeley, CA.
> Phone 510-841 5083, fax 510-841 5093.

> Digital Equipment Corporation, Maynard, MA.
> Phone 508-493 5111.

> First Byte, Torrance, CA.
> Phone 310-793 0610, fax 310-793 9611.

> Infovox, Solna, Sweden.
> Phone +46 8-764 35 00, fax +46 8-735 78 76

> Lernout & Hauspie, Burlington, MA
> Phone (+1) 617-238 0960

The following sections summarize some of the product offerings from some of these vendors.

Berkeley Speech Technologies

Berkeley offers Dialogic-compatible solutions under its BeSTspeech trademark. BeSTspeech is sold as firmware (software on a diskette which can be downloaded to the memory on an add-in card) for the Dialogic Antares platform and as stand-alone software. When running BeSTspeech, the Antares is a TTS generator but can not perform standard voice processing functions. A product designed to run on the

D/4x platform in a similar manner is planned for release by the time this book appears in print. This will have a minimum configuration of two D/4x boards connected via an AEB bus. A version of BeSTspeech is also available for Natural MicroSystems VBX DSP-based voice board line.

The currently available version of BeSTspeech running on the Antares offers 2 to 8 TTS generators. More boards may be added if more simultaneously active channels are required.

TTS generators are currently available for the following languages:

English
German
French
Spanish
Japanese.

Digital Equipment Corporation

DEC's Assistive Technology Division produces the DECTalk PC board, which is a full-length board containing one TTS generator. Up to four boards may be installed in a single PC- or AT-compatible computer. An RCA jack is provided for the audio output line: there is no voice bus connector on the board. To attach a DECTalk PC board into an AEB configuration such as a D/4x, an AMX board and an adapter device known as an AIA/2 is required. To incorporate a DECTalk PC into a PEB-based system, an MSI with at least one station card and an AIA/2 is required. An advantage of this board is that it does not require an IRQ.

DEC currently offers generators for:

English
French
German
Spanish.

First Byte

First Byte's ProVoice product, unlike the others described here, runs on the host PC's 80x86 processor. This eliminates the need for additional hardware, but adds a significant burden on the processor which may limit the capacity of the overall system, for example in the number of lines which may be serviced simultaneously. At the time of writing, ProVoice is distributed as an object code library for Microsoft Visual C++. The application must be re-written to call ProVoice's *getevent* API call instead of the Dialogic standard *gtevtblk* routine.

First Byte currently has generators for:

    English (North American)
    French
    Spanish.

Products under development include British English, Italian and Japanese.

Infovox

The Swedish company Infovox produces half-length expansion cards which are PC- and AT-compatible and carry a TTS generator. A single RCA jack is provided for the audio output. Infovox boards may be integrated with Dialogic AEB or PEB systems in a similar way to the DECTalk PC board.

Languages currently offered include:

    English (North American)
    English (British)
    French
    German
    Italian
    Spanish
    Swedish
    Danish
    Icelandic.

TTS Methods

This section provides a brief outline of the techniques used by a typical TTS generator in the order they are typically used in processing input; the details vary among the different generators.

Text Input.

The ASCII text to be spoken is first read from a file or memory buffer. An input rate of about 10 bytes per second is required to give the average of three spoken words per second which is typical for natural speech.

Text Normalization.

There are numerous special cases to be identified. For example, should "Dr." be spoken as "Doctor" (a person's name) or "Drive" (an address)? As this example shows, considerable intelligence may be required in the generator to make the determination. Numbers also require special attention. 1994 should probably be spoken as "Nineteen ninety four", unless it's the number of a taxicab, but $1994 should be spoken as "nineteen hundred ninety four dollars." Following text normalization, the phrase "Dr. Ron Smith won $1984" will be translated to an internal representation equivalent to "Doctor Ron Smith won one thousand nine hundred eighty four dollars".

Exception Dictionary Lookup.

Dictionaries are used to isolate exceptions to common pronunciation rules. The exact stage where the dictionary is used depends on the detailed strategy used by the generator. Berkeley Speech Technologies literature quotes the example of the English word "of", which is pronounced with a "v" rather than with an "f" sound. This exception is identified early because of Berkeley's reliance on rule-based translation of spellings to phonetic spellings, which takes place in the next step. The importance of the extent and quality of exception dictionaries varies greatly with the language and application. English has relatively many exceptions to spelling pronunciation rules, especially when proper names in North America must be pronounced: place names and peoples' names are derived from many languages other than English. A language like Norwegian, where spelling rules were defined much more recently and which is therefore more consistent, may require less attention from exception dictionaries.

Text To Speech

Conversion To Phonetic Spelling.

Webster's Collegiate Dictionary provides a phonetic spelling of every word: "voice", for example, is phonetically spelled as \'vóis\. A phonetic alphabet is used to describe in detail how the word is spoken. For example, hyphens are used to separate syllables. Since "voice" is a single syllable, no hyphens are used. The high set mark ' indicates that the stress is at the beginning of the word. (You say "VOiss", not "voISS", to use an informal phonetic spelling.) The various vowel and consonant sounds are assigned specific marks. The "s" in \'vóis\, for example, represents the hard "s" as in "kiss" rather than the soft "s" as in "noise".

The TTS generator must arrive at a phonetic spelling for each word in the phrase to be spoken. This can be done using rule-based and look-up techniques. Look-up is conceptually simpler: the generator has a table (Webster's Collegiate Dictionary would be one option) which converts conventional spelling to phonetic spelling. The main disadvantage is the large amount of memory required. Remember that each variant of each word must be included: it is not enough to store speak=\'spék\: the words speaks, spoke, spoken, and speaking must also be translated. The rule-based approach attempts to derive the pronunciation from the conventional spelling. For example, it would be a rule that "c" would become phonetic "k" (as in car, record), unless followed by "ei", in which case it becomes phonetic "s" (as in ceiling or receive). A combination of both techniques may be used.

Phonetic Modification.

The phonetic spelling derived may not create the precise sound which would be made by a person. For example, the "t" sound in "cat" is subtly different from the "t" sound in "tom", but this is not usually recognized by linguists. The phenomenon of co-articulation should also be considered. A person will not speak two "t" sounds when saying "test tube" or two "n"s in "seven nine".

Inflection.

The overall "shape" of the phrase determining the pitch and weight (emphasis) of the voice is known as the *inflection* of the phrase. One factor determining the inflection will be whether the phrase is an exclamation, question or statement. The three words "Hello!", "Hello?" and "Hello." have essentially the same phonetic components but different inflections.

Voice Generation.

All the information is now available to synthesize the voice. This may be done by concatenating together stored elements and/or by using algorithms which use parameter-driven rules to create the final waveform. Stored elements may include *phonemes*, which are the smallest units of identifiably different sounds, and *disyllables*, also called *diphones*, which are pairs of sounds such as \vó\ and \is\, which might themselves be concatenated to make \vóis\.

TTS Features

The first and most obvious TTS feature is the quality and clarity of the generated voice. While this is of course a highly subjective issue, a group of listeners would probably agree on "winners" and "losers" among today's offerings. Clearly, live demonstrations of the available generators are an essential first step in choosing the technology for a given application.

Other important considerations include the following.

Integration Method.

The TTS generator may require significant additional hardware, such as a D/12x board or board + AMX + AIA/2 or board + MSI + station card + power supply + AIA/2 combination. This can add to costs and require more slots. On the other hand, a solution running on the host CPU may limit the number of calls which can be processed simultaneously.

Number Of Simultaneous TTS Channels.

The expandability of the system in terms of the maximum number of recognizers which can be supported in one computer may be a significant limitation.

Voice Characteristics.

The generator may provide several options for determining the style of the spoken voice, including:

Text To Speech

Volume.
Speed.
Gender
  (male/female).
Character
  (there may be many male/female options, or tunable parameters determining the character of the voice).
"Excitement" level
  (a given voice may sound subdued and professional for giving stock quotes but generate more excitement when hosting a game show).
Language
  (ability to switch languages in mid-sentence or from phrase to phrase, c'est à dire).

Embedded Commands.

It may be possible to embed commands in the stream of text being spoken. A command may be identified by an unusual string of characters, such as "[:" which is used by DECTalk PC. The command itself will be used to control some aspect of the speech generation: for example, the string "[:punc all]" in the text would request that the generator speak punctuation marks, so that "then, as now" would be spoken "then comma as now". The embedded command "[:punc none]" could be used later to turn this option off. Embedded commands may be the only method offered to alter voice characteristics.

Asynchronous Commands.

A typical asynchronous command is "stop speaking now". The essential point is that the command is executed immediately, in contrast to an embedded command which is executed only when all the preceding text has been spoken. Asynchronous commands may be provided for controlling voice characteristics such as speed and volume. Asynchronous commands are essential if touch-tone commands are to be used to interrupt or modify the flow of speech.

Ability To Speak Proper Names.

Proper names present a special challenge to TTS generators due to the large number of special cases. I have a colleague with last name McCabe. Attempts by commercial products to pronounce her name have ranged from almost perfect to "Em Cee Cabe". If name or address pronunciation is an important part of your application, careful testing of representative samples for acceptable results is recommended.

Chapter 11

International Issues

Introduction

There are a number of issues to be considered when taking a voice processing system developed for the US to be installed in another country. The following sections cover some of the major issues related to the voice processing technology itself: the product developer will of course also need to consider translating manuals and screen prompts, training local distributors, marketing and other business issues.

Approvals

The Post Telephone and Telegraph (PTT) administrations, usually controlled by their governments, provide telephone and telecommunications services in most countries where these services are not privately owned. In CCITT documents, these are the entities referred to as "Operating Administrations."

It is not a simple thing to obtain approval from the PTTs to sell and use telecommunication equipment of any kind in their countries. The world is far from being one in the field of telecommunications. Meeting international requirements typically means providing hardware and software modifications to the product, unique to each country, and then going through an extremely rigorous approval process that can average between six to nine months. Products are required to meet both safety and compatibility requirements.

Regulatory approval (also known as homologation) may be possible at the level of the telephone line interface, of complete components such as boards, or may involve a complete computer system including chassis, power supply and add-in boards.

National safety and telecommunication standards exist for each country, some are similar to US standards and others are quite different. In the US, the FCC is responsible for the approval of telecommunication equipment. Canada's CSA and DOC standards are similar to those used by the FCC. However, in the United Kingdom, the BABT requirements are quite different.

Due to the relative youth of the voice processing industry, it is not always possible to obtain approval for just the voice processing card, and in that case it is necessary to approve a complete system. In Japan, where it is not possible to obtain card approval, Dialogic Systems KK, the Japanese subsidiary of Dialogic Corporation, provides a range of approved systems containing voice processing cards.

Naive Users

The users of your system may not be as familiar with automated call processing as in the US. They may never have been transferred into voice mail or used a banking by phone service. This may suggest more extensive explanation of the operation of the system, fewer menu options, and so on. A field trial with prospective users of the application is highly recommended.

Algorithms

Algorithms for constructing numbers, dates etc. from pre-recorded vocabularies of digits, months and so on will almost certainly require modification. It is not enough simply to re-record the component vocabulary files. For example, in Danish "25" is spoken as "Five and twenty", literally translated. Or worse (for the English-speaking programmer), in French, "82" is "Quatre vingt douze", literally "four score twelve". Other complications may arise from variations which just don't exist in English. For example, English has only one kind on "one": the "1" required for "one house" is the same as the "1" for "one dog". However, in Danish, you need two different kinds of "one": "et" (neuter), as in "et hus" (one house) and "en" (gendered) as in "en hund" (one dog). In French, there is "une" (female) as in "une maison" and "un" (male) as in "un chien". So even if you revise your algorithm for speaking a number, you will still need a facility in your application for specifying the gender or whatever of the thing that you are counting. This can have surprisingly broad consequences. For example, in an automated order-entry system, you will probably have a database of items which may be ordered. In implementing this database for France or Denmark, you would need to add a new field to the database indicating whether the word for the item is gendered/neuter or male/female. If your system needs to support more than one language, it may need such a field for each such language.

Of course, you can solve these linguistic problems by using the international standard Esperanto language for all your products (don't plan on selling many systems, though).

Telephone Numbers

If your application needs to store and interpret telephone numbers, it will need to adapt to the wide variation of numbering schemes used in different countries. The consistent US/Canadian system of a three-digit area code, and seven-digit number starting with a three-digit local code is not found in many other countries. For example, in England, area codes vary from two to five digits, and local numbers vary from three to seven digits. Methods of reaching an operator, making a local or long distance call by dialing a prefix, reaching directory information or emergency services also vary widely.

Bi- And Multi-Lingual Applications

Countries or regions which speak more than one language (examples: Canada, Belgium, Switzerland) may require systems to provide an initial greeting where the caller can select his or her preferred language. This complicates even further the algorithm and scripting considerations. Your choice of languages offered may even become a political issue in some regions.

Script Translation

This is perhaps the most obvious requirement: translation of menus and other pre-recorded messages to one or more non-English languages. As with English systems, questions of the gender of the speaker, local dialects and accepts, wording of the prompts and recording technology used will all need careful consideration. The issue of naive users and multi-lingual requirements may prompt a re-design of parts of your script and call flow.

Rotary Phones

Few callers may have touch-tone phones. Some major markets have touch-tone penetration as low as 5%. This may require the use of pulse digit recognition, voice recognition or other alternative input methods.

Trunk Types And Signaling Protocols

The type of trunk supported may require different hardware and different signaling technologies. For example, digital E-1 trunks vary widely in the details of the signaling used to accept and initiate calls, even within a given country. There may be little or no disconnect supervision, or disconnect supervision may require tone detection rather than loop current monitoring. You should examine each point in your application where the VRU interacts with a remote telecommunications device: accepting an incoming call, dialing an number, detecting hangup, transferring a call, supervising a transfer, analyzing call progress, collecting DID digits and so on; each of these operations may require modification to adapt to local protocols.

Digital trunks in particular provide significant challenges. E-1 trunks, while similar in principle to T-1, vary widely where the details of call protocols are concerned. An application needs to be aware of and control many low-level details of the protocols for accepting, making, transferring and disconnecting calls. Each country has its own variation on the theme and may require a significant development effort on the part of the systems developer.

ISDN appears to be gaining wider usage in Europe than within North America at the present time. ISDN is often provided on the same type of trunk as E-1: 2.048 MHz digital lines with 32 time-slots. ISDN also varies widely in its detailed implementation.

Regulatory Issues

There may be specific requirements for particular applications. For example, pay-per-call systems in the USA are required to play a 20-second explanation of the charges associated with the system; the caller may hang up in this period and not be charged. There are analogous requirements for different types of applications in different countries. Out-dialing is often subject to particular scrutiny. In the US, a business conducting automated dialing may be required to have a prior relationship with the called parties. In Belgium, the following restrictions apply to the repetition of a call by an automated out-bound dialing application. The application must wait for 5 seconds before retrying the number for the first time, and then must wait for at least 1 minute between subsequent attempts. No more than 4 attempts should be made within one hour of the initial attempt, although in the case of an alarm system up to 15 attempts may be made.

The application developer should ensure that the application fully complies with any other local regulations, particularly in the case of automated out-bound dialing.

Chapter 12

E-1

Introduction

Many countries outside of the US and Canada use E-1 rather than T-1 digital trunks. E-1 is similar to T-1 in principle, but quite different in the details.

The following table outlines the important features of E-1 with the corresponding T-1 features listed for comparison:

Feature	E-1	T-1
Number of time-slots	32	24
Number of voice channels	30	24
Where used	Europe, Asia	US, Canada
Tone signaling	R2/MF, DTMF	DTMF, MF
Digit dialing	R2, pulse, DTMF	DTMF, MF
Voice encoding method	8 KHz 8-bit PCM	(Same)
PCM scale	A-law	Mu-law

The "PCM scale" refers to the way voice samples are converted to volume. Each 8-bit binary value sent on a digital trunk is converted to a loudness or volume value by referring to a table:

8-bit Sample	Volume
0000000	0
0000001	3
0000010	5
0000011	8
...etc...	

The conversion tables used by T-1 and E-1 are different. The table used by T-1 is called the *Mu-law* or *µ-law* table, the table used by E-1 is called the *A-law* table.

Unfortunately, details of the E-1 protocol differ from country to country and even within a given country.

Compelled Signaling

Compelled signaling is often used to convey information between two devices connected via an E-1 trunk. To understand compelled signaling, let's begin by reviewing the more familiar (non-compelled) signaling on T-1 and analog trunks in the US. Digits and other information is sent by using DTMF or MF tones. ("Regular" MF tones as used in the US are sometimes known as *R1/MF tones*).

DTMF and MF tones are *dual tones*, sounds composed of two pure frequencies played at the same time for a short duration. If information is to be transmitted, one device gets ready to receive the digits (in the jargon, *enables a DTMF or MF receiver*). There are standards which determine how quickly digits may be sent, what the gaps should be between the digits, and so on. It is up to the receiving equipment to make sure that it can detect digits reliably under these conditions.

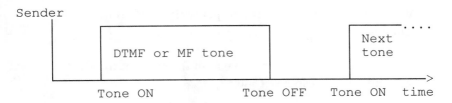

Sending DTMF or MF digits is done simply by sending a sequence of dual tones with a short gap between then.

Compelled signaling adds audience participation to the simple DTMF/MF signaling model. The receiver acknowledges each tone that is sent. The acknowledgment itself takes the form of a tone.

This may be visualized in the following diagram:

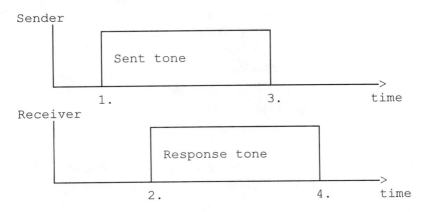

The basic element of compelled R2/MF signaling is the exchange of a pair of tones between a sender and receiver.

The sequence of events is like this (the step numbers refer to the above diagram):

1. The sender begins transmitting a tone.

2. The receiver detects the tone. In response, the receiver begins transmitting a response tone.

3. When the sender detects the response, it turns off the original tone.

4. When the receiver detects the end of the original tone, it stops sending the response tone.

The advantages of this signaling method are:

Since it waits for acknowledgment of each tone, the sender automatically adapts the speed of the transmission to the ability of the receiver to detect the tones and to the speed of the link. On an international link, for example, there may be a significant delay.

Since tones are sent in both directions, there is an opportunity for the receiver to communicate information back to the sender. For example, the receiver of a DID call may say "that extension is busy" back to the phone company, which can then send a busy tone back to the original caller.

Types of R2/MF Signaling

There are two main types of R2/MF signaling protocols:

CCITT Standard R2/MF, variants of which are used in most countries using E-1. (CCITT is an international standards body for telecommunications).

Socotel Signaling, variants of which are used in France and Spain. (Typically for R2 signaling, the protocol differs between these two countries).

E-1 197

The main differences between the two protocols are:

> In the CCITT standard, response tones themselves carry information. There are 15 different possible response tones, each one conveys a different instruction or piece of information back to the sender. The handshake is always in the same direction once started: the sender always initiates by sending a tone, the receiver responds with a "backwards" tone.
>
> On an in-coming call, the phone company is always the sender and the VRU is always the receiver. The call starts with the phone company sending a tone which is the first digit of the dialed number. The typical sequence for an in-coming DID call will be:

Phone Company	VRU
Sends tone=digit 1	Response tone=send next digit
Sends tone=digit 2	Response tone=send next digit
Sends tone=digit 3	Response tone=send next digit
Sends tone=digit 4	Response tone=got all digits, send caller category
Sends tone=category	Response tone=done, put call through

> In Socotel signaling, the response tone is always the same tone: a single tone (tone containing just one pure frequency) of 1900 KHz (France) or 1700 KHz (Spain). Since the receiver cannot convey any information other than "I got that tone" by sending the acknowledgment tone, the handshake has to "change directions" if information is to be sent back to the original sender. The change of direction happens by agreement on a pre-arranged sequence of tones which will be exchanged:

Phone Company	VRU
	Sends tone meaning "please send DID digits"
Sends tone=digit 1	
Sends tone=digit 2	
Sends tone=digit 3	
Sends tone=digit 4	
	Sends tone meaning "got those digits"
	Sends tone meaning "please put the call through".

Each tone sent by the phone company or the VRU is acknowledged by sending the standard response tone.

This is the typical sequence of tones which are exchanged to accept an incoming Socotel DID call. As this sketch shows, the VRU acts as a sender three times and as a receiver four times.

E-1 Signaling Bits

Signaling bits are used to carry the current state of the connection: is there a call in progress, or is the line idle. Changes in the signaling bits indicate:

- Start of in-coming call (ring)
- Start of out-going call (seize)
- Caller hang-up (disconnect)
- VRU hang-up (disconnect)

T-1 channels carry two signaling bits which can take the value 0 or 1. These bits are known as the A and B bits. Usually, the B bit is kept equal in value to the A bit and carries no additional information. Most T-1 installations use the following meanings for A bit values:

A=1   Connection is active; call in progress.
A=0   Connection is not active; line is idle.

E-1 channels carry four signaling bits, named A, B, C and D. The C and D bits are usually held at constant values, and an application can ignore them.

The exact meaning of the A and B bits may vary from region to region. Certainly the CCITT standard and Socotel meanings differ.

E-1: The Details

E-1 is a digital transmission standard which is similar to T-1 but differs in the details. There are 32 channels (time-slots) on an E-1 connection, these are time-division-multiplexed into a single 2048 Kbps bit stream.

R2/MF signaling is an international standard signaling protocol which is widely used on E-1 trunks in the Europe and the Far East to implement enhanced services such as ANI and DNIS. R2/MF may also be encountered in some analog services.

R2/MF is an elaborate protocol using a number of dual tones. The tones are similar to DTMF digits but different in detail. *Compelled signaling* refers to a method of sending tones where each tone sent results in a reply tone from the equipment receiving the tone.

E-1 And T-1 Trunks Compared

Time-division-multiplexing or *TDM* is the process used by both T-1 and E-1 to combine many channels into one stream of data. The data is transmitted as a series of *frames*. Each frame is divided into 24 (T-1) or 32 (E-1) different *time-slots* which are 8 bits long. A complete frame is then 24 x 8 or 32 x 8 bits long, together with a one or more extra bits for synchronization. The 24 or 32 channels are combined at one end of the trunk and split apart at the other end. This process happens so quickly that each conversation is seamless.

A typical T-1 or E-1 frame looks like this:

```
01010011  10001001  11110010  ...  11111101     1..
Slot 1    Slot 2    Slot 3         Slot 24/32   Sync bit(s)
```

Typical frame transmitted on a digital trunk consists of one set of samples from each conversation.

Each "slot" is an 8-bit sample of the amplitude of the sound in the conversation, encoded using PCM (Pulse Code Modulation). There are 8,000 samples transmitted per second. The number represented by the 8 bit sample is a direct measure of the loudness (amplitude) of the sound, converted on a non-linear scale. Unfortunately, E-1 and T-1 use different scales to do the conversion: the scale usually used by T-1 is called the Mu-law scale, E-1 generally uses A-law.

T-1 or E-1 data is simply a rapid succession of frames.

There are two streams of data in an E-1 or T-1 connection: transmitted data and received data, each of which is a series of frames like the one described above. Unlike an analog connection, where there is simply sound on the line represented by the amplitude of the current flowing, digital connections carry two completely separate channels. From the point of view of the VRU, the receive channel is usually the caller's voice and the touch tones he or she sends, the transmit channel is the one used to play messages or dial numbers to the phone company.

The Dialogic PEB and SC buses use a similar TDM system to transmit conversations between voice boards.

Digital trunks use signaling bits to play the role of loop current in an analog connection. If loop current is flowing, an analog connection is active. If no current is flowing, the subscriber is on-hook and there is no connection. If the CO applies ring voltage to the line, an incoming call is being signaled.

On T-1 trunks, a single bit, known as the A bit, plays a role similar to loop current. If the A bit is 1 (one), the connection is active. If the A bit is 0 (zero), there is no connection. (Remember that there are two A bits: the A bit received from the CO, and the A bit transmitted by the CPE. These two A bits do not necessarily have the same values). To signal an incoming call, the CO changes the A bit from 0 to 1. To terminate a call (hang up), the CPE changes the transmitted A bit from 1 to 0.

In fact, T-1 trunks usually carry two signaling bits: the A bit and the B bit, but the B bit is usually held to the same value as the A bit and carries no additional information.

The PEB bus only carries a single signaling bit, which is closely analogous to the T-1 A bit.

T-1 trunks generally use what is known as *robbed-bit* signaling to transmit the A and B bits. Every sixth and twelfth frame that is transmitted has a slightly different interpretation. The least significant bit from each time slot contains the A and B bits respectively. Of course, this changes the transmitted sound slightly, but the difference is too small to hear.

```
         A          A          A                     A
    00000111   00101001   00000000   ...   00010101       1
    Slot 1     Slot 2     Slot 3          Slot 24     Sync bit
    A=1        A=1        A=0             A=1
```

Typical sixth frame using T-1 robbed-bit signaling. The least significant bit position in each time-slot, marked A, is "robbed" from the audio to carry the A bit value for that channel.

```
         B          B          B                     B
    01010011   10001001   00000000   ...   11111101       1
    Slot 1     Slot 2     Slot 3          Slot 24     Sync bit
    B=1        B=1        B=0             B=1
```

Typical twelfth frame using T-1 robbed-bit signaling. The least significant bit in each slot carries the B bit value.

E-1 trunks carry four signaling bits, named A, B, C and D. In the T-1 tradition of only using half the available bits, C and D usually carry no useful information (they are kept at constant values), but the AB combination does carry important signals. E-1 time-slots 0 and 16 are reserved for signaling information. Since a separate channel is used to carry the ABCD bits, this is sometimes called out-of-band signaling.

The following table compares the major features of T-1 and E-1.

Feature	T-1	E-1
Data rate	1544 Kbps	2048 Kbps
Sampling method	8 KHz 8 bit PCM	(same)
PCM scale	Mu-law	A-law
Time-slots	24	32
Conversations	24	30
Signaling bits	AB	ABCD
Signaling method	Robbed-bit	Slots 0 and 16
Digit signaling	DTMF or MF	R2/MF, DTMF, pulse

R2/MF Tones

R2 signaling tones are sounds sent in the usual audio channel of a connection. First we'll introduce the terminology which is often used in discussing R2/MF.

R2 signaling takes place between two *registers*: an *outgoing register* and an *incoming register*. Typically the outgoing register will be the phone company (CO) and the incoming register will be the VRU (Customer Premise Equipment, or CPE).

Signals (tones) sent from the outgoing register are called *forward signals*.

Signals (tones) sent from the incoming register are called *backward signals*.

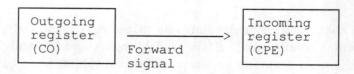

Forward signals are tones sent to the VRU by the CO.

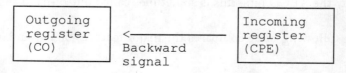

Backward signals are tones sent to the CO by the VRU.

Each forward signal from the CO results in a backward signal as a response from the CPE. This happens with the following steps:

1. The exchange of tones starts when the CO begins sending a forward tone.

2 As soon as the forward signal tone is detected by the CPE, it responds by sending a backward tone.

3. When the CO detects the backward tone, it stops the forward tone.

4. When the CPE detects that the CO has stopped sending, it stops also.

5. When the CO detects that the CPE has stopped sending the backward tone, it may begin sending the next forward tone.

This may be visualized in the following diagram:

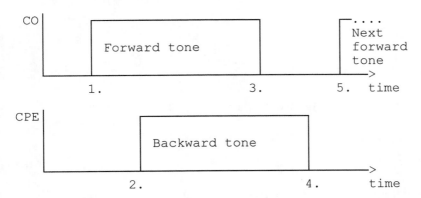

The basic element of compelled R2/MF signaling is the exchange of a pair of tones between the CO and CPE.

Typically, the forward tone will be one DID digit, and the backward tone will be a tone meaning "I got that digit, please send the next digit."

R2 Dual Tone Frequencies

R2 tones are dual tones, i.e. two different frequencies played at the same time in a similar fashion to DTMF. Forward and backwards tones are built from two different sets of frequencies from non-overlapping ranges. This avoids possible problems due to a device "hearing" its own tone reflected back from the other device, a situation which may be unavoidable if there is an analog device somewhere in the circuit. Tones are constructed from a set of six possible frequencies, making a total of $(6 \times 5)/2 = 15$ different possible tones:

Frequencies for forward tones (Hz)	Frequencies for backward tones (Hz)
	540
	660
	780
	900
	1020
	1140
1380	
1500	
1620	
1740	
1860	
1980	

Each tone is constructed from a pair of frequencies. For example:

```
Forward tone 1  =     1380 + 1500
Backward tone 1 =     1140 + 1020
```

The Dialogic Voice Software Reference Manual contains complete listings of the frequency assignments.

Interpreting R2 Tones

Each tone has two (occasionally more) possible interpretations depending on the context in the exchange of information between the CO and CPE.

Forward tones are interpreted from two sets of meanings: Group I and Group II. When a call starts, forward tones are understood to be Group I tones. A switch to Group II may be triggered by a backwards tone.

Similarly, backward tones are interpreted from two sets: Group A and Group B. When the call starts, Group A applies. Again, a switch to Group B may be requested by sending a backwards tone.

When the call starts, the CO will generally start by sending the first digit of the dialed number specification (DID or DNIS) as a Group I tone, the CPE will respond with a Group A backwards tone meaning "I received that digit OK, please send the next digit".

According to CCITT standard 1914-ANS 532 04, National R2 MFC Register Signaling, meanings are assigned to tones as follows:

> Group I refers to the *address* information, i.e. to a specification of where the call should be routed (DID, DDI, DNIS) and/or where the call originated (ANI).
>
> Group II refers to the *category* of the caller, i.e. to a specification of the priority of the call and how it should be billed.
>
> Group A are control signals for requesting the transmission of address digits. Provision is included for requesting re-transmission of digits not properly received.
>
> Group B are indications to the CO of the condition of the subscriber's line.
>
> (Group C is a third group of meanings which is used to transmit a two-digit tariff number and only applies to network signaling and will not normally be required by CPE).

A complete listing of the CCITT standard meanings of the tones follows. Those marked "Not CPE" will not be used by typical CPE applications.

Group I	Meaning (Address Specification)
1, 2 ... 10	Digits 1, 2, ... 0
11 - 14	(Not CPE).
15	End of address, or, Request not accepted.

Group II	Meaning (Subscriber Category)
1	Ordinary subscriber.
2	Subscriber with priority.
3	Maintenance equipment.
4	(Reserved for future use).
5	Trunk operator.
6	Data transmission.
7	International subscriber.
8	International data transmission.
9	International subscriber with priority.
10	International trunk operator.
11	Coin box.
12	Subscriber with priority with home meter.
13	Ordinary subscriber with home meter.
14	Interception service operator.
15	(Reserved for future use).

Group A	Meaning (Address Control)
1	Send next address (DID, DNIS) digit (n+1).
2	Send last but one address (DID, DNIS) digit (n-1).
3	Send Group II signal and change over to Group B interpretation.
4	Congestion. (Not likely to be used by CPE).
5	Send Group II signal. If two or more A5 tones are sent, this requests the first (next) digit of the originating caller's number (ANI).
6	Address complete, begin billing for call, open talk path. (The CPE must know how many address digits were expected and send this tone when all digits have successfully been received).
7	Send last but two address digit (n-2).

8	Send last but three address digit (n-3).
9	(Reserved for future use).
10	Send next address digit and change to Group C meanings. (Not likely to be used by CPE).
11	Send country code indicator. (Not CPE).
12	Send language or discriminating digit. (Not CPE).
13	Send circuit type. (Satellite links only, not CPE).
14	Request for information on use of echo suppresser. (Not CPE).
15	Congestion in international exchange. (Not CPE).

Group B	Meaning (Subscriber Line Condition)
1	Line free, start billing, last party released. (Not CPE).
2	Send SIT to indicate long-term unavailability. (Not CPE).
3	Subscriber's line busy. (Not CPE).
4	Congestion. (Not CPE).
5	Unallocated number. (Not CPE).
6	Subscriber's line free, open talk path, start billing. (This is the tone that will usually be used by the CPE to end the R2 transaction with the CO).
7	Subscriber's line free, no charge, open talk path. (Not CPE).
8	Line out of order.
9	Line marked for intercept service. (Not CPE).
10	(Reserved for future use).

Socotel R2 Signaling

Spain and France use a variant of R2/MF which uses a rather different system of tones. As with the CCITT standard method, there are two sets of dual tones, one used for forward signaling and one used for backward signaling. The main difference is that a single frequency acknowledgment tone (1700 Hz in Spain, 1900 Hz in France) is used to respond to a signal in either direction.

MF signaling starts with the CPE sending a backward tone to the CO, which responds using the acknowledgment tone. (This is the reverse of the CCITT standard, where the CO sends the first signal).

A Socotel forward tone is sent in the following way:

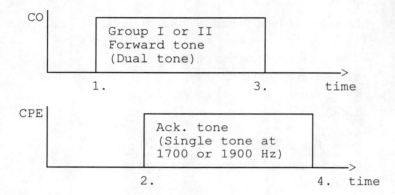

A Socotel forward tone. Sent from the CO to the CPE. The CPE acknowledges with a single tone, which causes the CO to stop sending.

A Socotel backward tone is sent in the following way:

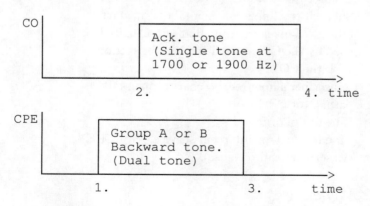

A Socotel backward tone. Sent from the CPE to the CO.

The Dialogic Voice Software Reference manual contains tables of frequencies for Socotel forward and backward signaling.

Socotel Signal Interpretation

Forward signals are interpreted in two ways, also called Groups I and II. Group I tones I-1, I-2 to I-10 represent the digits 1, 2, .. 0. Group II tones also are used to indicate the category of the caller:

Group I tone	Meaning (Address Digit)
1, 2 .. 10	Digit 1, 2 ... 0

Group II tone	Meaning (Caller Category)
1	Regular subscriber.
3	Special service (within the area).
4	National number, but the caller is located in a different area code from the calling party.
10	International number.

Group A tone	Meaning (Address Control)
1	Send digits.
2	(Reserved)
3	Send Group II tone for caller category
4	Send digits.
5	(Reserved)
6	Send all digits.
7	Send digits.
8	Switch to Group B meanings.
9	(Reserved)
10	Congestion.

Group B tone	Meaning (Line Condition)
1	Complete call, open talk path.
2	Congestion.
4	Busy.
8	Line dead.
9	End of selection.

Tone A1, A4, A6 and A7 all request a series of digits from the CO. The exact digits to be sent depend on the class of service and caller category.

ABCD Signaling Bits: CCITT Q.421

Unfortunately, ABCD signaling bit usage—like the details of the compelled signaling—vary from country to country and even from region to region.

In this section we will look at a concrete example from the CCITT Blue Book recommendation Q.421 for System R2.

Recommendation Q.421 only uses the A and B bits, C and D values are held constant at CD=01. Signaling bits transmitted by the CO and received by the CPE are called *forward* signals, as for R2 tones.

Signaling bits transmitted by the CPE to the CO are called *backward* signals.

For consistency with the usual PEB and T-1 usage, we will refer to forward signals as *receive bits*, and to backward signals as *transmit bits*, looking at the connection from the CPE point of view.

Under normal conditions, the significance of the bits is as follows:

Bit	Meaning
Receive A bit	Identifies the condition of the CO line.
Receive B bit	Indicates CO failures.
Transmit A bit	Identifies hook state of CPE line (on- or off-hook).
Transmit B bit	Indicates idle or seized state of CPE equipment.

The following table summarizes the values of the bits for usual states of the E-1 circuit:

E-1

State	Receive (fwd.)		Transmit (bckwd.)	
	A	B	A	B
Idle	1	0	1	0
Seized	0	0	1	0
Seizure acknowledged	0	0	0	1
Answered	0	0	0	1
Clear-back	0	0	1	1
Clear-forward	1	0	0	1 or 1 1
Blocked	1	0	1	1

Clear-back is a hang-up by the CPE.

Clear-forward is a hang-up by the original caller.

Note that the receive B bit is always zero under normal conditions. The CPE therefore only has to watch for A bit transitions, in a similar way to T-1.

Note also, however, that the meaning of the A bit is reversed relative to T-1: 0 corresponds to a call in progress, 1 to an idle line.

For example, a typical incoming call progresses in the following way:

Receive		Transmit		Action
1	0	1	0	Both CPE and CO idle.
0	0	1	0	CO signals ring by seizing
0	0	0	1	CPE acknowledges ring

(R2/MF signaling follows)

If the caller hangs up, this will be signaled by the CO going into the clear-forward state:

Receive		Transmit		Action
1	0	0	1	CO signals hang-up by returning to idle.
1	0	1	0	CPE acknowledges by also going to idle.

If the CPE wishes to terminate the call, this is signaling by the CPE going into the clear-back state:

Receive		Transmit		Action
0	0	1	1	CPE signals hang-up.
1	0	1	1	CO acknowledges clear-back by going to idle.
1	0	1	0	CPE returns to idle also.

ABCD Signaling Bits: Socotel

Socotel is a signaling standard based on E-1 which is used in France and Spain. A form of compelled R2/MF signaling is used, but it does not conform to the CCITT standard described in detail earlier. The use of the signaling bits is also different. As before, however, the CD bits are kept at the fixed values CD=01. In addition, the B bit is held fixed at value B=1, with one possible exception: the CO may use the B bit to send charging pulses.

Normal states for a connection are:

State	Receive A	Transmit A
Idle	1	1
Seizure (=ring)	0	1
Seize acknowledge	0	0
Clear-back	1	0
Clear-forward	0	1
Blocked	1	0

A typical out-bound call starts as follows:

State	Receive (bckwd.) A	Transmit (fwd.) A
Idle	1	1
Seize (=go offhook)	1	0
Seize acknowledge (R2/MF signaling)	0	0

Note that for the out-bound call the CPE becomes the out-going register and is this responsible for forward signaling, and the CO becomes the in-coming register and is responsible for backward signaling. The examples in this chapter generally assume that the CPE is the in-coming register, which is the more usual case for automated voice response equipment.

During the call, the B bit received from the CO may be used for charging pulses, where the B bit is set to zero for an interval of 50 to 200 ms. If charging pulses are not to be counted, the B bit may be ignored.

A typical in-bound call starts as follows:

State	Receive A	Transmit A
Idle	1	1
Seize (=ring)	0	1
Seize acknowledge (R2/MF signaling)	0	0

DID Example: CCITT

The most commonly used service on E-1 is probably DID, where the last few digits of the dialed number are transmitted to the CPE. The following traces the progress of a typical 4-digit DID call on an E-1 service using CCITT protocol. The DID digits used for this example are 3276.

In these examples, AB represents signaling bits A and B. Group A and B tones are represented by A-n and B-n where n is the tone number, similarly Group I and II tones are represented by I-n and II-n.

State/Action	CPE	CO
1. Idle.	AB=10	AB=10
2. Seize (=ring). CO changes AB to 00.		AB=00

State/Action	CPE	CO
3. Seize acknowledge. CPE changes AB to 00. MF signaling now starts.	AB=11	
4. CO sends 1st digit. CO sends Group I tone I1 .. I10.		I-3
5. CPE responds with A1 meaning "send next digit".	A-1	
6. CO sends 2nd digit.		I-2
7. CPE responds with A1	A-1	
8. CO sends 3rd digit.		I-7
9. CPE responds with A1	A-1	
10. CO sends 4th digit.		I-6
11. CPE responds with A3 meaning "all digits received, switch to Group II/B meanings.	A-3	
12. CO sends tone II-1 meaning Normal Subscriber		II-1
13. CPE responds with tone B6, meaning complete call.	B-6	

MF signaling ended by category tone/B6 response.

DID Example: Socotel

The following example traces a typical 5-digit Socotel DID call. The DID digits used for this example are 46189.

E-1

State/Action	CPE	CO
1. Idle.	AB = 11	AB = 11
2. Seize (=ring). CO changes AB to 01.		AB => 01
3. Seize acknowledge. CPE changes AB to 01. MF signaling starts.	AB => 01	
4. CPE sends A4 meaning send DID digits	A-4	
5. CO responds with DID digits.		I-4 I-6 I-1 I-8 I-9
6. CPE sends A-8, meaning switch to Group B/II.	A-8	
7. CPE sends B-1, meaning line ready, complete call.	B-1	

Protocol Outlines For Selected Countries

In this section we will outline some of the details of the signaling protocols in some selected protocols. These outlines will necessarily omit many of the details: a complete exposition of each including all error handling would require a book in itself. The selected countries illustrate much of the variation which an equipment vendor must take into account, including ABCD bit usage and types of DDI digit transmission (CCITT R2, Socotol R2, R1/MF, and decadic pulse where one of the signaling bits is switched on and off to play a role analogous to loop current in a rotary dial).

If four 0/1 digits are shown, this refers to the ABCD bit states. "DTMF" indicates that a string of DTMF digits is sent. "MF" indicates that a string of MF (R1) tones is sent. An MF string will start with a KP tone (represented as *) and will terminate with an ST tone (represented as #). "Ring tone" indicates that the CPE will send audio corresponding to the standard ring-back tone in that country. Typically two ring cycles are required. "An" or "Bn", for example "A4" indicates that the an R2 tone from Groups A or B is sent. "I_n" or "II_n" indicates that an R2 tone from Groups I or II is sent. The states are listed in the order they typically occur in a call.

Belgium R2 (inbound).

State	CO	CPE
Idle	1001	1001
Ring	0001	
Acknowledge		1101
R2 signaling:		
send 1st digit, Group I	I_n	
acknowledge 1st digit		A1
send 2nd digit	I_n	
acknowledge 2nd digit		A1
...		
send last digit	I_n	
number complete		A6
ring back		Ring tone
Hook up (make talk path)		1101
Clear forward (hang up)	1001	
Clear back (hang up)	1001	1101

E-1 **217**

Netherlands ALS70D (inbound).

State	CO	CPE
Idle	1001	1001
Ring	0001	
Acknowledge		1101
Ready to receive DDI digits		0101
DDI digits	DTMF	
Number complete		1101
Ring back		Ring tone
Hook up (make talk path)		0101
Clear forward (hang up)	1001	
Clear back (hang up)		1101

Sweden SS 63 63 33 (inbound).

State	CO	CPE
Idle	1001	1001
Ring	0001	
Acknowledge		0001
DDI digits	DTMF	
Number complete		1001
Ring back		Ring tone
Hook up (make talk path)		0101
Conversation		0001
Periodic B pulse (stay alive)		0001/0101
Clear forward (hang up)	1001	
Clear back (hang up)		1001

Turkey R1 (inbound).

State	CO	CPE
Idle	0101	0101
Ring	1101	
Acknowledge		1101
Ready to receive DDI digits		0101
DDI digits	MF	
Ring back		Ring tone
Hook up (make talk path)		1101
Clear forward (hang up)	0101	
Clear back (hang up)		0101

Spain Socotel KD3/MFE (inbound).

State	CO	CPE
Idle	1101	1101
Ring	0101	
Acknowledge		0101
Socotel-type R2 signaling:		
Request 1st digit		A4
Send 1st digit	I_n	
...		
Send last digit	I_n	
Switch to B		A8
Ready for call		B1
Ring back		Ring tone
Hook up (make talk path)		1101
Clear forward (hang up)	1101	
Clear back (hang up)		0101

UK BT Call Stream (inbound).

State	CO	CPE
Idle	1111	0111
Ring	0011	
Acknowledge		1111
Send DDI by pulsing A bit	1011/0011	
Ring back		Ring tone
Hook up (make talk path)		0011
Clear forward (hang up)	1111	
Clear back (hang up)		1111

Chapter 13

Voice/Data Protocols

Introduction

The vast majority of telephone calls fall into one of two categories: voice or data. Voice calls are generally person-to-person, but also include person-to-machine applications like many of those discussed in this book: banking by phone, audiotex and so on. Data calls are modem-to-modem, the prototypical example is a call to a bulletin board or on-line service. Data calls are characterized by the use of sound modulation to transmit digital information as sound (hence the word "modem" for *mo*dulator/*dem*odulator).

An important category of data calls is fax, where images are transmitted between modems specifically designed for this task (remember that a conventional fax machine is just a modem with a printer built in).

Many future applications will include the ability to combine, or switch between, voice and data in a single call. One relatively familiar example today is fax on demand. The caller dials a service from a fax machine and selects a document by dialing touch-tone digits. The system prompts the caller to press the Start button, and switches from interactive voice to fax transmission mode. This works relatively easily because the Group 3 fax protocol enables an exchange of tones to begin the call. Switching back to voice once the document has been transmitted may, however, be difficult or impossible depending on the fax machine design since the fax protocol was not designed with this voice/fax mode switching in mind.

New applications, made possible when both caller and called party have compatible voice/data equipment, will enable data to be transmitted as part of a voice call. For example, a technical support person might be able to download and examine your CONFIG.SYS file while you are discussing a problem, or a graphic designer could send sketches for a new ad layout directly to the client's screen while discussing a concept.

To address this need, a family of protocols is under development, commonly known as Simultaneous Voice and Data (SVD). The best-known of these is currently Radish Communication Corp.'s (Boulder, CO 303-443 2337) VoiceView protocol. VoiceView is planned for inclusion in Microsoft Windows '95 (not shipping at the time of writing). In fact, VoiceView is more properly described as an Alternating Voice and Data (AVD) protocol since it switches between voice and data but does not allow simultaneous transmission.

Most protocols are *point-to-point*, meaning that they involve only two parties as in a regular telephone call. However, advanced *multi-point* protocols involving two, three or more parties are also under development.

Alternating Voice And Data (AVD)

This type of protocol allows for switching from voice to data and from data to voice. VoiceView is the primary contender for becoming an industry standard following Microsoft's decision to license this protocol for Windows '95.

The Hayes AT command set provides a primitive example of an AVD-like protocol. When in command (voice) mode, issuing at ATDT command will cause the modem to establish a data connection. Once in data mode, sending data consisting of three consecutive ASCII plus characters "+++" followed by a pause with no data will switch the modem back to command mode. (Hayes claims a patent on this technique). If you are ever in a communications program and are trying to break out of a file transfer, try typing "+++" and waiting a few seconds. If you are using a Hayes or Hayes-compatible modem, you will probably be able to type an AT command, such as ATH0 to hang up the line. If the first time doesn't work, it's probably because your escape sequence got mixed up with data being sent by the modem, another try might succeed.

This is based on the idea that a data connection is highly unlikely to contain exactly this sequence of events. Transmitting this chapter by modem might easily include the "+++" sequence, but it would be protected from a mode switch by the immediately following characters.

Simultaneous Voice And Data (SVD)

SVD protocols allow for voice and data to be transmitted at the same time. This naturally may involve degraded speech quality and/or reduced transmission speed on a standard voice connection.

There are two main types of SVD: analog (*ASVD*) and digital (*DSVD*).

Analog SVD has the advantage that sound can be transmitted without encoding and decoding (modulation), and therefore with higher fidelity and without additional processing power in the modem. This simplification means that voice/data modems based on a single Digital Signal Processor (DSP) are well within reach of current technology. However, minor modifications to current data protocols (V.34 and V.32bis) were required to implement ASVD.

Digital SVD typically requires two DSPs per channel, because of the extra demands of digitizing voice. DSVD expands the existing V.34 and V.32bis data carrying protocols by adding further negotiation to the capabilities exchange phase where a connection is established and an switching protocol to signal the transition from voice to data and back.

Available Protocols

Some protocols currently available to developers of voice hardware include the following.

ADSI

The *Analog Display Services Interface*. This is an AVD protocol which requires the caller to have an enhanced telephone with a screen which can display an array of characters. ADSI therefore offers very little flexibility in terms of the contents of the data which can be transmitted, and has a fixed transmission speed of 1200 baud. ADSI has been used in very few commercial applications, and its lack of features makes it appear probable that it will be superseded by more flexible protocols such as VoiceView.

VoiceView.

Radish Communications Corp.'s point-to-point AVD, to be included in Windows '95.

VoiceSpan.

AT&T Microelectronic's point-to-point ASVD protocol.

Talk Anytime.

Point-to-point DSVD protocol from Multitech.

Digital SVD Specification 1.1.

A competing point-to-point DSVD standard from a consortium including Intel, US Robotics, Hayes and Creative Labs.

Voice/Data Protocols

V.DSVD

A multi-point DSVD protocol under consideration by the Telecommunications Industry Association. No products are available supporting this protocol at the time of writing.

Dialogic And Voice/Data

Dialogic has announced that the FAX/40 daughter-board for the D/4x series will support Radish's VoiceView protocol. This will enable Dialogic applications to switch between voice and fax when communicating with VoiceView-compatible modems.

Since Dialogic products do not currently support traditional data modem transmission, the ability of a Dialogic-based application to exploit the full potential of voice/data protocols will require third-party or proprietary firmware solutions.

VoiceView

As of May 1995, Radish Communications Systems claimed the following list of VoiceView "supporters", although the exact definition of "supporter" was not provided.

AT&T
Ameritech
Analog Devices
Boca Research
Cardinal Technologies
Creative Labs/Digicom Systems, Inc.
Cirrus Logic, Inc.
Dialogic Corporation
Diamond Multimedia
Digital Sound Corporation
GVC
Hayes Microcomputer Products, Inc.
IBM
Intel Corporation
Innovative Trek Technologies, Inc.
Microsoft Corporation
Motorola

MultiTech Systems, Inc.
Octel Communications Corporation
Practical Peripherals
Radish Communications Systems, Inc.
Rockwell International
Sierra Semiconductors
U.S. Robotics, Inc.
Zoom

The VoiceView Protocol is defined in the VoiceView Protocol Specification which is provided to vendors upon licensing the technology from Radish. The base license fee at the time of writing is $20,000. The specification details the information necessary to incorporate VoiceView technology into a wide variety of hardware products such as modems, modem chips, DSP code, screen-based telephones, desktop fax machines, PDAs, voice response systems, and voice mail systems.

Two VoiceView users can exchange files, business cards, or play and interactive game during a call. VoiceView-enabled service providers handle client calls using call center agents equipped with VoiceView-enabled modems or a VoiceView-enabled call processing system.

The VoiceView technology includes several major areas of functionality. Most importantly, it specifies the mechanism for switching between different information transfer modes during a telephone call.

VoiceView Modes

The protocol currently defines four modes of information transfer: voice, modem data mode, facsimile data mode, and VoiceView data mode.

In voice mode, the modem or *Data Circuit Terminating Equipment (DCE)* typically provides a direct connection between the user's telephone and the telephone network interface. This enables a conversation to take place between users. During voice mode, the DCE must monitor the telephone line to detect an incoming signaling sequence (start tone) which requests it to switch to one of the data transfer modes.

Voice/Data Protocols

When the DCE switches to a data transfer mode, the local telephone interface is typically muted so that the user does not hear the data transmission and so any ambient room noise does not corrupt the data. In the case of a simultaneous voice/data mode, the phone would only be muted for a few seconds while the data link is established and the compressed voice transmission protocol initiated. Information Transfer Modes In order to promote easy implementation, the protocol makes use of several existing data modulation schemes to transfer information across the telephone network. The protocol is extensible to allow the use of new schemes as they become available. VoiceView allows an application to switch between voice mode and any of the following three data transfer modes during a call:

> VoiceView Data Mode, which uses facsimile modulation schemes including V.17, V.29, V.27ter, and V.21 and is optimized for most switched voice and data applications.

> Modem Data Mode, which utilizes existing modem protocols such as V.22 bis, V.32, V.32 bis and V.fast (V.34) and is compatible with existing data communications products.

> Fax Data Mode, which transfers data via T.30 fax protocols used in existing fax devices Applications can invoke multiple types of data transfer modes during the course of a call to convey different types of information between users.

VoiceView is designed to make it possible to select the most appropriate mode and to switch between modes. VoiceView Data Mode is a unique information transfer mode defined by the VoiceView Protocol. It is the information transfer mode that has been specifically optimized for exchanging the typical kinds of information involved in telephone transactions. It is optimized for speed in completing typical transactions and designed to provide reliable and efficient data communications over the telephone network. VoiceView Data Mode is built on the foundation of traditional half-duplex facsimile data modulation schemes such as V.17, V.29, V.27ter, and V.21. These protocols provide reliable data transport with minimal startup overhead. VoiceView Data Mode also includes an HDLC-based error correcting data link layer to insure error-free transmission of the data over the telephone network.

DTE Interface

VoiceView also includes the definition of the message communications across the DTEDCE interface for VoiceView-enabled modems which communicate with a PC or other type of computing equipment. The message structure is an extension of the Hayes "AT" commands used in modems today. New commands have been defined to initiate each data mode, to accept a request to switch to a data mode, and to send and receive the Capabilities Query. Event reports notify the DTE when a request to switch to data mode has been received or when capabilities information or an ADSI CPE response has been received. Additional parameters have been defined which control the DCE's response to a Capabilities Query, allow the DTE to read or change the capabilities information in the local DCE, allow the DTE to set the VoiceView Data Mode transmission speeds and enable the DCE to provide detailed error information to the DTE.

VoiceView was designed to operate with existing data and fax communications modes in the DCE. VoiceView functionality is implemented in a new service class (+FCLASS) which is modular and can be added to existing products. The transport and format of data messages exchanged between two VoiceView applications is specified in the VoiceView Information Exchange Protocol. It is specified a separate document, the VoiceView Software Developers Specification. This protocol provides a common standards-based method for exchanging information over a phone connection using the VoiceView technology. It includes a transport layer protocol and a binary file file transfer format using the standard T.434 protocol.

Implementing VoiceView

Since the VoiceView Protocol has been designed to make maximum use of existing capabilities in data and fax modems little or no changes are required in a modem's hardware design. VoiceView requires that certain capabilities be provided by a DCE's data pump, controller, and DAA (Data Access Arrangement). The VoiceView Protocol requires that the data pump, either a modem chip or DSP, implements the appropriate tone detection and generation capabilities as well as the specified data modulation protocols. The modem chip or an additional microprocessor controller must also provide the support for the HDLC error correcting protocol. In addition, the DCE design must accommodate the switching between modes. DAA modifications may require the inclusion of a second relay in addition to the one typically found in most modems for controlling the switching between voice and data modes. However, modem designs which separate the voice signal into transmit and receive paths via DSP/CODEC technology can naturally perform the switching function without the additional relay. A monitor circuit must be included in the design to allow the modem chip or DSP to "listen" to the telephone line while the DCE is in voice mode. This circuit enables the DCE to detect the inband start tone used to switch from voice mode to a data transfer mode.

VoiceView implementations also experience two challenges that may be unfamiliar to most modem developers: talk-off and talk-down. These two phenomena are very familiar to designers of voice mail and call processing systems. Talk-off occurs when audible energy received from either the near-end or the far-end results in a false interpretation of the start tone signal. Talk-down occurs when speech or background noise from the local telephone causes the near-end DCE to miss a valid start tone. All VoiceView DCEs should incorporate a means for reducing the effect of these two phenomena. The least cost (no hardware change) design will utilize the DSP capabilities in the modem. VoiceView Protocol implementations which make use of generic DSPs have the opportunity to program these devices to detect the VoiceView signaling tones.

Chapter 14

Voice Boards

Introduction

The PC-based voice processing industry is built around one fundamental component: the *voice board*. A voice board, also called a *speech card*, is an IBM PC- or AT-compatible expansion card which can perform voice processing functions.

A voice card has several important characteristics:

Computer Bus Connection

In the PC world, there are four main variants of the computer bus connection:

1. 8-bit PC/XT.

2. 16-bit AT, also known as *ISA* or *Industry Standard Architecture*.

3. PS/2-compatible 32-bit *Micro Channel* or *MCA*.

4. The EISA 32-bit superset of the ISA: an alternative to the MCA.

Telephone Line Interface

Again, there are three common variations:

1. The most obvious way: RJ-11 or RJ-14 jacks, which allow one or two analog two-wire lines to be connected directly to the board.

2. Digital interface, such as T-1.

3. Bus interface. It may be that the voice board itself does not have a telephone line interface, requiring an additional board in order to connect with a phone line.

Voice Bus Connection

A *voice bus* is a bus independent of the computer controlling the card which allows audio and signaling information to be passed between different voice processing components. At the time of writing, the great majority of PC-based voice processing installations probably make no use of a voice bus, but it is anticipated that this feature will become increasingly important as features such as fax integration, voice recognition and text-to-speech, which often are implemented by add-in boards connected via a voice bus, become more common. Voice bus options include:

1. None.

2. Dialogic's *Analog Expansion Bus (AEB)*. The AEB is similar to a bundle of two-wire analog connections.

3. Dialogic's *PCM Expansion Bus (PEB)*. The PEB is similar in character to a digital T-1 connection. The PEB is now obsolete, having been replaced by the SC Bus, but there is still a large number of installed PEB-based systems.

4. The *Multi-Vendor Integration Protocol (MVIP)*, currently supported by Rhetorex and Natural MicroSystems among voice board manufacturers. MVIP is based on Mitel's ST bus specification and is more or less a head-to-head competitor with Dialogic's SC Bus.

5. *SC Bus.* A vital part of the SCSA architecture (see the chapter on SCSA), the SC Bus is Dialogic's next generation voice bus.

Operating Systems Supported

While not strictly speaking a hardware component, the device drivers provided for the PC operating system by the voice board manufacturer will be a critical resource for most systems integrators. The PC-compatible operating systems most widely used in the voice processing industry are:

1. MS-DOS.

An operating system with major drawbacks for software developers, but so widely used and understood, with so many connectivity options and software development tools available at very affordable prices, that it still manages to dominate. A significant advantage of DOS is that it requires very little overhead, unlike traditional pre-emptive multi-tasking operating systems which take significant resources whenever the PC's clock ticks. Disadvantages are the difficulty of writing C programs to control multiple channels using state machines, and the problems of memory management (640Kb may not be enough).

2. Windows

At the time of writing, Windows NT is rapidly gaining in popularity with computer telephony developers, although DOS is still very widely used. Windows 95, with its inferior robustness and less stable multi-tasking, is generally used only for a few low-density solutions and for clients in client/server applications such as call centers, where the computer telephony server is likely to be running NT. Dialogic currently offers only a limited range of Windows 95 drivers for its products.

3. OS/2.

OS/2 has clearly lost the battle for the corporate desktop in the US. However, a few niche computer telephony developers, especially those integrating with IBM systems, continue to use this operating system. Most major vendors who used to concentrate on OS/2 have moved to NT. Dialogic does provide OS/2 drivers for its most popular boards.

4. UNIX.

UNIX comes in many flavors. UNIX, like OS/2 and NT, is not a real-time operating system (which means that response times on any given channel may be uneven), and also requires significant memory and processor resources. However, like OS/2, UNIX provides a multi-tasking environment which makes software development conceptually simpler than state-machine C programming for MS-DOS. UNIX is used to control most central office switches in the US, and for this reason alone has a stable niche in the computer telephony market. SCO UNIX, the most popular PC UNIX flavor, currently lacks support for multi-threading, which gives NT a big advantage in terms of performance.

Voice Boards

5. *QNX.*
Should perhaps be classified as a UNIX flavor; this operating system has won a small but loyal following in the voice processing community. A major issue is a lack of drivers for Dialogic boards.

Telephony Functions

At a minimum, a voice board will usually include support for:

1. Going on- and off-hook (answering, initiating and terminating a call).

2. Notification of call termination (hang-up detection).

3. Sending flash-hook.

4. Dialing digits. Touch-tone (DTMF) will be the most common, also supported may be rotary (pulse) digits detection, rotary digit recognition (recognizing the clicks as sounds over a distant connection rather than the loop pulses themselves) and MF.

Other common telephony functions include:

5. Call progress analysis. Determining the result of a transferred or out-dialed call by "listening" for tones such as ringing, busy or operator intercept; detecting the cadence of the voice answering the call, and so on.

6. Disconnect tone monitoring. For example, disconnect monitoring may detect "re-order" or dial tone from a PBX when the called extension hangs.

7. General tone detection. This means detection of user-defined tones. For example, it may be desired to detect an in-coming fax call by detecting the standard tone made by a fax machine.

Voice Processing Functions

"Standard" voice processing functions include:

1. Recording audio to a file.

2. Playing audio from a file.

Playing and recording can be done in several different ways. Features to consider when deciding which model of voice board to use include:

a) Digitization method (PCM, ADPCM, other).

b) Sampling rate (samples/second measured from the audio waveform), 6 kHz, 8 kHz and 11 kHz are common. Faster sampling means better quality but more data transfer and therefore load on the disk and a smaller maximum number of lines which can be handled.

c) Bits per sample (4 and 8 are common). More bits again means better quality but more disk space and heavier loading.

d) Automatic Gain Control. Best quality is obtained by setting the level high enough that background noise is suppressed, but not so high that distortion results. The board may be able to adjust the recording level automatically to match the line conditions.

e) Silence compression. For dictation and other applications, this option removes periods of silence from a recording to save disk space and make transcription more efficient. This feature is not used for most consumer applications such as voice mail because the message playback often sounds very "choppy".

f) "Special Effects" such as speed-up and slow-down without change of pitch.

3. Detecting tone digits. Touch-tone (DTMF) is standard, MF may be included for special applications including detecting ANI and DNIS.

4. DTMF cut-through, also known as touch-tone type-ahead. This allows the caller the ability to press a touch tone while a prompt is being played, optionally interrupting the prompt. While this feature is offered by almost all voice boards, it is not as simple as you might think. On an analog phone line, the audio sent out on a line "loops back" through the caller's telephone and an "echo" is received back at the voice board together with an any speech, background noise or touch-tones coming from the caller. You can hear this type of echo by blowing into the mouthpiece of a telephone which is off-hook: if there is circuit, you will hear a hissing sound in your ear. (Blowing in the mouthpiece is a quick check used by telephone equipment engineers to see if there is loop current on a line which is not producing dial tone).

5. Talk-off and play-off protection. Talk-off is a "false positive" detection of a touch tone. Typically talk-off happens when a person is recording an audio file (say, leaving a voice mail message) and the frequencies in the voice happen to match those of a touch tone digit. The voice board reacts as if a touch tone were pressed, and may terminate the recording immediately and move to another menu selection in the application. Protection against talk-off can be achieved in several ways:

> a) By restricting the list of touch tones to be detected when recording. For example, by only allowing a zero as in "Please begin recording at the tone, press zero when finished".

> b) By specifying how much the touch tone frequencies must stand out from background frequencies before a detection is reported.

c) By specifying a guard time, i.e. that the touch tone frequencies must be detected for a certain minimum length of time before a detection is reported. By specifying a longer time, false positives will be reduced, but the caller may have to hold down the touch tone button longer. Some telephones don't allow the caller to hold a tone: the sound is produced for a preset duration irrespective of how long the button is pressed.

d) By specifying the allowable range of "twist" (relative strengths of the two frequency components).

e) By specifying the range of allowed deviations from the precise frequencies specified for the touch tone. By narrowing the range, false positives are reduced, but there will be a greater chance of failing to detect equipment which does not produce exactly correct frequencies.

Play-off is similar to talk-off, except that the touch tone is detected in the audio being played by the voice board. This can be extremely puzzling: the system reacts as if a ghost had pressed a touch tone. Features to suppress play-off are similar to those for talk-off.

Other voice processing functions may include:

1. Rotary digit detection. Rotary digits are much harder to detect than tone, since the drops in loop current which count the digits are not transmitted across telephone switches. This means that algorithms must be developed to analyze the audio signal to "listen" for the characteristic clicks made by pulse dialing. Consider, for example, the problem of distinguishing a pulsed "1" digit (one click) from static on the line.

2. Voice Recognition. Also known as speech to text, the most common usage is to recognize Yes/No or single digit spoken responses to prompts.

Voice Boards

3. Text-to-Speech. This feature converts any given string of words, specified as ASCII text, to a synthesized spoken voice.

4. Conferencing/switching. In other words, the ability to connect two or more callers in a multi-way conversation.

Fax Integration

Important fax functions include:

1. Record fax data to a file.

2. Send fax data from a file.

3. Fax synthesis, i.e. the ability to take ASCII text and/or graphics format files and convert them to fax format for transmission "on the fly".

The Dialogic D/4x Series

The D/4x series manufactured by Dialogic Corp. of Parsippany, NJ (201) 993-3000 has effectively set the standard for multiple-channel PC voice processing cards, and is without doubt the most widely used component in the industry at the present time.

A D/4x board looks something like the following diagram.

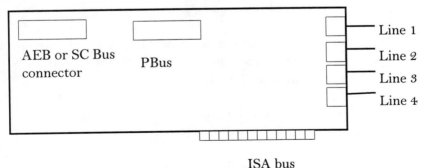

The Dialogic D/4x board. The de-facto industry standard.

ISA bus. The PC bus connector. The D/4x is an 8-bit, full-length card designed for the IBM PC/XT, AT and compatible computers.

Line 1 .. Line 4. The classic Dialogic boards (D/41B, D/41D, D/41E) support up to 4 phone lines, hence the "4" in D/4x. The telephone line connectors are standard modular RJ-11 jacks as would be found in any US home. Other Dialogic boards support different numbers of phone lines. For example, the D/21ESC supports two lines, and the D/160SC-LS supports 16.

AEB or SC bus connector. The *AEB* (*Analog Expansion Bus*) connector allows the D/4x to be connected to other voice processing boards which may take input from or create output for the analog voice signals processed by the D/4x. Examples of these boards are the AMX/81 switching matrix card, which can be used to "conference" D/4x channels, or the VR/40 speaker-independent voice recognition board. An SC bus connector allows a connection to other SC bus boards, which may provide services such as voice recognition or text-to-speech. The D/41ESC board has both AEB and SC bus connectors. Older boards, including the D/41D, have only an AEB connector. Boards built for lower cost, including the Proline/2V, have no expansion bus connector.

PBus. The on-board processor bus. This is used on certain older D/4x model boards, including the D/41B and D/41D to connect daughter-boards such as the MF/40, which adds MF digit detection capability, and the VR/40, which adds voice recognition capability.

It may be in order at this point to take a brief digression and comment on Dialogic board naming conventions (such as they are). Some Dialogic boards are named "B/nnxr", where:

B is the board type, eg D for voice boards, VR for voice recognizer, FAX for fax and so on,

nn is the number of phone lines supported,

x is usually 0 or 1, maybe 2, indicating the presence or absence of a major feature — for example, the D/40B does not support call progress analysis, while the D/41B does.

Voice Boards

r indicates a revision level: "A" for the D/40A, "B" for the D/40B, "D" for the D/41D. Some boards are named without a revision level.

Most boards with an SC bus connector are named B/nnrSC or B/nnrSC-i. The -i indicates if there is a trunk interface built into the card, so for example a D/160SC-LS has a built-in loop start interface for 16 analog lines, a D/240SC-T1 has a built-in T-1 interface. An exception to this naming rule is the D/41ESC, which has a loop start interface for four analog lines and should therefore be named D/41ESC-LS (but isn't). SC bus board types include:

D/
Voice boards, may include an interface if there is a -i in the name. Thus, a D/240SC has just 24 voice channels, the D/240SC-T1 has both 24 voice channels and a T-1 interface.

LSI/
Board with a loop start (analog) interface. If the board is named with x=1, then there are also voice processing channels on the board which have all standard capabilities for voice except playing and recording voice channels. Thus, a LSI/160SC is an interface-only card, the LSI/161SC also has 16 of these limited voice channels.

DTI/
Board with a T-1 or E-1 interface. They can be distinguished because a T-1 board will have 24 channels, an E-1 board will have 30 channels. As with the LSI, if x=1 then there are limited voice channels included on the board. So, the DTI/240SC is a T-1 interface only card, the DTI/300SC is an E-1 interface only card, the DTI/241SC has a T-1 interface and 24 limited voice channels, and the DTI/301SC has an E-1 interface and 30 limited voice channels. A D/240SC-T1 and DTI/241SC are therefore very similar except that the voice channels on the D/240SC-T1 are fully-functional.

Other Dialogic boards have been named with no apparent regard for logic (we didn't say any of this would be easy).

Several D/4x boards can be installed in one PC, allowing the designer to create systems supporting 8, 12, 16 ... phone lines. Using four-channel boards, however, one can quickly run out of available slots in the PC chassis, and this where higher-density speech cards can be preferable.

Using our earlier list of voice board functions, the D/4x scores as follows.

Feature	Supported by D/41ESC
Operating System	
MS-DOS	Yes
UNIX	Yes
OS/2	Yes
QNX	No
Windows NT	Yes
Windows 95	Yes
Telephony	
Go on-,off-hook	Yes
Hang-up detect	Yes
Send flash-hook	Yes
Dial digits	Yes
Call progress	Yes
Call monitor	Yes
Tone detect	Yes
Tone generate	Yes
Switching	Yes (2 calls can be connected via SC bus)

Feature	Supported by D/41ESC
Voice Processing	
Record audio	Yes
Playback audio	Yes
Get DTMF, MF	Yes
Recognize pulse digits	Yes (additional charge for firmware option)
Voice recognition	No
Text-to-speech	No
Fax integration	No

Some of the "missing" functions are provided by additional boards from Dialogic or third-party manufacturers which add features through a voice bus.

Chapter 15

Installing Boards

Introduction

In this section we will discuss some general points about installing voice card hardware and software. Since there are many different models of voice board, and the details of their installation are continually changing as new revisions of the hardware and driver software are released, we cannot discuss each board in detail; we refer you to the manufacturer's documentation for specifics. However, installing expansion cards in IBM-compatible PCs is a minefield for the unwary, so we'll offer some general advice on hardware installation. Most of this information is not specific to voice card hardware but to installing any expansion card. That's good, though—you will need to understand how the voice card might be conflicting with other hardware in your PC.

The general rule is simple: make sure that no two devices are using the same resource, except in the rare situation where they have been designed to do so (for example, some multi-serial boards can support COM 1 and COM 3 sharing IRQ 4).

While we've tried hard to explain everything in this section, whole books have been written on the architecture of the PC and how to install, maintain and troubleshoot PC hardware. You may find it helpful to invest in some of these books if you are not familiar with the technical terms and troubleshooting techniques described here.

We recommend that you use a high-quality memory manager such as Quarterdeck's QEMM. You will probably want to use EMS memory for storing Dialogic voice buffers, and QEMM also provides a useful utility called MFT (Manifest) which shows the memory layout and software interrupts used in your configuration. Utilities packages can also provide useful information listings, for example Norton Utilties' SYSINFO. Windows ships with a diagnostic utility called MSD (Microsoft System Diagnostics), go to your Windows directory and type MSD /? to get more information. PC quality control and maintenance software such as Touchstone's CheckIt may also provide invaluable information as you are trying to complete a successful installation. It is a good idea to try multiple utilities when trying to get system configuration information since any given utility may sometimes give incomplete or incorrect listings on any given topic.

Resources

Expansion cards take software and hardware resources when they are installed. Almost every expansion card or programmable device mounted on your motherboard will use one or more of the following resources:

Installing Boards 245

Hardware interrupts, commonly known as IRQ (interrupt request) lines,

Software interrupt numbers,

I/O port addresses,

Shared RAM addresses,

Direct Memory Access (DMA) channels.

Hardware Interrupts (IRQs)

Usually called IRQs. All IBM compatibles have IRQs from 0 to 7. AT-compatibles also have IRQs from 8 to 15. The following table summarizes the IRQs found on AT-compatibles. You must, of course, find an IRQ which is free on your PC to use for your installed voice card or cards. Note that Dialogic voice cards can share a single IRQ, so you only need to find one free IRQ.

IRQ	Usage on AT-compatible PC
0	Timer 0, never available for expansion cards.
1	Keyboard interrupt controller, never available for expansion cards.
2	Available for expansion cards (see the following caution re. IRQ 2/9).
3	Serial port COM 2, may be used by internal modem or serial port
4	Serial port COM 1, may be used by internal modem or serial port
5	Second parallel port (LPT 2), usually available for expansion cards because few PCs have LPT 2 installed.
6	Diskette controller, never available for expansion cards.
7	First parallel port (LPT 1).
8	CMOS clock interrupt.
9	Used by bus to cascade interrupts, not available for expansion cards (see the following caution re. IRQ 2/9).
10-13, 15	Available for expansion cards.
14	Used by hard disk.

Note that the comment "available for expansion cards" in the above table may not be strictly true. Additional devices not anticipated in the AT specification, such as sound cards, may be included in the base configuration of your PC or PC motherboard, these devices may use these IRQs listed above as available.

Some Dialogic voice boards are shipped with a default IRQ of 3. As the above table shows, IRQ 3 is used by an internal serial port or modem on COM 2. Make sure that that any COM 2 device provided by an serial communications card or your motherboard is disabled. If you have serial ports on your motherboard, you can often disable these ports in the CMOS setup which can be selected when you power-on reset your computer.

IRQ 2 and IRQ 9 are not as simple as other levels, you will sometimes see documentation refer to IRQ 2/9. (Some Dialogic hardware and software installation documentation refers to IRQ 9, this does not follow the industry-standard notation for interrupts). To explain this, we must touch on some of the complexity that lies "under the hood" of the interrupt hardware. The IBM XT had 8 interrupt lines, IRQs 0 through 7. In the AT, a new set of 7 IRQs was introduced by using a hardware engineering trick. IRQ line number 2 is shared by all devices which use the new interrupt numbers 10 - 15. An XT-style board which generates a hardware interrupt number 2 will actually produce a hardware interrupt 9 on the AT, as will devices using IRQs 10 - 15. The AT BIOS traps interrupt 9 and sends it to driver software as if an IRQ 2 had happened on an XT. In this way, software and hardware developed for the XT will run unchanged on the AT. New hardware developed for the AT must check the second interrupt controller to distinguish between an interrupt 9 which came from IRQ 2 or from IRQ 10 - 15. (Yes, this is confusing.) You will sometimes see this shared line referred to as IRQ 2/9, this is the line that Dialogic has sometimes called IRQ 9 its hardware manuals. You may need to experiment with setting 2 or 9 in your software or hardware setup to make a board work at this shared IRQ.

Installing Boards 247

Hardware Interrupt Priorities

A hardware interrupt is generated when a device such as the hard disk controller or a Dialogic voice card needs attention from the CPU. It is possible that more hardware interrupts will happen before the CPU has finished taking care of the previous interrupt. The interrupt controller hardware on the PC motherboard is responsible for deciding the order in which interrupts are serviced in this situation. The rule for the IBM XT is simple: the smaller the IRQ, the higher the priority. In other words, IRQ 0 is guaranteed to be serviced ahead of any other interrupts, IRQ 7 will be serviced last. On the AT, this is complicated by the "cascading" mechanism used for IRQ 2/9. IRQs 2/9 and 10-15 come between IRQ 1 and IRQ 3 in priority, as shown in the following table.

IRQ	Priority
0	Highest
1	
2/9 and 10 - 15	↑
3	Higher
4	Lower
5	↓
6	
7	Lowest

If you have a network adapter, it is important that the Dialogic board be installed at a higher priority IRQ than the network, otherwise audio data may not be transferred quickly enough to the voice channels, giving audio "dropout" effects as Dialogic buffers underflow. For example, if the network adapter is at IRQ 4, then the Dialogic board could be installed at IRQs 2, 3 or 10 - 15, but not at IRQ 5 or 7.

Software Interrupts

Software interrupts are numbered from 0 to 255 (00 to FF hex). The interrupt number is often called the interrupt vector number. When you read about "interrupts", make sure you know whether the author is talking about software or hardware interrupts. One way to tell the difference is that hardware interrupts can't have values larger than 15, and are generally specified as decimal numbers, software interrupts have values up to 255 (FF hex) and are generally specified in hex.

The Dialogic driver D40DRV is a TSR which uses a software interrupt, default value 6D hex. Parity Software's FREEVECT utility can be used to look for free interrupt vectors or to check if a given vector is in use. (FREEVECT is provided, for free, with all Parity Software products). Type FREEVECT [Enter] for a brief summary of how to use the utility. Look for a free vector before loading D40DRV. If 6D is in use, try to find another vector in the range 60 to 6F hex or 40 to 4F hex.

Use Parity Software's FREEVECT utility to check for software interrupt usage before loading D40DRV or other voice driver software. If you use it after loading D40DRV, it will show that the vector is in use but will not be able to report if the vector was used by something else before D40DRV was loaded.

Most expansion cards do not require a software interrupt. Dialogic voce cards use a software interrupt because the driver software is provided in a TSR, rather than in a true MSDOS device driver which would be installed in CONFIG.SYS.

The default Dialogic software interrupt vector number 6D is used by many types of VGA BIOS. Be sure to check that the vector you use is free. The vector is specified by the D40DRV -I command-line option. If 6D is taken, we recommend trying 60 or 48 as your next two choices. Note that D40DRV does not check that the vector is free before installing itself.

I/O Port Addresses

An I/O port is a way for the Intel CPU to communicate with an expansion card in the bus. Port numbers range from 200 to 3FF hex. I/O port addresses in common use on AT-compatibles are shown in the following table. Dialogic D/xx voice cards available at the time of writing do not use I/O port addresses, however some additional cards such as the FAX/120 fax card, do use a range of I/O ports.

Installing Boards

I/O range (hex)	Usage
200	Available (Used by SoundBlaster card)
201	Game adapter
202-277	Available
278-27F	Second printer port (LPT 2)
280-2F7	Available (COM 4 may use 2E8-2EF)
2F8-2FF	COM 2 (serial port or internal modem)
300-377	Available
378-37F	First printer port (LPT 1)
380-3AF	Available
3B0-3BF	Monochrome video/parallel port card (MDA)
3C0-3CF	Available
3D0-3DF	VGA video
3E0-3EF	Available (COM 3 may use 3E8-3EF)
3F0-3F7	Disk controller
3F8-3FF	COM 1 (serial port or internal modem)

Shared Memory

We will first briefly review the terminology of PC memory, which can certainly be confusing and complicated even for PC veterans. The notes here may help you understand the manufacturer's documentation.

Segment Address

Memory addresses are often specified as segment addresses. The address of a byte in memory can range from 0 (the first byte) up to the number of bytes of installed memory. The Intel chips have a strange system called segment:offset addressing where software refers to data in memory by giving a segment and offset. The CPU calculates the true address of the data by calculating address = segment×16 + offset. If a memory address is given as a segment, you can find the true address by multiplying by 16. For this reason, segment addresses are almost always given in hexadecimal notation, since multiplying by 16 in hex is as easy as multiplying by 10 in decimal notation: just add a zero. For example, segment address A000 hex is true address A0000.

Conventional Memory, also known as Lower Memory

Memory with addresses from 0 to 640 Kb (segments 0000 to A000 hex). Sometimes the term "conventional memory" is used to include both lower and upper memory.

Extended Memory

Memory with addresses greater than 1 MB (segment FFFF hex). This memory cannot be accessed directly from a DOS application unless it knows how to do really fancy tricks like switching to protected mode some of the time (an example of a program which can fully utilize extended memory is Parity Software's 32bit VOS call processing engine for MSDOS).

Upper Memory

Memory with addresses from 640 Kb to 1 MB (segments A000 to FFFF hex).

Upper Memory Block (UMB)

Extended memory which has been "mapped" by a memory manager to appear in upper memory. Upper Memory Blocks are used to fill areas which are not used by expansion cards such as display adapters and voice boards. Having upper memory blocks means that MSDOS programs have more memory available for storing data. It may also be possible to load TSRs and device drivers in Upper Memory Blocks if they are large enough and if the developers of the TSR or driver software have not made assumptions about where in memory they will be loaded.

Real Memory

Sometimes called DOS memory. Real memory includes Lower + Upper Memory: i.e. memory with addresses from 0 to 1 MB (segments 0000 to FFFF hex). This is the range of memory addresses which can normally be accessed by MSDOS programs..

Expanded Memory (EMS)

EMS is additional RAM, up to 8 MB. This memory is accessed through a window, called the EMS Page Frame, which is provided as a fixed 64 Kb range in upper memory. An EMS device driver is used to move the window to make different areas of EMS accessible through the page frame. Often, extended memory is used to provide expanded memory by using a device driver such as MSDOS EMM386 or as part of the functionality of memory managers such as QEMM and 386MAX. Pro: makes more than 1 MB available to a real mode application. Con: it's a

Installing Boards

slow way to access memory and complex to program. The Dialogic driver can store buffers in EMS, which can free lots of DOS memory.

High Memory Area (HMA)

A block of just less than 64 Kb above upper memory in segment hex FFFF. The HMA is the first 64 Kb of extended memory. The HMA is only available if the CPU is an 80286 or later.

Shared Memory

Also called shared RAM, this is memory mounted on an expansion card rather than the PC motherboard which is shared by the card and the PC's CPU. For example, the 4 Kb of memory starting at segment B800 is on the VGA adapter and is used to hold the 25 row x 80 column screen visible in character mode.

The Dialogic voice cards have shared RAM which is used to copy digitized audio data to and from the hard drive when recording and playing audio files. By default, this memory starts at segment D000. It is important to check that nothing else is using memory in the Dialogic shared RAM range. Points to check include:

The Page Frame address.
The page frame is created in upper memory by the EMS driver, which may be EMM386, QEMM or some other driver Some memory managers are able to detect the Dialogic board automatically, others, including EMM386, require a command-line parameter or configuration file entry specifying regions of shared memory in order that an alternative address is used. Typically the option is called something like EXCLUDE=xxxx-yyyy or X=xxxx-yyyy where xxxx to yyyy is the range of segment addresses, QEMM versions at the time of writing use ARAM=xxxx-yyyy on the DEVICE= QEMM386.SYS entry in CONFIG.SYS. We recommend that you explicitly exclude the memory region used by the voice cards even if the memory manager is able to identify it automatically.

Upper Memory Block (UMB) addresses.
UMBs are created by your memory manager, so again you must be sure that any required option is in place to specify where installed RAM is placed.

BIOS shadowing.
Some PCs have the ability to copy BIOS code into a re-mapped area of upper memory RAM. This can give a performance improvement because the ROM chips which store the BIOS are usually much slower than RAM. BIOS shadowing is enabled and disabled through the CMOS setup which is available when you do a power-off reset of the PC. If you have difficulties getting a Dialogic board to function and you don't know the address used for BIOS shadowing on your PC, make sure shadowing is disabled.

The Dialogic driver D40DRV by default will scan all of upper memory looking for voice cards. This may cause problems, including a lockup of the PC, when there are UMBs containing data used by TSRs or when shared RAM is present on other cards. The solution is to use the S command-line option, which restricts the range of addresses which D40DRV will scan. For example, if you are installing a D/21D with base address at segment D000 the card uses memory from D000-D0FF and the command line argument SD000/D0FF should be used. A four-channel D/41D card uses twice as much shared memory, the option might then be SD000/D1FF.

Windows 3.1, Windows for Workgroups and Windows 95 include built-in upper memory management software. You should make sure that Windows is informed about the shared memory regions used by your voice card. This is done via the EMMExclude= setting in the [386Enh] section in SYSTEM.INI. This should specify the range used by the Dialogic board(s) in a similar way to the S option to D40DRV. For example, if you are installing a D/21D board at base segment D000 you should specify EMMExclude=D000-D0FF.

To learn more about Windows SYSTEM.INI settings and troubleshooting memory management and expansion card issues, we

Installing Boards 253

highly recommend the inexpensive Windows Resource Kits available from Microsoft.

DMA Channels

Direct Memory Access (DMA) channels are high-speed data paths between expansion cards and your motherboard memory which bypass the CPU. Dialogic boards available at the time of writing do not use DMA channels. However, many multimedia sound cards do require DMA. Channels are classified as 8 or 16-bit. The following table lists the DMA channel assignments on a standard AT-compatible PC.

DMA Chan	8/16 Bit	Usage
0	8	Available
1	8	Synchronous host communications adapter
2	8	Disk controller
3	8	Available
4	N/A	(Cascade from first DMA controller)
5	16	Available
6	16	Available
7	16	Available

Trouble-Shooting Checklist

If you have trouble with your voice card, these are some important items to check:

Voice card IRQ conflicts with another device,

Voice card shared RAM region conflicts with the EMS page frame, BIOS shadowing or a UMB,

D40DRV software interrupt conflicts with the VGA BIOS or other software,

Run diagnostic software provided with the board, such as Dialogic's D40CHK.

Trouble-Shooting Strategy

If you are unable to identify an installation problem then the best approach is to start with the simplest possible hardware and software configuration and check if the voice board works in that configuration. If it does not, you may have a defective board. If the board does work, then gradually add the components you removed until you get a failure. Then you know that there is a conflict with the last component added. To get a simple starting configuration:

Remove all expansion cards except the voice board.

Rename startup files such as AUTOEXEC.BAT and CONFIG.SYS files so that the PC "clean boots".

Disable all devices and options in your CMOS setup including communications and printer ports, BIOS shadowing etc.

Don't load any resident software (including DOSKEY, disk cache software etc.) after the PC boots.

Chapter 16

Dialogic Product Line

Introduction

To understand the full Dialogic product line, we will first summarize the different types of connection that voice processing systems may make with the outside world, the different connections that may be made between voice boards, and the different capabilities which such systems may incorporate.

A block diagram of a generic voice processing system looks like this:

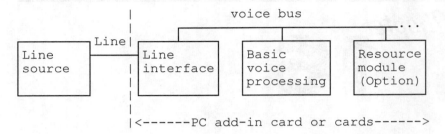

Generic voice processing configuration. Everything to the right of the vertical line is inside the PC, the "Line" is the trunk line connecting the PC to the outside world. Typically, "Line source" will be the telephone company or a phone system.

The elements of the diagram are as follows:

Line source
No general term for this seems to exist in the industry, so we made one up. By *line source*, we mean "the thing that's at the other end of the phone line". Typical examples will be:
    The phone company
    A business phone system (PBX).

Line
This is the physical connecting wire from the line service to the PC. Another name for this might be a *trunk*.

Line interface
This is the component which has a connector to the phone line.

Voice bus
The voice bus transfers audio and signaling information between different boards which together make up the voice processing system. Most D/xx boards offer a voice bus connector of some type. Some simple configurations will not make use of a voice bus.

Basic voice processing
There are some voice processing features which are always present in a Dialogic-based voice processing system. Which features are included

(eg. DTMF recognition) and which features are omitted (eg. voice recognition) is really a historical and technological accident; the feature list is based on the original and highly successful D/4x product line. These features will be provided by one of the members of the D/xx product line, and will include:

- Recording digitized audio
- Playing back digitized audio
- Dialing DTMF and pulse digits
- Recognizing DTMF digits
- Detecting and answering an incoming call
- Initiating an outgoing call
- Disconnecting a call
- Detecting a disconnect.

Other features, such as call progress analysis or speed/volume control, may or may not be present depending on which model D/xx board is being used. Some older models may lack some features which are standard in current models.

Resource

Additional voice processing capabilities may be provided by additional components connected through a voice bus. Examples include:
- Voice recognition
- Fax
- Text-to-speech.

The resource part of the configuration is optional: many applications need no more than the basic voice processing functions provided by the D/xx series.

Types Of Trunk

There are several different types of telephone line, which we summarize here. Dialogic provides boards which support one or more variants of each of these types.

Two-wire analog

This is the most familiar type of phone line, the type you probably have at home. It is generally carried on two-wire, *twisted-pair* cables. One end, generally a switch such as a PBX or phone company central office switch, provides power for loop current and ring voltage. The other end, typically a telephone, draws its power from the line and provides for dialing numbers.

Proprietary Analog
Most business phones have buttons and lights which communicate with the PBX. These lines are generally two-wire or four-wire analog lines which carry additional signals not supported by "normal" telephones.

DID analog
Electrically, DID analog lines are similar to "regular" analog lines: sound is carried by varying loop current. However, the connection is "backwards": the power is provided by the customer's equipment, digits are provided by the phone company switch.

Digital
Digital trunks send information as a stream of bits rather than by varying current. The best-known type of digital trunk in the US is T-1, which carries up to 24 conversations on two twisted-pair lines: one for received audio and one for transmitted audio. Unlike analog lines, which just carry sound, digital trunks separate receive and transmit channels. Outside the US, many countries provide digital service on E-1 trunks, which carry up to 30 conversations. ISDN service may be provided on trunks which are very similar to T-1 or E-1 service, except that one or more channels (*D-channels*) are dedicated to carrying signaling information such as ANI or DID digits for all conversations.

Voice Buses

(In case the spelling "buses" looks wrong to you, we should point out Webster's New Collegiate Dictionary allows both "buses" and "busses" for the plural of "bus").

A voice bus looks physically like a flat ribbon cable with two or more connectors. This cable will be used to connect two or more Dialogic cards.

Dialogic Product Line 259

There are four types of voice bus supported by the Dialogic product line: the Analog Expansion Bus (AEB), the PCM Expansion Bus (PEB), the SC bus and the SCX bus.

The AEB might be visualized as a bundle of four analog phone lines carrying audio and signaling information.

The PEB is a digital bus which carries 24 or 32 digital channels in a similar manner to a T-1 or E-1 trunk.

The SCSA specification includes two digital buses. The SC Bus is a "super-PEB" for connecting different voice boards within one PC. Unlike PEB-based boards, SC bus-based boards all include switching and routing functions. Instead of a maximum 32 channels, the SC bus carries 1024 channels, which may optionally be combined into "super channels" capable of carrying higher data rate information such as video or high-fidelity audio. A second bus, the SCX Bus, connects two or more PCs allowing for configurations exceeding the capacity of a single PC platform.

The AEB and PEB buses are completely incompatible: a PEB-based board cannot be connected to an AEB board, and vice versa.

The D/4x Board

The last sections got pretty complicated, so just to put things in perspective we'll describe the D/4x board. The D/4x includes everything required to create a voice processing system in a PC: a line interface and basic voice processing features. The board looks something like this:

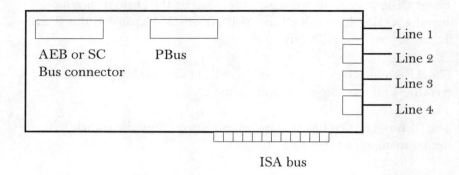

The Dialogic D/4x board.

The connectors on the board are as follows:

ISA bus. The PC bus connector. The D/4x is an 8-bit, full-length card designed for the IBM PC/XT, AT and compatible computers.

Line 1 .. Line 4. The classic Dialogic boards (D/41B, D/41D, D/41E) support up to 4 phone lines, hence the "4" in D/4x. The telephone line connectors are standard modular RJ-11 jacks as would be found in any US home. Other Dialogic boards support different numbers of phone lines. For example, the D/21ESC supports two lines, and the D/160SC-LS supports 16.

AEB or SC bus connector. The *AEB (Analog Expansion Bus)* connector allows the D/4x to be connected to other voice processing boards which may take input from or create output for the analog voice signals processed by the D/4x. Examples of these boards are the AMX/81 switching matrix card, which can be used to "conference" D/4x channels, or the VR/40 speaker-independent voice recognition board. An SC bus connector allows a connection to other SC bus boards, which may provide services such as voice recognition or text-to-speech. The D/41ESC board has both AEB and SC bus connectors. Older boards, including the D/41D, have only an AEB connector. Boards built for lower cost, including the Proline/2V, have no expansion bus connector.

PBus. The on-board processor bus. This is used on certain older D/4x model boards, including the D/41B and D/41D to connect daughter-boards such as the MF/40, which adds MF digit detection capability, and the VR/40, which adds voice recognition capability.

D/4x product line boards are designated D/NCR where:
N is the number of lines, which will be 2 or 4.
C is 0 or 1, indicating support for call progress analysis (0=no, 1=yes).
R is A, B or D, indicating a major release level.
A and B boards carry firmware in EPROM. D and later are DSP-based, firmware is downloaded from a disk file. New features may be added by firmware releases without a hardware upgrade.

For example, the D/40B has C=0, meaning no call progress analysis, and is a B board, meaning that it does not support downloaded firmware.

A DSP (Digital Signal Processor) is a microprocessor specially optimized for processing signals such as audio or video information in real-time.

To a large extent, D/4x models are backwards-compatible: software written for older boards and older device driver releases will work with later versions.

Dialogic has several versions of the D/4x series approved for use in many different countries outside the US. The list of approved products is continually growing, check with your hardware vendor for the latest list.

The D/2x Board

The D/2x is a two-line version of the D/4x. Instead of two RJ-14 jacks there are two RJ-11 jacks: one for each line. The D/2x is very similar in capabilities to the D/4x, except that the VR/40 daughter-board is not compatible with the D/2x. Only 1000 hex (= 4096 decimal) bytes of shared RAM are used by the D/2x.

The D/42-xx Boards

D/42 cards support proprietary analog interfaces used by certain common PBXs. Those currently included are:

    D/42D-SX   For the Mitel SX.
    D/42D-SL   For the Northern Telecom SL-1.
    D/42-NS   For the Norstar KSU.

These products support most D/41D features, and several capabilities related specifically to the supported PBXs.

The AMX/8x Board

The AMX might be described as a PBX card for the PC. It provides an 8x8 switch: channels to be switched may be AEB devices, such as D/4x cards, or telephones or headsets connected directly to the AMX. In this analogy, the D/4x functions as a trunk card for the PBX, the AMX itself the switch and the station interfaces. The card looks something like this:

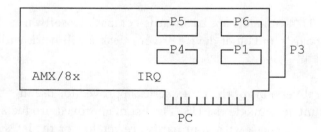

The Dialogic AMX/8x card.

The connectors are:

P1, P4, P5 and P6
These are AEB connectors. An AEB cable may be run from each one of these connectors to a D/4x or other AEB device. P5 and P6 are on the X axis of the 8x8 switch, P1 and P4 are on the Y axis.

P3

This connector provides loop current and may be connected to up to 8 telephones, headsets, fax machines or other telephone-like devices. A ring voltage may be applied to a device attached to P3. P3 is also on the Y axis of the switch.

PC

The AMX is a full-length, 8-bit card which may be installed in an XT- or AT-compatible PC. The AMX requires no shared RAM or I/O addresses.

IRQ

If the AMX is to detect seizure and disconnect (on- and off-hook) on a P3 station, an interrupt must be assigned to the board. This interrupt may not be the same as any other board installed in the PC (except for other AMX boards, which may share the same interrupt). Most AMX applications may be run without using an interrupt.

As a switch, the AMX might be visualized as follows:

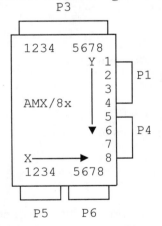

The AMX is an 8x8 switch. The P connectors, except for P3, are AEBs. P3 is a station interface for telephones etc. P1 and P4 are the Y axis, both P3 and P5,P6 are the X axis.

A significant problem with the AMX is a loss in gain. There is between 3dB and 6dB loss in gain (volume, amplitude) when a conversation is passed through the AMX. This may make the sound unacceptably faint.

Dialogic recommends that normally no more than two or three channels be connected (conferenced) through AMX switch-points. This problem may be rectified with line amplifier devices, but these add significant cost to the system.

Typical AMX configurations are as follows.

1 AMX and 1 D/4x
This allows any two or three lines on the D/4x to be connected together.

1 AMX, 1 or 2 D/4x's and 1 to 8 devices connected via the P3
The external devices could be operator headsets or Intel SatisFAXtion boards, for example. This allows any channel on a D/4x to be connected to the external device. You can't add another D/4x to this configuration since the remaining two AEB connectors on the AMX are on the same axis as the external devices.
An SA/102 adapter will usually be needed in this type of configuration. The SA/102 attaches to the P3 and provides a cable running to 8 RJ-11 jacks to which the external device(s) can be connected.

1 AMX and 2 to 4 D/4xs
This allows any two, three or four lines to be connected, whether they are on the same or different D/4x boards.

More than one AMX may be installed in one PC: they may even be connected via an AEB to extend the dimensions of the switching matrix.

The MF/40

The MF/40 is a daughter-board for older versions of the D/4x line, such as the D/41B, which does not support MF detection. The MF/40 is mounted on the μ-bus of the D/4x and is also connected via an AEB cable. When attached, the D/4x will be able to recognize MF digits as an alternative to DTMF digits. The MF/40 requires no additional IRQs, shared memory addresses or I/O ports.

The VR/40

The VR/40 is a daughter-board for the D/4x lines which is mounted on the μ-bus of the D/4x is also connected via an AEB cable. The VR/40 adds voice recognition capabilities. Typically, a vocabulary will be able to recognize spoken digits "zero" to "nine" and a few simple words such a "yes" and "no". The recognizer is implemented in firmware and vocabularies developed by Voice Control Systems (VCS). The VR/40 requires no additional IRQs, shared memory addresses or I/O ports.

The DID/40

The DID/40 is an external unit which provides a connection between D/2x or D/4x boards and analog DID trunks.

The DTI/124

The DTI/124 provides an interface from a T-1 trunk to 6 AEB connectors. Typically, these will be used to connect 6 D/4x boards.

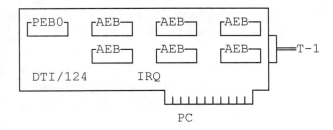

The Dialogic DTI/124 card.

Any T-1 conversation may be routed to any AEB channel. This allows sharing of resources. For example, if only one of the six D/4x boards has a VR/40 daughter-board, an incoming call may be routed to this board if there is a free voice recognition channel. If required, an AEBs could be connected to an AMX board. This would allow sharing of other types of resources: perhaps a live operator attached through a headset, or a third-party AEB-compatible voice recognition board such as a VPro-4 from Voice Processing Corporation (VPC).

The main function of the DTI/124 is to allow systems integrators to take advantage of existing inventory of D/4x boards. New T-1 configurations would usually be designed with PEB-based D/12x boards. Another use is to exploit resource boards, such as the Gammalink fax board, which only offer AEB interfaces. The DTI/124 allows such resources to be attached directly to a T-1 trunk.

The PEB0 connector allows two DTI/xx boards to be connected in a *drop and insert* configuration where two T-1 trunks are connected together through the PC:

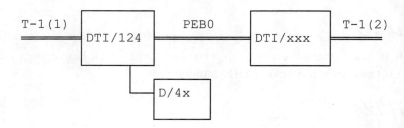

Drop and insert configuration. An incoming call can be "dropped" to the D/4x, providing voice processing services, and subsequently "inserted" into an out-going call from the second DTI/xx. Note that a D/4x on the T1(1) side cannot communicate with T-1(2); additional D/xx boards would be needed on the second DTI.

In a drop and insert configuration, a call on T-1(1) on the DTI/124 can be connected (*dropped*) to the D/4x for voice processing services, or connected through the PEB0 (*inserted*) into a call on T-1(2) attached to the other DTI/xx board.

A typical drop and insert application might be *talking yellow pages*. A caller can query a database of phone numbers to find, say, a florist in the local area. A menu option could provide the caller the option of being connected immediately with the florist. This could be accomplished by dialing out on the second T-1 and connecting this call with the original caller). In order to dial out, the second DTI/xx might also be a DTI/124 with one or more D/4x boards attached. Since a D/4x board might only needed for a few seconds when dialing and analyzing call progress, a full T-1 span on the out-bound side might be serviced with only one or two D/4x boards, depending on the details of the application.

Note that the PEB0 connector is not compatible with PEB connectors found on most other boards, such as the D/12x. "Normal" PEB connectors are occasionally called PEB1 to distinguish them from PEB0. PEB0 connectors are only found on DTI/100, DTI/101, DTI/124 and DTI/211 boards.

No switching is available on the PEB0: channel 1 on T-1(1) is "hard-wired" to channel 1 on T-1(2), etc.

The DTI/124 requires an IRQ (which may be shared with other DTI, DMX and MSI boards) and an I/O port address, though no shared RAM addresses.

The DTI/100

The DTI/100 is like a DTI/124, except that there are no AEB connectors. Its purpose is to provide a connection to a second T-1 in a drop and insert configuration. Its only active capability is to pulse dial numbers, which might be utilized to complete an out-going call through a PBX or phone company switch (although most do not support pulse dialing on T-1).

The LSI/120

The LSI/120 provides a connection between up to 12 analog phone lines and a PEB bus.

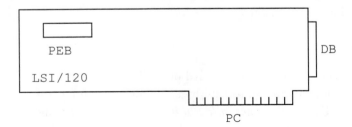

The Dialogic LSI/120 board

There is an 8-bit, AT- (not XT-) compatible PC bus connector. No IRQ, I/O port address or shared memory is required by an LSI/120.

The connector marked DB in the above diagram is a 25-pin DB connector for 12 lines (there is therefore one unused pin). An SA/120 adapter and cable combination attaches to DB and provides 12 RJ-11 sockets for telephone lines.

The LSI/120-JP is a variant of the LSI/120 approved for use in Japan (as part of a complete approved platform).

The DID/120

The DID/120 provides a connection between up to 12 analog DID trunks and a PEB bus.

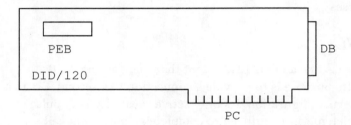

The Dialogic DID/120 board.

In most respects, the DID/120 is very similar to the LSI/120 board, except that a power supply is needed to provide loop current. One power supply can serve one or two DID/120 boards.

A typical DID configuration would support 12 DID trunks with a DID/120 and one D/12x card.

The D/12x

The D/12x series is a 12-channel board providing basic voice processing services and a connection to a PEB bus. It is important to realize that the D/12x does *not include a line interface*: at least one additional PEB-based board must be included to connect the D/12x to the outside world. The board looks something like this:

Dialogic Product Line **269**

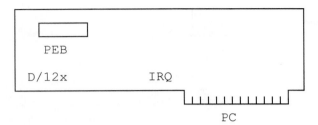

The Dialogic D/12x board

The D/12x series uses a DSP processor to provide essentially the same features as three D/4x boards. DSP-based voice processing platforms such as the D/41D and D/121B are generically called *SpringBoards*, the DSP firmware is called *SpringWare*.

One IRQ is needed by the D/12x (which may be shared with other D/xx boards), 6000 hex (24576 decimal) bytes of shared RAM in the D000 or A000 segment, and one I/O port address. The PC bus connector is 8-bit, but AT-compatible only (not XT).

There are currently three members of the D/12x series: the D/120, D/121A and D/121B. The D/120 is a now obsolete board which lacks support for call progress analysis. The D/121B supports the same firmware capabilities as the D/41D. The D/121A was similar to the D/121B but had less memory and a slower processor, and therefore lacked support for speed/volume control and could not handle as many global tone detection templates at one time.

A minimum configuration involving a D/12x is to use an LSI/120 and a D/12x in combination to provide 12 lines of analog voice processing capacity:

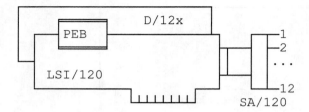

Minimum configuration with a D/12x and LSI/120 connecting to 12 analog phone lines. The SA/120 adapter converts from the LSI connector to 12 RJ-11 jacks. A PEB cable connects the D/12x and the LSI.

An LSI/120 and D/12x combination provides almost identical capabilities to three D/4x boards, but uses one fewer slot.

The FAX/120

The FAX/120 is a PEB resource board providing 12 channels of fax store and forward. It takes a full-length slot. Unlike most resource modules, it is not independent: it effectively adds fax capabilities to a D/121A or D/121B channel, it cannot be installed without being "tied" to a D/12x board. It may not, for example, be installed on a PEB by itself and shared through a DMX, as other resource boards may.

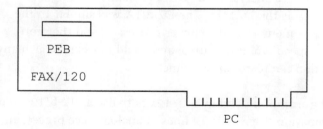

The Dialogic FAX/120 board

The FAX/120 is capable of recording and playing back fax data in the same way that a D/xx can record and play back audio data. The current firmware does *not* support conversion of ASCII or graphics files to fax format; planned future releases will add this capability.

The FAX/120 requires a range of I/O port addresses, but no additional IRQs or shared RAM.

If a FAX/120 channel is playing or recording fax, the corresponding D/12x channel is "busy" and may not be used for other purposes.

The FAX/120 currently does not support 32-channel PEBs or A-law PCM encoding. This means that in an E-1 installation, a DMX will be needed to convert PEB types, and the FAX/120 must be installed on a different PEB than the DTI/212.

A minimal FAX/120 configuration would be: 1 LSI/120, 1 D/121B and 1 FAX/120, which would provide 12 channels of simultaneous voice and fax.

The VR/121

The VR/121 is a PEB resource board which provides 12 independent voice recognition channels.

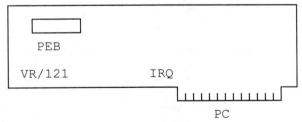

The Dialogic VR/121 board

It requires an IRQ, which will be shared with D/xx boards. A range of I/O port addresses is needed, but no shared RAM.

The recognition firmware and vocabularies are based on technology provided by Scott Instruments. Like other voice recognition boards, the vocabularies typically consist of the digits "zero" through "nine" and a few simple words such as "yes" and "no".

The VR/121 is an independent resource board which may be installed on a separate PEB and shared through a DMX if required. Any one PEB channel can be routed to any voice recognition channel.

The VR/xxp

The VR/xxp provides an alternative PEB-compatible voice recognition platform.

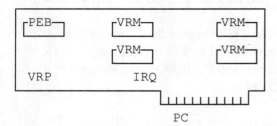

The Dialogic VR/xxp card.

The VR/xxp starts with a baseboard, the VRP, to which can be added from one to four VRM daughter-boards, each of which has four voice recognition channels. The recognition technology is based on the same Voice Control Systems firmware as the VR/40. Otherwise, the board provides similar capabilities to the VR/121.

The MSI

The MSI is used to connect a PEB to up to 24 operator headsets.

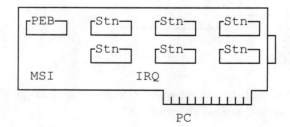

The Dialogic MSI card. "Stn" indicates an MSI station card; from one to six may be installed on the MSI baseboard.

From one to six daughter-boards, known as MSI Station Interfaces, may be installed, each one for servicing up to four headsets. An MSI Power Module (an external device) is needed to provide loop current to the Station Interfaces, one Power Module can serve one or two MSI boards. An SA/240 adapter will usually be used to convert between the MSI connector and 24 RJ-11 jacks for the headsets.

Any PEB channel can be connected to any station: the MSI acts as a 24 x 24 switch.

Note a very important limitation of the MSI: it cannot ring a station. Regular telephones are therefore not usually suitable for attaching to an MSI since the MSI cannot signal an incoming call by ringing the phone. As an alternative, the MSI can announce a call to a headset by playing a message or a by a special tone, called a "zip" tone

The MSI uses an IRQ, which may be shared with other DTI-class boards such as DTI/xx and DMX boards, and a range of I/O port addresses. Shared RAM is not needed.

The MSI/C (MSI Revision 2)

The MSI rev. 2, also known as the MSI/C, provides the same capabilities as the MSI with conferencing added. A conference is a conversation between three or more parties. A conference may be between parties connected to the MSI via the PEB, via headsets, or both. With the current release of firmware (8/93), the MSI rev. 2 may be configured in two ways, allowing for a maximum conference of 4 or 8 parties respectively. If a maximum size of 8 is selected, this reduces the total number of conferences that can be in progress simultaneously.

Since the MSI does provide gain control, the MSI provides an alternative to the DMX (see next section) when drop and insert is required on a single PEB with an analog interface card such as an LSI/xx or DID/xx. The MSI rev. 2 conferencing feature can be used to connect an inbound caller to an outbound caller with adjustment of the gain. The MSI rev. 2 might then be configured without station cards, in which case it would not require a power supply or SA/240.

The MSI/SC

The MSI/SC is similar to the MSI/C except that it provides an SC bus interface rather than a PEB interface. The MSI/SC-R additionally is able to ring stations.

The DMX

The DMX (Digital Matrix Switch) provides switching capabilities for from one to four PEBs.

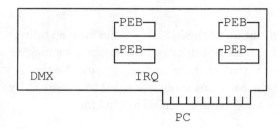

The Dialogic DMX card.

The DMX has only one function: to connect and disconnect conversations. The DMX can route the receive and transmit half of a conversation separately. A receive channel can be connected to one or more transmit channels on the same or on a different PEB. A typical example of transmitting a receive channel to more than one transmit channel would be for a listen-only type of application such as a lecture, where many callers could listen to one speaker but not respond, or for music on hold. Usually, one caller will be connected to one other caller, making a call which is passed through the PC. The DMX is not able to create conferences of two or more callers where all parties can talk to all other parties, this requires an MSI rev. 2 or third-party conferencing device.

The DMX has three major applications:
1. Connecting two callers in drop and insert configurations.
2. Sharing resources.
3. Converting from T-1 to E-1 type PEBs.

A typical drop and insert configuration would have a DTI/211, two D/121Bs and one DMX, all of which would be on one PEB. The D/121Bs would be able to answer incoming calls, the DMX could be used to connect the caller to a second outbound call made on another channel.

Dialogic Product Line

A typical application of sharing resources might have four LSI/120s, four D/121Bs, one VR/40p and one DMX. Four LSI/120s (48 lines) cannot be installed on one PEB since the maximum number of conversations on a T-1-type PEB is 24. If voice recognition were required as an occasional option, and the DMX were not available, a VR/40p would have to be installed on each PEB. Using the DMX, two LSIs would be attached on one PEB, the second two LSIs on a second PEB, and the VR/40p on a third PEB, all of which would be attached to the DMX:

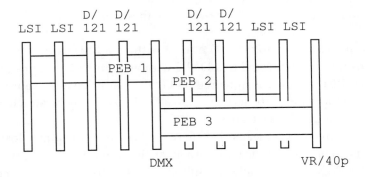

DMX used to share a VR/40p between four LSIs. This view (we are looking down at the top of the PC rather than from the side) shows the three PEB cables.

This configuration allows a call on any of the four LSIs to be routed to the VR/40p, sharing the four voice recognition channels between 48 lines.

There are several important limitations of the DMX to be aware of when designing a configuration:

1. The DMX cannot detect changes in signaling bits on an attached PEB.

2. The DMX cannot generate signaling bits on an attached PEB. The exception is that it can generate a fixed A=1 value, which it will do automatically when there is no other source of signaling, there is no application software control over this feature.

3. The DMX does not pass signaling bit values through when a connection is made.

4. The DMX does not have any way to correct the gain (volume) of the audio on a connection. In particular, it cannot compensate for the loss caused by making connections through two analog interfaces such as LSI/xxs.

Since the DMX will be made obsolete when the SC Bus becomes available (see the SCSA chapter), it appears that Dialogic is not planning to improve the DMX to overcome these limitations.

The consequences of limitations 1, 2 and 3 is that an LSI/xx or DID/xx cannot be connected to a DMX without a D/xx board present on the *same* PEB to control signaling. DTI/xx boards have the capacity both to detect and generate signaling, so a DTI board may be attached via a PEB without a D/xx board.

The lack of gain control means that there may be an unacceptable loss of gain if a call is passed twice through an analog interface card (LSI, DID) via a DMX. For example, a configuration with one LSI/120, one D/121B and one DMX could be used for drop and insert to connect an inbound call to an outbound call, but there will be a loss in gain in making the connection which may not be acceptable in some environments. (However, some Parity Software customers have been able to install this configuration with satisfactory results).

The DTI/211

The DTI/211 is a PEB interface board for a T-1 trunk.

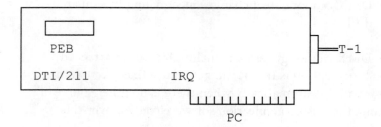

The Dialogic DTI/211 board

An IRQ is required, which may be shared with other DTI, DMX and MSI boards, and a range of I/O port addresses, however no shared RAM is required.

A typical PEB configuration for a T-1 would have one DTI/211 and two D/121Bs.

The DTI/101 is an older revision of the DTI/211. Unlike the DTI/211, it does not require that the DTI driver be loaded when simple configurations are used. This may save DOS RAM in the lower 1MB, which may be important. The DTI/101 also supports pulse dialing using the transmit A bit, which the DTI/211 does not support. If two DTI boards are connected via a DMX, the DTI/211 must be used. The DTI/101 is compatible with the DMX as long as only one DTI board is present.

The DTI/212

The DTI/212 is a PEB interface board for an E-1 trunk.

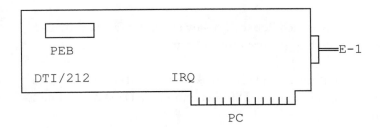

The Dialogic DTI/212 board

An IRQ is required, which may be shared with other DTI, DMX and MSI boards, and a range of I/O port addresses, however no shared RAM is required.

A typical configuration for a E-1 would have one DTI/211 and four D/81As.

The D/81A

The D/81A is very similar to the D/12x, except that there are 8 instead of 12 voice processing channels. It is a DSP-based board run by downloaded firmware.

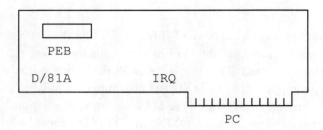

The Dialogic D/81A board

One IRQ is needed by the D/81A (which may be shared with other D/xx boards), 4000 hex (16384) decimal) bytes of shared RAM in the D000 or A000 segment, and one I/O port address. The PC bus connector is 8-bit, but AT-compatible only (not XT).

The D/81A can be switched between A-law encoding (as used by E-1 trunks) or μ-law encoding (as used by T-1 trunks).

A typical D/81A configuration will consist of one LSI/80-xx and one D/81A for 8 analog lines, or one DTI/212 and four D/81As for one E-1 trunk (30 channels).

The LSI/80-xx

The LSI/80-xx provides a connection between up to 8 analog phone lines and a PEB bus.

Dialogic Product Line

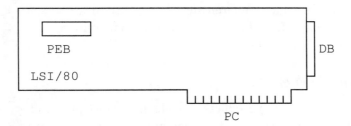

The Dialogic LSI/80 board

There is an 8-bit PC bus connector. No IRQ, I/O port address or shared memory is required by an LSI/80.

The connector marked DB in the above diagram is a 25-pin DB connector for 8 lines. An SA/120 adapter and cable combination attaches to DB and provides 12 RJ-11 jacks for telephone lines (8 of which will be used).

There are several members of the LSI/80-xx series approved for use in different countries, at present including:

LSI/80-DK	Denmark
LSI/80-CH	Switzerland
LSI/80-UK	United Kingdom

There may be additions to this list as more approvals are obtained.

The DCB (Spider) Board

The *DCB* (previously known as the *Spider*) is a pure conferencing board.

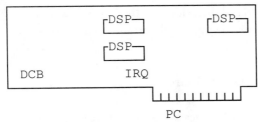

The Dialogic DCB board

The DCB board is used for conferencing. Three DSP processors are included on the board. Each processor can create one or more conferences of up to 32 parties.

HD SC Bus Products

Dialogic *High Density* (*HD*) products provide more voice processing channels and optional trunk interfaces combined into a single card taking one slot. The HD series features an SC bus interface (see the SCSA chapter).

With the introduction of a new product line, Dialogic is introducing a new naming convention for the boards:

   c/nnvSC-ii

where:

c/	indicates the class of board: D/ for voice processing, LSI/ for analog interface, DTI/ for digital interface. Note that D/-type board may also include an interface, if so this is indicated by a –ii suffix to the board model name.
nn	is the number of channels.
v	is the major release level, starting at zero.
SC	indicates that there is an SC Bus connector
-ii	indicates the interface type:

   -LS   Loop start
   -T1   T-1
   -E1   E-1

If there is no -ii in the product name, this indicates that there is no trunk interface on the board (i.e., the interface must be provided through the SC Bus connector).

Among the most popular HD boards are:

D/160SC-LS	16-channel voice with loop start interface.
D/240SC	24-channel voice.
D/320SC	32-channel voice.

Dialogic Product Line **281**

  D/240SC-T1 24-channel voice with T-1 interface.
  D/300SC-E1 30-channel voice with E-1 interface.
  D/320SC   32-channel voice (no interface)
  LSI/161SC   16-channel limited voice (no play or record) with loop start interface.
  DTI/241SC   24-channel limited voice with T-1 interface.
  D/240SC-2T1 24-channel voice with two T-1 interfaces.

The HD boards will share several characteristics:

Common Shared Memory.

An HD board requires 32 Kb shared memory. However, this 32 Kb is shared between all HD boards installed in the system (older board architecture require each channel to have a 1 Kb region permanently assigned). This permits configurations with more than the older limit of 64 voice processing channels. The base address of the shared memory is selectable anywhere from hex 8000 to E800 on a 32 K boundary. 32 K = 8000 hex, so the permitted base addresses will be 8000, 8800, 9000, 9800, A000... and so on.

Common IRQ.

All installed HD boards share a single IRQ selected from 2/9, 3, 4, 5, 6, 7, 10, 11, 12, 14 or 15, selected by software.

Maximum 16 Boards Per System.

The HD architecture allows a maximum of 16 HD boards per system. Combined with the common shared memory, this allows systems with much larger capacities than the older limit of 64 voice processing channels per chassis.

PEB Or SC Bus Mode.

Most HD boards are configurable to run in either an SC Bus or PEB mode. Using PEB mode will enable PEB resource boards such as the VR/160p to be utilized in an HD configuration. SC bus routing features not offered by PEB are not available in PEB mode.

The D/160SC-LS

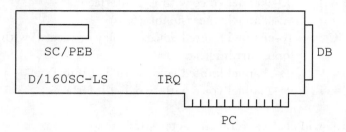

The Dialogic D/160SC-LS board

The D/160SC-LS offers 16 channels of voice processing and a DB37 connector (shown as DB) which can be broken out into 16 RJ-11 interfaces for loop start phone lines.

The D/240SC

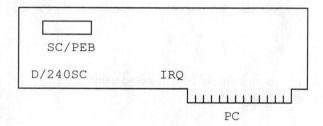

The Dialogic D/240SC board

The D/240SC offers 24 channels of voice processing. A trunk interface must be provided via the SC Bus connector.

The D/320SC

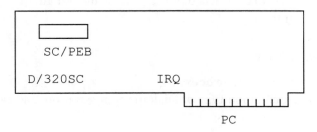

The Dialogic D/320SC board

The D/320SC offers 32 channels of voice processing. A trunk interface must be provided via the SC Bus connector.

The D/240SC-T1

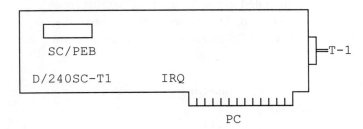

The Dialogic D/240SC-T1 board

The D/240SC-T1 offers 24 channels of voice processing and an interface to a T-1 trunk.

The D/300SC-E1

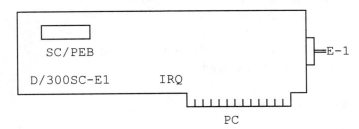

The Dialogic D/300SC-E1 board

The D/300SC-E1 offers 30 channels of voice processing and an interface to an E-1 trunk.

Typical Configurations

Analog lines with basic voice processing
This is by far the most common type of Dialogic configuration, and can be achieved in the following ways:

>   One or more D/2x boards.
>   One or more D/4x boards.
>   One LSI/120 one D/12x, SA/120.
>   One LSI/80-xx and one D/81A (in approved countries), SA/120.

These boards may be mixed: for example, a D/21B, D/41D, D/121B and LSI/120 may all be installed in the same PC. The differences between using D/2x and D/4x or D/12x are in the number of slots required and whether AEB or PEB buses are provided for future expansion with voice recognition, fax etc.

Digital lines with basic voice processing
A T-1 installation would be:

>   One D/240SC-T1,

An E-1 installation would be:

>   One D/300SC-E1.

Analog lines with optional drop to live operator
An AEB configuration would be:

>   From one to four D/4x boards and one AMX board.
>   SA/102 to connect the AMX to a headset or phone.

A SC bus configuration would be:

>   D/160SC-LS with MSI/160SC.

Dialogic Product Line 285

T-1 with drop to live operator (inbound/outbound call center)
This could be done in a PEB configuration as follows:

D/240SC-T1 with MSI/240SC.

Dialogic Product Line Summary

This table summarizes the most popular boards in the Dialogic product line, it is not intended to be complete.

Board	Function	Channels	Bus	Comments
D/2x	Basic VP	2	AEB	
D/4x	Basic VP	4	AEB	
D/8x	Basic VP	8	PEB	Needs LSI, DTI or other PEB interface board.
D/12x	Basic VP	12	PEB	Needs LSI, DTI or other PEB interface board.
LSI/120		12	PEB	PEB interface board for 12 analog lines. Needs SA/120.
DID/120		12	PEB	PEB interface board for 12 analog DID lines. Needs power supply, SA/120.
DID/40	DID	4	---	Connects D/2x or D/4x to analog DID lines.

Board	Function	Channels	Bus	Comments
MF/40	MF	4	AEB	Adds MF detection to D/4xA and D/4xB.
DTI/101		24	PEB	PEB interface board for T1. Has PEB0 for hard-wired drop-and-insert with second DTI.
DTI/211		24	PEB	Like DTI/101, newer revision.
DTI/212		30	PEB	PEB interface board for E1.
AMX	Switch	8x8	AEB	Can switch up to 4 AEB devices (typically D/4x boards), and up to 8 stations. Gain loss may be a problem. Needs SA/102 if connected to stations.
VR/40	Voice recog.	4	AEB	D/4x daughter-board. VCS recognizer.
VR/xxp	Voice recog.	4 to 16	PEB	Consists of VRP baseboard and 1 to 4 VRM daughter-boards. VCS recognizer.

Dialogic Product Line

Board	Function	Channels	Bus	Comments
VR/121	Voice recog.	12	PEB	Scott Instruments recognizer.
FAX/120	Fax	12	PEB	Needs D/121A or D/121B board in parallel. No A-law support, cannot convert ASCII or graphics to fax.
DMX	Switch	4 PEBs	PEB	Connect two channels on up to 4 PEBs. No gain control, may be a problem if two analog interfaces. No more than one DTI/101 may be attached. Cannot detect or control signaling.
MSI	Switch	4 to 24	PEB	Connects a PEB to 4 - 24 operator headsets. Cannot ring stations. Needs power supply, SA/240.
MSI/C	Switch	0 to 24/PEB	PEB	As MSI, but can also conference two or more parties on the PEB and/or operator stations.

Board	Function	Channels	Bus	Comments
LSI/80-xx		8	PEB	Like LSI/120, but with 8 channels. Approved for certain countries.
D/42	Basic VP	4	AEB	As D/4x, but for proprietary PBX interfaces.
D/160SC-LS	Basic VP+ LS interface	16	SC/PEB	Voice processing and loop start interface.
D/240SC	Basic VP	24	SC/PEB	Voice processing. Needs interface.
D/240SC-T1	Basic VP+ T-1 interface	24	SC/PEB	Voice processing and T-1 interface.
D/320SC	Basic VP	32	SC/PEB	Voice processing. Needs interface.
D/300SC-E1	Basic VP+ E-1 interface	30	SC/PEB	Voice processing and E-1 interface.
DTI/241SC	Limited VP+ T-1 interface	24	SC/PEB	Voice processing (no play/record) and T-1 interface.
D/240SC-2T1	Basic VP+ 2 T1 interfaces	24/48	SC/PEB	Voice processing and T-1 interfaces.

Chapter 17

Antares

The Dialogic Antares Series

The Dialogic Antares boards provide text-to-speech and voice recognition capabilities to SC bus and PEB-based systems.

At the time of writing, Antares is the only available hardware platform which adds voice recognition or text-to-speech for SC bus systems.

Antares is quite different from other Dialogic boards. An Antares board can be envisaged as a computer in itself, a complete platform which is capable of running third-party firmware products. It has four DSP processors which run under Dialogic's *SPOX* firmware operating system. A complete *SDK* (*Software Development Kit*) is provided so that developers can write their own firmware.

While Antares firmware products currently concentrate on text-to-speech and voice recognition features, if you are a firmware developer you can create any type of call processing product you like: fax, conferencing, Internet voice gateway or anything else you can imagine.

The following diagram illustrates the principal hardware elements of the Antares boards.

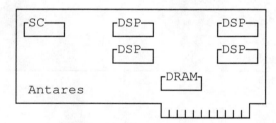

The Antares board.

The main components are:

- SC Bus chip. This can be run in SC or PEB mode.

- Four DSP processors. These are Texas Instruments TI TMS320C31s. Each DSP has SRAM memory which is dedicated to that one processor.

- Global DRAM memory which is shared by all DSPs.

Antares Board Models

There are currently three different Antares board models: the 2000/50, 3000/50 and 6000/50. The /50 in the model number refers to the clock speed of the DSPs, which is 50 MHz in all models. The difference

Antares

between the models is in the amount of SRAM (per-DSP) and DRAM (shared) memory.

The following table shows the amount of each memory type in the three board models.

Model	Per-DSP SRAM	Shared DRAM
Antares 2000/50	512 kb	4 MB
Antares 3000/50	512 kb	8 MB
Antares 6000/50	2 MB	8 MB

Antares Firmware Products

An Antares board does nothing useful until firmware is downloaded to the DSPs. Several vendors offer firmware products for text-to-speech and voice recognition.

In most cases, you will need a *dongle* (copy protection device) mounted on the Antares board in order for a given firmware product to run. These dongles are sold in some cases by Dialogic and in some cases by the firmware vendors.

The following vendors offer Antares firmware products.

Company Name	Contact Information
Voice Control Systems (VCS)	(214) 386-0300
Lernout & Hauspie (L&H)	(617) 238-0960
Voice Processing Corp. (VPC)	(617) 577-8422
PureSpeech	(617) 441-0000
Telefónica	Dialogic Telecom Europe (DTE) +32 2 712-4311
CSELT	DTE (as above)
France Telecom/CNET	DTE (as above)
Centigram	(408) 428-3796

Most of these vendors offer demo lines which you can call to try out their products. Contact the vendors for details.

The following table summarizes the products currently available from these vendors. ASR refers to Automated Speech Recognition, which covers speaker-independent voice recognition and/or voice verification, and TTS which is Text-to-Speech. The "max chans" column indicates the maximum number of simultaneous channels supported on a single Antares card. You may be able to load more than firmware type on a single board, depending on the vendor.

Vendor	Product name	Type	Max chans	Comments
VCS	VCS ASR	ASR	32	Offers a wide range of languages and features, including discrete and continuous numeric, alphanumeric vocabularies and voice cut-through.
L&H	L&H.asr 1500/T	ASR	12	Languages include English, French, German, Spanish, Dutch and Italian. Features include word spotting and continuous speech.
VPC	Vpro	ASR	8	Several languages supported, with discrete and continuous digit recognition.
PureSpeech	ReCite	ASR	8	Large vocabulary technologies.
Telefónica		ASR	16	Spanish vocabularies, including Castilian, Catalan and Galician.
CSELT	AURIS	ASR	8	Italian vocabularies.

Vendor	Product name	Type	Max chans	Comments
CSELT	FLEXU S 1000	ASR		Word spotting.
CNET	PHIL90	ASR	8	Word spotting.
L&H	L&H.tts 2000/T	TTS	24	English, German, Dutch, French, Castilian Spanish and Italian.
Centigram	TruVoice	TTS	12	
Telefónica		TTS	32	Castilian, Catalan and Galician Spanish.
CSELT	Eloquens 2000	TTS	16	Italian.

Programming For Antares

Unfortunately, Dialogic did not standardize application programming functions or interfaces for any types of Antares product. This means that each vendor provides its own proprietary API for each firmware type. The application developer is therefore forced either to choose one particular vendor's API, or to go through the extensive work of supporting multiple APIs. It is beyond the scope of this book to describe each vendor's functions in detail, contact the individual companies or Dialogic for more information.

Chapter 18

PEB Boards

Introduction

The PEB is a voice bus: a transmission highway which conveys audio and signaling information between different components in a voice processing system. Physically, it is a ribbon cable with two or more connectors which attach to voice boards.

The PEB is conceptually similar to a T-1 span. Information is transmitted digitally, divided into several conversations. Each conversation is called a time-slot. The number of time-slots is 24 or 32, depending on the clock speed selected for the bus. When the PEB is connected to a T-1 trunk, or a network device designed for North America, then the clock speed of the PEB is 1.554 MHz and there are 24 time slots. When the PEB is connected to an E-1 trunk or a device designed for most European and Asian countries, the clock speed is 2.048 MHz and there are 32 time-slots. On a 2.048 MHz PEB two of the time-slots are actually used for signaling and only 30 time-slots are available for conversations.

Every PEB configuration starts with one or more *network modules*, a board which connects the audio bus with the outside world. The network module is responsible for converting between the PEB format for audio and signaling information and the transmission standard of the external connection. Network modules include:

LSI/120.
The LSI/120 connects twelve two-wire analog lines to a single PEB. There may be one or two LSI/120s on one PEB, making a total of 12 or 24 channels.

DTI/xx.
A board from the DTI/xx family connects one digital trunk, T1 or E-1, to one PEB.

Aculab.
A board from a third-party vendor supporting 2.048 MHz ISDN in several European countries.

Dianatel EA24.
A board from a third-party vendor supporting 1.554 MHz ISDN and T-1 in the US.

MSI.
The MSI (Modular Station Interface) board connects up to 24 external devices (telephones, headsets, fax machines or other devices with a telephone-type two-wire analog interface) to a PEB. The main difference between the LSI and MSI is that the

PEB Boards

MSI provides loop current to the lines, and is thus electrically like a switch rather than like a telephone. The MSI board itself will be equipped with up to six daughter-boards, each one of which connects the MSI with four devices. Note a significant limitation of the MSI — the daughter-boards which apply loop current cannot apply ring voltage.

In addition to network modules, the PEB configuration will usually include one or more resource modules. A resource module processes conversations on the PEB. Examples of resource modules are as follows:

D/12x.
The D/12x adds twelve channels of the usual voice processing features: playing and recording digitized audio, touch-tone detection and generation, and so on. A PEB time-slot can be routed to any channel on a D/12x connected to that PEB.

VR/121.
The VR/121 adds twelve channels of voice recognition features. A PEB time slot can be routed to any channel on a VR/121 connected to that PEB.

FAX/120.
The FAX/120 adds twelve fax reception and transmission capabilities. While it is a separate board with a PEB connector, is not quite an independent resource module in the same way as the VR/121. Each channel on the FAX/120 follows a paired channel on a D/12x board. In effect, the FAX/120 is a daughter-board of a D/12x. This means that if a D/121 channel is routed in a particular way, the FAX/120 channel follows the same routing.

MSI.
The MSI can be a resource module as well as a network module. When configured as a resource module, the MSI can be thought of as a switch which can connect any in-bound call arriving on the configuration's network module to any MSI station set.

Example Configurations

The simplest configuration has an LSI/120 and a D/12x board:

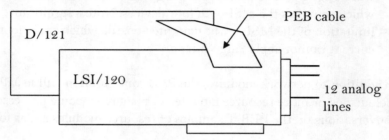

Simplest PEB configuration: one LSI/120 and one D/12x. Here, the LSI functions as the network module, taking up to twelve analog lines from the CO switch, PBX or other source and converting the conversations to digital format on the PEB. The D/12x card is able to perform voice processing functions on these twelve channels via the PEB connection.

For 24 lines, a second LSI/120 and D/12x could be added to the same PEB.

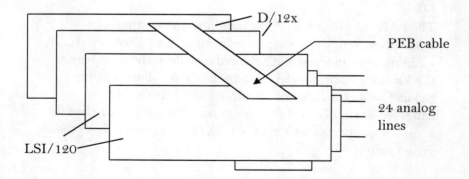

24-line configuration with two LSI, two D/12x boards. In this configuration, one D/12x board will be connected to PEB time-slots 1 through 12 (the LOW slots), the other will be connected to time-slots 13 through 24 (the HIGH slots).

For a single T-1 span, the following configuration would be used:

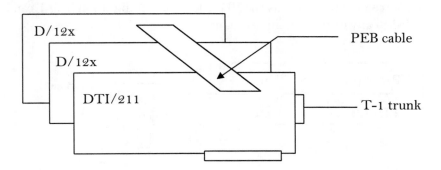

PEB configuration for T-1. In this setup, the DTI/211 card is the network module, the D/12x cards are the resource modules providing voice processing services.

If the MSI is used as a network module, telephone-like devices (phones, headsets, fax machines etc.) can be connected directly to the PEB via an MSI daughter-board:

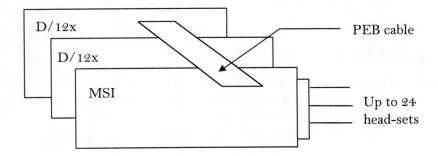

Using an MSI to connect phones to D/12x boards. This configuration allows a stand-alone PC to connect telephones and equivalent devices such as fax machines directly to voice processing services provided by D/12x speech cards. Unlike the LSI, the MSI provides loop current to drive the analog stations.

Resource modules, such as the VR/160p voice recognition board, can be added to any of these configurations.

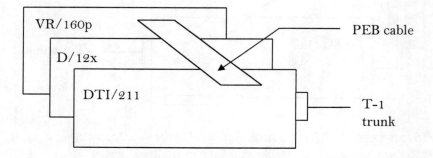

24 T-1 channels sharing 16 voice recognition channels.
This configuration has 24 channels of voice processing capacity provided by the D/12x boards and 16 channels of voice recognition provided by the VR/160p. Using the time-slot routing capabilities of the VR/160p, any one of the 24 active PEB channels can be sent to any of the 16 voice recognition channels.

Drop And Insert Configurations

Configurations such as those shown in the last section, which include just one network module for each time-slot, are known as *terminating* configurations. A configuration which includes a second network module for each time-slot is known as a *drop and insert* configuration.

Drop and insert applications may be visualized as having a "T" shape:

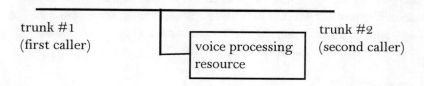

Drop and insert configuration. With drop and insert, there are two network modules. An in-coming call from one network trunk may be "dropped" to a voice processing resource or "inserted", i.e. passed through the VRU to an out-going call on the second network trunk.

An in-coming call may be initially "dropped" to the voice processing resource. It may be later "inserted" into an out-going call on the second network module, freeing the voice processing resource to process another channel. If only one side of the call (in-bound or out-bound) may be dropped, i.e. use a voice processing resource, this is known as a *one-way drop and insert*. If both the in-bound and out-bound call may be dropped, this is called *two-way drop and insert*.

An example of a drop and insert application takes in-coming calls from a T-span, collects digits from the caller, then connects the caller to an out-bound call. This is used, for example, for services which take advantage of cheaper toll rates in the US to offer reduced rate billing between Europe and Japan.

The configuration could be as follows.

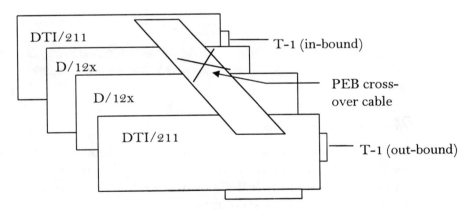

Two-way drop and insert example. This setup uses two T spans and a two D/12x boards. The 24 incoming time-slots share the 12 voice processing channels on one D/12x card, the 24 out-going time-slots share the 12 channels on the other D/12x. Because of the way the PEB connections work, a D/12x can only talk to one of the network modules. Since the voice processing channels are only required for a short time — to collect the digits from the caller, then to dial out and complete the out-bound call — a 2:1 ratio between calls and voice processing channels may be sufficient. The "×" symbol in the diagram shows where the PEB cable "crosses over" to make a hard-wired connection between receive and transmit time-slots.

An example of a one-way drop and insert configuration would be a call center design using a single PC. An in-bound T-1 span is connected to a

DTI/211 with two D/12x boards available to play menus and interact with in-bound calls. As an out-bound network module, the MSI board could provide connections for up to 24 live agent stations. The PC will be able to switch any of the 24 in-bound time-slots to any agent via the PEB.

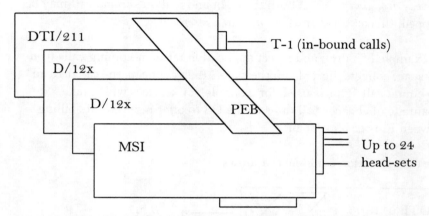

Call center design using MSI. The D/12x channels provide voice processing services to the in-bound calls, which may be switched to any available operator station through the MSI.

The DMX Switching Card

The DMX (Digital Multiplexer) card provides a switching matrix for up to four PEBs. The data on the receive half of any time-slot can be routed through to the transmit half of any other time-slot connected to the DMX. Only one source of data is permitted, which means that multi-way conferences are not possible with this device. However, it is possible to connect one source to many transmit time-slots, for example to broadcast "music on hold" to any number of channels.

PEB Boards

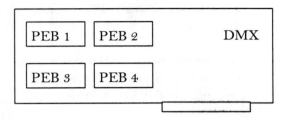

The DMX card.
The DMX has four PEB connectors. It can make a full- or half-duplex connection between any two PEB time-slots. The DMX is used as a switching card and also to convert between T-1 and E-1 clock speeds, the PEBs connected to the DMX may be run at different speeds.

Once two time-slots are connected through the DMX, the resulting set-up is in effect a drop-and-insert configuration where two network modules will be tied together.

The simplest configuration with a DMX is a 12-line system which has the ability to make a one-on-one conversation between any two channels.

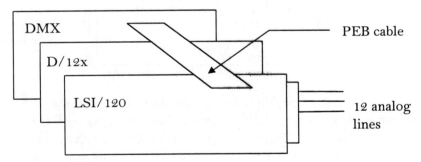

12-line DMX configuration.
With this configuration, a caller on any of the twelve in-coming lines can be "conferenced" with a caller on any other line using the switching capabilities of the DMX card. The DMX can switch between channels on up to four PEBs.

The DMX board cannot generate signaling information, nor can it detect transitions in signaling bits (hang-up, cease, wink, or flash-hook). Robbed bit signaling is not transmitted across a DMX connection. A resource module will therefore be required on any PEB channel which is routed through the DMX.

Note also that the DMX does not compensate for the loss in gain across an analog interface. This means that a conversation between two LSI/120 channels may have an unacceptably low volume.

As a further example of the use of the DMX, we can greatly improve the flexibility of the earlier "hard-wired" drop and insert configuration between two DTI/211 by replacing the PEB cross-over connection with a DMX card.

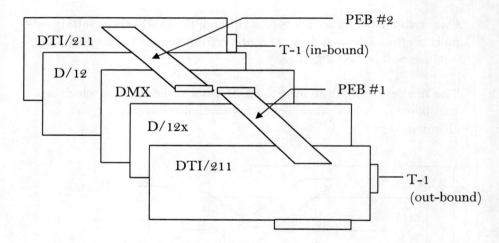

Two-way drop and insert using DMX. By using a DMX board, the flexibility of a drop and insert application can be considerably improved. There are now two independent PEBs, one for each DTI/211. Each PEB is connected to the DMX board. This enables a call arriving on a given T-1 to be connected to any call on the same T-1 or the opposite T-1.

Chapter 19

SCSA And The SC Bus

What Is SCSA?

The easy thing to explain about *SCSA* is what the acronym stands for: *Signal Computing System Architecture*. The hard part is to explain what the architecture really means. That's not because SCSA (pronounced "scuzza") involves more rocket science and brain surgery than anything else in this business.

No, the problem is that SCSA is just big: it covers a broad swathe of hardware and software specifications and has ambitions to solve a wide range of current and future technology challenges.

SCSA is an open standard. A consortium of leading telecommunications and computing technology players, led by Dialogic and including companies such as IBM and Seimens are, at the time of writing, cooperating in development of the SCSA specification. The specification documents will be available to anyone who wants them for a nominal fee. There will be no technology license fees charged to developers who wish to create SCSA-compatible products.

The most important components of SCSA for the voice processing developer are a *bus* and a uniform *API* (Application Programming Interface).

The primary bus is the *SC Bus*. SC Bus is a high capacity bus which is designed to be the "next generation PEB". Where the PEB has up to 32 time-slots, the SC Bus will have up to 2,048 time-slots: enough bandwidth for high-fidelity audio, full-motion video and other demanding applications of the future. The standardized API should make SCSA components such as voice processing, voice recognition, speech synthesis, video boards and others accessible independent of the component manufacturer.

When I think about SCSA, the analogy which comes to mind is the familiar "PC-compatible architecture". For hardware or software to be "PC-compatible" involves many layers from the chip level up to the operating system and applications programming interfaces. Since so many vendors respect and follow these compatibility levels, hardware and software components can be "plugged and played" on a wide range of computers. At the chip level, the PC-compatible computer must have a processor compatible with the Intel 8086, interrupt controllers which work like (or are) Intel 8259As, and so on. The expansion bus must be an ISA bus, i.e. work like the AT bus. The operating system must function like MS-DOS or PC-DOS. Device drivers for add-in products like expanded memory or CD-ROMs must conform. All of these levels operate according to (more or less) published standards and are offered by more than one vendor. AMD, for example, has "cloned" several of

Intel's 80x86 chip family. Digital Research offers DR-DOS as an alternative operating system. As a result, companies like Dell and Compaq can offer complete computer systems with few or no components from IBM, who created the standard. Most would agree that the open standards created for the IBM PC and AT were the driving force behind the explosive growth of the desktop computer industry. The objective of SCSA is to create an open set of standards for the next generation of computer/telephone integration products in order to fuel a similar growth of this market segment.

SCSA Components

The SCSA specification consists of the following components:

1. SC Bus
2. SCx Bus
3. SC Bus hardware
4. SCfirmware
5. SCdriver
6. Configuration Database
7. Device programming interface
8. SCapi
9. Server API

SC Bus

The SC Bus itself is a ribbon cable connecting components all mounted in one host computer. Think of it as a "super PEB". The bus can run be run at data rates of 2.048 to 8.192 MHz (32.768 Mbps to 131.072 Mbps) with a default of 4.096 MHz (65.536 Mbps). The 16 data lines of the bus give a maximum of 2,048 possible time-slots. The default data rate supports 1,024 time-slots. Each time-slot is bi-directional and carries 8-bit, 8 KHz (64 Kbps) data, just like a T-1, E-1, ISDN, PEB or MVIP time-slot. The term *bi-directional* refers to the independent data streams usually referred to as transmit and receive time-slots respectively.

Signaling is handled on a separate data line, in contrast to the standard PEB signaling which uses a single robbed bit. The use of a separate signaling line is referred to as *clear channel* data transport to contrast with robbed bit, where bits are occasionally "stolen" from audio data to represent the signaling state of the line. With clear channel transport, data is never altered, however slightly, between transmitter and receiver.

Time-slot bundling, resulting in what are sometimes called *virtual time-slots* or *virtual channels*, is supported. This refers to combining two or more 64 Kbps channels into a single logical channel which has a higher data rate, allowing for higher fidelity audio or other types of data such as video to be transmitted.

The SC Bus is a *multi-master* bus with *automatic clock fall-back*. This means that any SCSA board in a system can be the "master" used to coordinate the activities of the other boards connected to the bus. In current voice buses, one board (typically a network module such as an LSI or DTI) is designated as the master. If the firmware or clock chip on the master board fails, other SCSA boards in the configuration will detect the failure and "agree" on a new master. Since the implementation is at the hardware rather than firmware or driver level, the switch to a new master and recovery can be achieved quickly enough to prevent a loss of signal. The end result? Fault tolerance. SCSA systems will be better able to tolerate component failures and take corrective or remedial action than those we are building today.

Time-slot switching capability is built into SC Bus hardware connectors. This means that the ability to connect the receive half of one time-slot to the transmit half of a second time-slot (i.e. to create a talk-path between two channels) which now requires a DMX will be included in every SCSA board.

A comparison of major features with the PEB and Mitel ST bus (which is the basis for the MVIP bus) follows. The features shown for the ST bus may not be simultaneously achievable in one system.

SCSA And The SC Bus

Feature	SC Bus	ST	PEB
Max time-slots (channels) per box	2048[1]	512	128[2]
Max T-1 or E-1 trunks per box	64[1]	16	4
Time-slot bundling	Yes	Yes[3]	No
Time-slot switching	Yes	Yes	No
Clear-channel	Yes	Yes	No
Out-of-box expansion	Yes[4]	No	No
Clock fall-back	Yes	Yes	No
Multiple master	Yes	Yes	No

Notes:
[1] At highest 8.192 MHz rate
[2] With four E-1 trunks, four T-1s would give 96 channels.
[3] Requires additional circuitry
[4] Requires separate SCx bus adapter

The SC Bus is designed to be electrically compatible with the PEB and ST (MVIP) buses. This means that an SCSA board could be connected directly into a PEB or MVIP system. However, simply having an SCSA connector will not guarantee this capability, and several SCSA features will not be available in PEB or ST compatibility mode. Think of the 3½" diskette drive on your PC as an analogy. This drive is physically able to read a diskette from an Apple Macintosh computer, but cannot do so without additional software on the PC. SCSA will be physically compatible with these older buses, but SCSA component designers will need to take additional steps if this potential is to be realized in practice.

SCx Bus

The *SCx Bus*, or *SC expansion bus*, is a cable which can be run between different computers containing SCSA boards. The usual jargon is that SCx Bus is an "out-of-box expansion bus". The SCx Bus is the realization of an architecture Dialogic has been promoting for some time, which was earlier termed *DNA*, for *Dynamic Node Architecture*. This will have several benefits. Larger systems will be possible by tying multiple computers together. Redundant systems can be built so that, for example, voice mail messages can be recorded on more than one hard drive, or critical resource modules can be duplicated on more than one computer. If one computer fails, the message or resource will still be accessible. This is analogous to "disk mirroring" within a single chassis. Expensive resources such as voice recognition or video processing can be shared between more channels.

Up to 16 computers ("nodes") may be connected via the SCx Bus.

SC Bus Hardware

The details of the hardware are of interest mainly to board designers. Several approaches to building the SC Bus interface hardware will be possible, using discrete or integrated components. This allows the board designer to choose from more than one chip vendor.

SC Firmware

The firmware is software which directly controls the SC Bus hardware and other board-level components. It will implement primitive functions such as switching and routing of time-slots. At the time of writing, the specification of the firmware is at an early stage. It is anticipated that the firmware will be designed and implemented in a modular fashion that will allow designers of new boards will be able to take pre-written firmware and "link" it to firmware specific for their new board, relieving them of the burden of writing standard SC Bus functions in each new design.

SC Driver

The SC driver specification has changed considerably as the SCSA specification has matured. As the fourth edition of *PC Telephony* nears the presses, the latest incarnation is the S.100 API specification from *ECTF*, (the *Enterprise Computer Telefony Forum*, www.ectf.org). There are as yet no released products, but Dialogic is nearing the end of a beta cycle and expects to be shipping a product in the near future.

For the applications programmer, the SC APIs should at last provide real hardware independence. This means for example that voice recognition boards from different manufacturers will respond to the same programming commands and generate the same event codes.

SCSA And The SC Bus

For the board developer, this means that device driver development is simplified because all that has to be written is a relatively small number of routines which are specific to the board under development, the rest is handled by the universal functions.

The driver will eventually support four different programming models:

Synchronous
Calls to functions like "play a speech file" or "wait for the caller to enter a touch-tone" put the program to sleep ("block") until the operation is completed or interrupted. Conceptually, this is the simplest model. It is the model usually used by UNIX and OS/2 programmers with the current Dialogic drivers, and by developers using Parity Software's VOS environment.

Asynchronous polled
This is the model used by the current Dialogic MS-DOS driver. The application polls the device driver for events using a function call like *gtevtblk* and uses a state table to track the current state of each call.

Asynchronous call-back
In this model, a program notifies the driver of a function to be called whenever a given type of event is received. Structurally, this is similar to asynchronous polled, but has less overhead since there is no polling loop in the applications program. UNIX programmers will recognize an analogy with the *signal* function.

Asynchronous interrupt
This is a similar model to asynchronous call-back where a main thread of execution is added to the application: this thread is interrupted whenever an event is reported.

Single-tasking operating systems such as MS-DOS will of course only support the asynchronous programming models.

Programming models will be discussed in detail in a later chapter.

Configuration Database

Currently, there is no way for a Dialogic-based device driver or applications program to determine the complete configuration of a system.

The SCSA configuration database will provide a mechanism where devices and applications can automatically determine all relevant details of a system configuration. Today, each application designer who needs to a flexible range of system configurations must define a proprietary scheme to achieve the same purpose. A Configuration API will provide function calls to allow the application to access the configuration database information.

Device Programming Interface

The *Device Programming Interface (DPI)* is the lowest level of SCSA accessible to an application program. These functions will be similar in style to the C-callable functions currently used to control Dialogic and third-party boards. Use of the functions will be specific to particular board types and will require a fairly detailed understanding of the operation of the board and the installation environment. However, using the DPI should still simplify the software development process considerably compared with current technology. Today, the function calls and event processing which must be done to utilize a Dialogic VR/40 are quite different from those for a Voice Processing Corp. VPro-4x, despite the functional similarity of these two boards. Using DPI, the programming interfaces for voice processing boards from different vendors should be similar or identical, and a common subset of calls should be sufficient to incorporate either into an application.

Server API

The SCSA Server API provides a remote access, client-server API for applications programs who wish to utilize voice processing functions on components in a remote voice server connected via a LAN or other communications link. The server API, currently available in a product from Dialogic called AppServer, is a bi-directional, byte-stream protocol implemented as a sequence of messages to and from the server. A message might be a request from the application client to start playing a message, or an unsolicited notification from the server to the client that the caller hung up.

Chapter 20

The SCX Bus

Extending The SC Bus

The SCX bus extends the SC bus to two or more PCs. This allows large, fault-tolerant switching equipment to be built based on PC platforms.

The fundamental components is an SCX bus adapter, the SCX/160, which is an SC bus board with an RS-485 connector for the SCX bus itself which allows connection to a second PC.

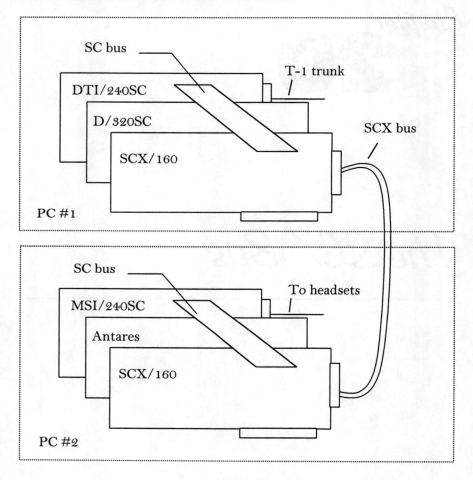

An example SCX bus configuration. The SCX bus extends the SC bus to two or more PCs.

The SCX bus adapter copies SC bus time-slot data between PCs. The time-slot numbers are the same on all PCs. For example, if the first channel on the Antares board is assigned to transmit to time-slot 90 on the SC bus of PC#2, the same data will appear on time-slot 90 of PC#1.

The SCX Bus

The SCX bus gives a number of advantages:

- Expensive devices such as Antares resources can be shared across more than one system.

- Larger switches can be created by combining multiple PCs.

- Systems can be made more robust and fault-tolerant through the clock-fallback features of the SC and SCX buses.

To explain exactly how the SCX bus works, we must first digress for a brief technical note on the SC bus.

In our simplified descriptions of the SC bus, we claimed there was a single wire which carried all time-slots. This is not true. In fact, there are 16 separate wires each carrying 64 time-slots. All these lines are driven by the same clock. The SCX bus copies data from one or more of these 16 lines. You can't choose to take some time-slots from a line but not others, it's an all or nothing choice.

An SCX Example

The SCX bus can be quite confusing, so let's consider the above example in detail. The configuration is as follows:

PC#1:
    DTI/240SC   connection to a T-1 trunk
    D/320SC   32 voice channels to be shared across both PCs
    SCX/160

PC#2:
    MSI/240SC   connection to 24 head-sets
    Antares   8 channels of voice recognition
    SCX/160

Each PC must be configured so that devices transmit to separate groups of SC bus time-slots. This is done with a configuration utility which is supplied by the SCX bus adapter. The result might be something like the following table.

PC	Devices	SC bus time-slots
1	DTI 1 .. 24	0 .. 23
1	Voice 1 .. 32	24 .. 55
1	SCX#1	64 .. 127
2	MSI stations 1 .. 24	64 .. 87
2	Antares voice rec. 1 .. 8	88 .. 95
2	SCX#2	0 .. 64

In PC#1, the first SCX adapter is transmitting to time-slots 64 to 127 on SC bus #1. This data originates in PC#2 with the MSI and Antares devices.

In PC#2, the second SCX adapter is transmitting to time-slots 0 to 63 on SC bus #2. This data originates in PC#1 with the DTI and Voice devices.

The result of this scheme is that the data in any given time-slot is the same on both SC buses.

Programming The SCX/160 Adapter

The SCX bus adapter does not have a switching API which the application can use while it is running. For the most part, the SCX bus is a purely static device. The few API calls which the adapter does support are for monitoring alarms (to notify when a bus appears to be out of synch) and for managing the clock mastering hierarchy in the system.

Programming An SCX Bus System

If your system uses an SCX bus, you have a new level of complexity to take care of. Neither the SC bus nor the SCX bus can carry programming commands, they carry only audio data. This means that an API command for a given device must always be issued on the PC where the device is installed, even if data from the device is routed to a different PC.

The SCX Bus

Consider our example system. Suppose we want to play a voice message to a headset on the MSI board. This requires the following steps:

- Route the MSI device listen channel so that it is taking data from the SC bus time-slot for a voice device on the D/320SC board.

- Play the voice message on that voice device channel.

  To execute the first step, you must issue an *ms_listen* call on PC#2 for the MSI device to connect it to the voice device's transmit time-slot. Since the D/320SC is in the other PC, there is no way to query the Dialogic drivers to find the appropriate time-slot number. It is your responsibility in writing your application to have an configuration file of some kind (or hard-coded data in your source code) which associates time-slots with devices. You could create this data by running a test program on each PC which used the *xx_getxmitslot* APIs to query the transmit time-slot numbers for each device type.

  To play the voice message, you need to execute a play API call on PC#1 to the chosen voice device. To implement this type of system, you will need a method of communicating between application code running on PC#1 and code running on PC#2. This is needed in order to keep track of which devices are available to be assigned and which are already in use in calls. The Dialogic API does not assist you in this area.

Chapter 21

Switch Cards

Introduction

There is a healthy niche market for switching and conferencing components which fill needs not met by products such as Dialogic's AMX, DMX and MSI cards. Independent vendors have created cards which interface through the AEB, PEB, MVIP and SC buses.

Third Party Switch Card Vendors

The three major independent vendors of switching cards are:

Amtelco, McFarland, WI.
Phone 608-838 4194, fax 608-838 8367.

Dianatel, San Jose, CA.
Phone 408-428 1000, fax 408-433 3388.

Excel, Sagamore Beach, MA.
Phone 508-833 2188, fax 508-833 2188.

The following sections will outline the capability of some of the products offered by these vendors.

The Amtelco XDS Switch

The XDS is a 512 x 512 port digital switch, permitting up to 256 full-duplex (two-way) conversations offering both PEB and MVIP connections and seven connections to the proprietary APIB bus (Amtelco PCM Interface Bus).

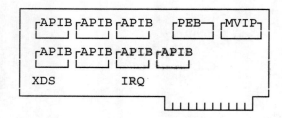

The XDS has interfaces to an E-1 type PEB or an MVIP and to seven APIB buses.

Each APIB can support up to 30 ports of XDS Line Interface Cards. The XDS itself offers both switching and conferencing capabilities.

The PEB bus operates at 2.048 MHz, and thus supports E-1 type PEBs only. Up to 32 channels of voice processing can be added using D/81A cards.

Switch Cards

A maximum of 21 conferences can be established with participants from up to 64 ports. This places a limit of 21 three-party or 16 four-party conferences.

The XDS Line Interface card may be visualized as follows.

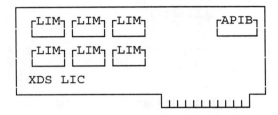

The XDS Line Interface Card can interface up to six analog trunks or stations. LIM indicates an appropriate Line Interface Module for the type of trunk or station.

Line interface modules are available for the following types of line:

Loop start
Ground start
E&M
Battery feed
DID
Station (applies battery for telephone or headset).

Amtelco also offers two types of AEB interface card. One is partly similar to an AMX, including four AEB interfaces, an interface with battery for up to eight telephones, headsets etc, and an APIB bus connector for attaching and XDS switch. The other interface card provides six AEB connectors and one APIB connector: this allows D/4x and other AEB compatible boards to be connected directly to an XDS.

The Dianatel SS96, SS192 And SB

Dianatel's SS96 (*SmartSwitch* for 96 channels) is a board which is similar in outline to the Dialogic DMX. It has four PEB interfaces, and has the capability to connect any two PEB channels in the same way as the DMX.

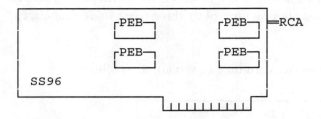

The Dianatel SS96, like the DMX, has four PEB interfaces. An RCA input jack is provided for music on hold.

The SS96 does not require an IRQ. It offers largely the same capabilities as the DMX, except that as of the time of writing it does not support 2.048 MHz PEB buses, as required by E-1. Additional features include the following.

Support For Signaling.

The PEB can carry a single signaling bit, analogous to the A bit on a T-1 trunk, encoded using the robbed bit technique. The DMX does not support this bit: it is not passed through, signaling bit changes cannot be reported to an application, and the signaling bit value cannot be set except to a default "A bit high" when no other device is transmitting a signaling bit. By contrast, the SS96 is able to perform all these functions. This allows a PEB connected to the SS96 to control an LSI without requiring a D/xx. When the DMX is used, a D/xx must be present on an LSI PEB to answer calls and detect caller hang-up by monitoring and setting the PEB signaling bit.

Music On Hold.

The SS96 has an RCA input jack for music on hold.

Expandability To 8 PEBs.

An SS192 daughter-card may be mounted on the SS96, adding a further four PEB interfaces.

Conferencing.

An SB96 (*SmartBridge 96*) daughter-card may be mounted on the SS96 adding conference capabilities.

Switch Cards

Conference Size.

Conferences may be from 3 to 24 parties, exceeding the current 8 party limit of the Dialogic MSI/C. Only one of the SS192 or SB96 daughter-cards can be mounted on one SS96.

Out-Of-Box PEB Expansion.

Special PEB cables may be used to connect SSxxx boards in two different PCs.

To illustrate the possible types of configuration, we will sketch a 96 by 96 switch which can be built with Dianatel components for performing ISDN to T-1 conversion:

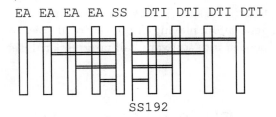

A configuration for ISDN to T-1 conversion.
EA indicates a Dianatel EasyAccess 24 board, which can be interfaced to a PRI ISDN trunk. A SS96 with SS192 daughter-board provides 8 PEB connections (each PEB is shown as a double line ═). DTI is a Dialogic DTI/211 for connecting to a T-1 trunk. Music on hold can be provided through the SS96, and DTMF detection and dialing through an optional daughter-card (the TG24X) of the EA. D/121B boards could be added for further voice processing services.

The Dianatel CO24

The Dianatel CO24 is an out-of-box device which may be connected to the SS96 via a PEB cable. No PC slot is needed. It can support up to 24 analog stations such as headsets or telephones. Unlike the MSI, it does have the ability to ring a station.

The Excel PCX512 Series

Excel's PCX512 series of PC cards, like Amtelco's XDS line, communicates using a proprietary digital bus. The range of cards offered includes the following.

MX/CPU-512
This is a switching matrix and controller card which can service up to 512 channels. A proprietary digital bus provides communications with other cards, and an RS-232 serial port allows for off-line diagnostics.

MX/CPU-16
This is a 96-channel version of the MX/CPU-512.

PCMFDSP
A less than elegant acronym, "PCMFDSP" stands for "PC-based multi-function digital signal processor". The card provides voice processing resources for the MX/CPU card. SIMM modules carrying DSPs are mounted on the baseboard to provide specific services, which may include tone detection and generation, playing of pre-recorded voice prompts which have been encoded into software, call progress analysis, and conferencing.

Network interface cards.
Interfaces are available for T-1, PRI ISDN, and E-1 including R2 signaling capabilities.

PCULC-6 and PCULC-12
This is a "universal line card" with 6 or 12 ports which can support several types of trunk or station under software selection, including:

- Loop start
- Ground start
- DID
- E&M
- Station (applies battery for telephone, headset etc).

DTMF receivers are included on the board.

PCTGR-24
This is a tone generator and receiver card.

PCRBI

This is a "resource bus interface" card which allows a configuration to be attached to a PEB or MVIP bus.

Chapter 22

AEB Switching

Introduction

The Dialogic AMX/8x board provides a switching matrix which can be used to connect D/4x speech card channels to other speech card channels or to external devices with analog two-wire telephone interfaces. Functionally, the card is an 8x8 matrix which has five connectors P1, P3, P4, P5 and P6:

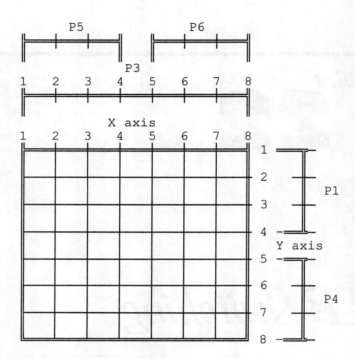

The card looks like this:

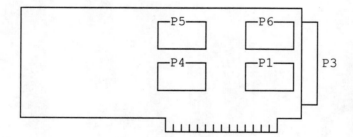

The AMX is an 8x8 switching matrix. The four connectors P1, P4, P5 and P6 can be connected using flat cables to D/4x boards or to additional AMX boards.

The P3 connector on the back of the AMX card connects to exactly the same channels as P5+P6, and also provides loop current so that external devices which are electrically like telephones can be connected to a D/4x channel — these devices may be telephones, operator headsets, fax machines etc.

AEB Switching

By convention, P5 and P6 are considered to be the X axis of the matrix, P1 and P4 are considered to be the Y axis. The P3 connector is also connected to the X axis.

The SA/102 unit from Dialogic connects to P3 on the back of the AMX card and converts this to 8 RJ-11 jacks which can be used for headsets and other devices:

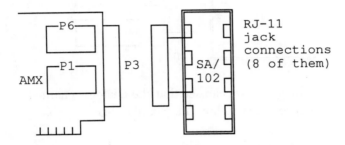

If the local operators plug their headsets in and out of the SA/102, the ESD unit, also available from Dialogic, should be used to prevent damage to the AMX card due to surges.

If it is desired to connect an audio unit with RCA jacks, such as a tape recorder, to the AMX, the Dialogic AIA/2 device can be used. The AIA/2 converts audio from the RCA jack to an RJ-11 which can be connected to the SA/102.

D/4x boards are connected to the AMX by using a flat cable which goes from the D/4x P3 connector to the AMX P1, P4, P5 or P6 connector.

Any switch-point in the 8x8 matrix can be "on" (made) or "off" (broken). The Dialogic driver function *sw_on(x, y)* turns on a given switch-point, the function *sw_off(x, y)* turns off the switch point.

The following sections provide some example configurations, other configurations are possible.

One AMX And Four D/4x Boards

This configuration has a single AMX board and two D/4x boards on each AMX axis:

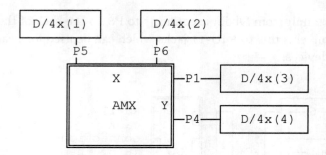

Looking at the AMX diagram, we can see that the Dialogic channel numbers are connected to the switching matrix like this:

D/4x Board	Channel	AMX Connection
(1)	1 - 4	X1 - X4
(2)	5 - 8	X5 - X6
(3)	9 - 12	Y1 - Y4
(4)	13 - 16	Y5 - Y6

The D/4x board number is assigned by the D40DRV driver when the system is started, and is determined by the base address of each board (set by DIP switches). If the board numbers are not connected in the order shown, this table must be adjusted accordingly.

By making switch-points, calls can be "conferenced". For example, with the command:

```
sw_on(2, 7);
```

the switch-point in the matrix at X=2 and Y=7 is turned on, and a call on D/4x channel 2 will be connected to a call on D/4x channel 15.

With the command:

```
sw_on(2, 2);
sw_on(6, 2);
```

AEB Switching

calls on D/4x channels 2, 6 and 10 will be conferenced.

NOTE
Dialogic recommends that no more than two X-axis channels are connected in a conference, and no more than one Y-axis channel.

NOTE
There is a significant loss of gain (gain is a measure of the amplitude of the signal, i.e. how loud the sound is) when a call is conferenced through the AMX. Two local calls may sometimes be conferenced acceptably, but two long-distance calls may suffer an unacceptable loss of quality when conferenced through the matrix. This may be cured by the use of "line amplifiers".

Two D/4x Cards And One External Device

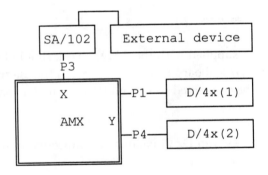

We suppose that the external device is connected to RJ-11 jack number 1 on the SA/102, and thus to the X-1 connection on the matrix. The D/4x channels are connected as follows:

D/4x Board	Channel	AMX Connection
(1)	1 - 4	Y1 - Y4
(2)	5 - 8	Y5 - Y6

This configuration allows the D/4x cards to "share" the device (up to eight devices, if required). For example,

```
sw_on(1, 3);
```

will connect D/4x channel 3 to the device, in general,

```
sw_on(1, line);
```

will connect D/4x channel number *line* to the device.

One AMX And One D/4x

An AMX card can be used to conference channels on the same D/4x card.

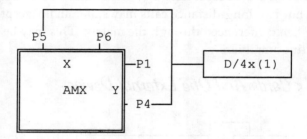

A "terminator assembly", supplied with the AMX board, should be placed on P6 if only one D/4x board is used as shown in the diagram. A second D/4x board can be connected to P6 and P1 in a similar manner if required.

In this configuration, the a given D/4x channel is connected both to a Y axis point and to an X axis point:

D/4x Channel	AMX Connection
1	X1 Y5
2	X2 Y6
3	X3 Y7
4	X4 Y8

Thus, for example,

```
sw_on(1, 6);
```

will conference D/4x channels 1 and 2.

AEB Switching

NOTE
A D/4x channel may not be conferenced with itself — a command with x and y having equal values will be invalid in this configuration.

Four D/4x Boards Sharing Eight Devices

By using two AMX boards connected to each other, four D/4x boards can share up to eight local devices:

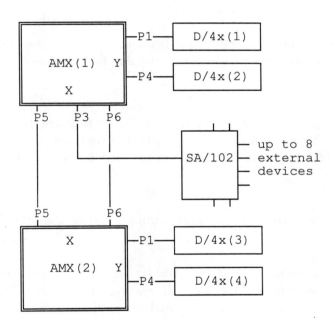

With this configuration, channels are connected to the two AMX cards as follows:

D/4x Board	Channel	AMX Connection
(1)	1 - 4	Y1 - Y4
(2)	5 - 8	Y5 - Y8
(3)	9 - 12	Y9 - Y12
(4)	13 - 16	Y13 - Y16

SA/102 external devices are connected to AMX board (1) as follows:

External Device	AMX Connection
1 - 8	X1 - X8

AMX board (1) is connected to board (2) along the X axis:

AMX (1) Connection	AMX (2) Connection
X1 - X8	X9 - X16

Channels on D/4x boards (1) and (2) can be connected directly to an external device by setting one switch-point, for example:

```
sw_on(2, 5);
```

will connect channel 5, the first channel on D/4x board (2), to external device 2. Since the X axes of the two boards are connected, D/4x boards (3) and (4) can connect to the external devices in a similar way,

```
sw_on(10, 13);
```

will connect channel 13, the first channel on D/4x board (3), to external device 2.

By adding one more AMX card, this type of configuration can be extended to two more D/4x boards, making a total of 24 D/4x channels sharing up to eight external devices.

Making And Breaking Switchpoints

Matrix switchpoints are controlled by the functions *amx_off*, *sw_on* and *sw_off*. The *amx_off* function, which takes no arguments, clears (turns off) all switchpoints on all AMX boards,

Connections are made at matrix switchpoints using the *sw_on* function, which takes the x and y coordinates as arguments:

```
code = sw_on(x, y);
```

AEB Switching

Coordinate values x and y range from *1* up to the number of installed ports. Both x and y must be on the same AMX board. The return value *code* will be zero if successful, or a positive error code.

Connections are broken at matrix switchpoints using the *sw_off* function, which takes the x and y coordinates as arguments:

```
code = sw_on(x, y);
```

The return value *code* will be zero if successful, or a positive error code.

Switching For The PEB

The AMX lives in the analog world of the AEB. Two newer cards offer switching matrix options for the PEB digital bus: the *MSI* (Modular Station Interface), and *DMX* (Digital Multiplexer).

The MSI connects PEB channels to analog station lines. The MSI applies loop current to station lines so that telephones and telephone-like devices (headsets, fax machines etc.) can be attached to PEB conversations:

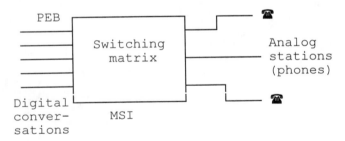

The MSI board.
The MSI provides a switching matrix which can switch a conversation on the attached PEB to an analog station such as a telephone or headset. Unfortunately, the MSI cannot apply ring voltage, so other means of signaling an incoming call must be used for the stations.

The DMX connects conversations on up to four PEBs. Any caller on any PEB can be connected to any caller on the same, or a different, PEB:

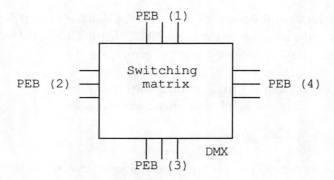

The DMX board. The DMX board can connect conversations between up to four PEBs. Only one-on-one connections are possible, the DMX cannot connect three or more callers in one conversation.

The DMX can create a conversation between any two callers, or broadcast audio from one conversation to many callers. It cannot, in its version current at the time of writing, conference three or more callers in one conversation.

The MSI and DMX are described in more detail in the chapter on PEB-compatible devices.

Chapter 23

Telephony Sound Files

Introduction

First some high-school physics. The human ear is a pressure-wave detector: what we hear as sound is the result of variations in air pressure. Sound waves propagate from a disturbance much like water waves spreading from a stone dropped into a lake. Corresponding to the height of the water is the pressure of the air: the air pressure varies from the prevailing average just as the surface of the water washes above and below the mean water level.

An object such as a violin string which can be made to vibrate will produce a sound. The volume of the sound as perceived by the ear is related to the size of the pressure variation, in other words, to the heights of the peaks and depths of the troughs as the pressure varies from the average. The greater the variation, the louder the sound. The amount of the variation, i.e. the difference between the current pressure and the average pressure, is called the *amplitude* of the wave at that time. The amplitude is directly related to the volume, although the relationship is closer to being logarithmic than linear — this is due to the way the ear and brain work together in perceiving the sound.

A violin string is special because it tends to vibrate a set number of times per second, producing a smoothly varying air pressure wave looking like a sine function from trigonometry. This sound is close to being a pure tone containing just one frequency. (In practice, of course, the sound made by a violin contains many frequencies, but these are of lower volume and tend to be *harmonics*, frequencies such as double or treble the dominant frequency which blend in harmoniously. It is the presence of these harmonics and the strength of the mix that give the violin its distinctive sound; a piano string, for example, will produce a different mix).

As a rule of thumb, the human ear can hear frequencies from about 50Hz to 20KHz, however, a much narrower range suffices to transmit speech acceptably: a range from a 300Hz to 2KHz to 3KHz is widely considered to do the job.

Digitization

Digitization is the process of converting an analog signal, such as the height of a sound wave, into numbers. A given number of times per second (the *sampling rate*), a digitization device measures the amplitude of the signal and registers the amplitude as a number, called a *sample*. The circuit performing this operation (often a *codec*, for coder/decoder) selects the binary sample value closest to the measured amplitude, the error introduced by this approximation is called *quantization noise*; it turns out that the ear is surprisingly sensitive to these errors.

In the case of sound digitization, samples will, on average, add up to zero: half will be negative, indicating pressure less than average, half will be positive, indicating pressure above average.

When digitized audio is saved in a buffer or file, a series of samples will be stored one after the other, the position of the sample in the file will be directly related to (usually proportional to) the elapsed time.

Reconstructing the signal reverses the process: a decoder takes samples as input, interpolates between the values, and produces a new waveform as output. Each storage method has its own characteristic problems which result in frequencies being generated which are artifact of the digitization process. These spurious frequencies are removed by using *anti-aliasing* filters designed for the given digitization scheme. Typically, anti-aliasing filters will remove low-frequency hum below about 60Hz and high frequency quantization noise from above 4KHz.

Preparing Digital Audio Files

The menus and prompts of your system are, of course, the most significant part of the system to the caller. There are several factors to consider when designing the prompts:

1. Wording.
How to explain in clear, unambiguous, concise language the instructions or information to be conveyed by the prompt.

2. The speaker's voice.
Should prompts be recorded by a man or a woman? Some researchers have expressed the view that instructions are better accepted from a woman, while information is better conveyed by a man. There is also the question of accent: a system serving a local area may be perceived as more friendly and welcoming if prompts are spoken in the local dialect, callers from outside the area may find the accent less pleasant or even hard to understand.

3. Special effects.
Prompts can use fades, echoes, sound effects, be mixed with music or natural sounds (birdsong, rockets taking off...) to sound more interesting, attractive or attention-grabbing.

4. Technology.
The fidelity and flexibility of the recording and editing technology will have a major impact on the quality of the resulting audio files.

Technologically, there are two main issues to be considered when preparing digital audio files: how to get the sound into digital format on a hard drive, and how to edit the resulting files.

The PC world offers four main options for recording a "first draft" digital audio file:

1. Using a *CO simulator*, a box which provides loop current and enables a regular two-wire telephone to be connected directly to a standard telephone interface on, for example, a D/4x board.

2. Recording over a telephone line from a telephone.

3. Using a standard sound board, such as SoundBlaster.

4. Using a proprietary sound board custom-built for call processing applications, such as the Audio Work Station.

The CO simulator is a useful tool for creating, testing and demonstrating voice processing applications. Typically it is a small box with two RJ-11 jacks, one for a line to the phone and one for a line to the voice card. A button may be provided which will assert ring voltage on the voice card side, some models will assert ring when one end goes off-hook (a blessing when doing repeated testing). Often an input jack for connection to a tape deck or other standard audio equipment is also included. CO simulators are great for developers and testers who need to run many calls through a system — it saves tying up two CO lines or PBX extensions at the desk, and saves on the toll for local calls, which can add up rapidly at business toll rates.

Telephony Sound Files

Dialogic provides a CO simulator box, the AC/101, which does not provide ring voltage. It provides a low-fidelity connection. Skutch Electronics, Inc. of Roseville, CA 916-786-6186 provides a line of low-cost range of telephone line simulators which have been third-party labeled and re-sold by several vendors.

Recording over a phone line can produce quite acceptable results, and may be the only solution for information lines which are updated remotely. It will help if the phone call is local, and if a specialized telephone with a high-quality moving-coil microphone is used rather than the cheap and cheerful carbon microphones which are found in most telephones.

Once your prompt has been recorded, there is still much which may need to be done. For example, you may wish to trim the periods of silence at the beginning and end of the recording. You may wish to "cut and paste" between voice files, mix voice with music, add custom tones, fade in and out, and so on. These functions are performed by a *voice editor*. A leading voice editor product is VFEdit from Voice Information Systems of Bryn Mawr, PA 215-747-5035. Note that VFEdit, at least at the time of writing, uses the Dialogic board to record and play back sound. An alternative is Parity Software's VOX Studio editor and converter product, which can record to the much higher-fidelity Wave format and subsequently convert to Dialogic format. In many cases, this will give significantly better sound quality.

The top-of-the line solution for recording quality prompts to a computer hard disk is the custom PC board product from Bitworks, of Thornhill, Ontario 416-881-2733, which includes a custom board and software which can take input from a high-quality microphone or stereo units such as CD players. This Audio Work Station package includes editing and digitization conversion software incorporating special effects and mixing capabilities. The results from using this product will be significantly higher fidelity than any solution which uses the digitization circuitry on current Dialogic products and for the most demanding applications may offer slightly better results that VOX Studio. The drawbacks of this product are its complex, character-mode

DOS interface and arcane command-line utilities, which are powerful but take some time to learn, and a relatively high cost.

Digitization Methods

There are many different storage schemes for digitized voice, which provide different answers to the following questions:

1. How should the number assigned to the sample be related to the amplitude? A simple linear relationship may not be the best choice, perhaps a logarithmic representation should be chosen.

2. How many bits per sample should be used? The more bits that are stored, the more accurately the sample is represented and the greater the maximum and minimum amplitude that can be stored.

3. Should each sample be stored as a complete binary number? You might think that the answer should be yes, but the widely-used alternative is a *differential* scheme, which exploits the fact that a given sample is likely to be close in value to the preceding sample, so that the difference between the two is stored — which may be a smaller number which can be represented accurately using fewer bits.

Some of the more widely used digitization schemes are as follows.

Pulse Code Modulation, PCM.

PCM stores a complete representation of each sample. *Linear* PCM stores a binary number directly proportional to the amplitude of the sample. Research has shown that the ear is more sensitive to variations at low amplitude, so non-linear conversion charts (scales) have been defined which allow fewer bits to store the same range in amplitude by providing more points at the quieter end of the scale and fewer at the louder end. Two non-linear scales are in wide use: *mu-law* within the US and Japan, and *A-law* in the rest of the world. US T-1 digital trunks use 8-bit mu-law PCM, European E-1 trunks use 8-bit A-law PCM. As you would expect, 8-bit refers to the number of bits per sample. Eight bits allows for one sign bit (is the amplitude greater than or less than "zero", the average air pressure?) and seven bits for the magnitude, which represents one of

128 possible points on a non-linear scale converting samples to amplitudes.

Differential Pulse Code Modulation, DPCM.

Voice signals vary relatively slowly. The largest component tends to be around 1KHz, well under the maximum or 3KHz or 4KHz required for a reasonably high-quality voice link. DPCM simply starts at zero, then transmits the difference between the preceding sample and the current sample each time. DPCM can get into trouble with quickly-changing signals, which leads to the more sophisticated ADPCM method.

Adaptive Differential Pulse Code Modulation, ADPCM.

The ADPCM scheme extends DPCM by allowing for a changing *step size*. To calculate the amplitude, the following equation is used:

> Amplitude = (Amplitude of last sample) +
> (difference x step size).

The "difference" is the binary number stored for each sample. The trick is to vary the step size depending on the current volume: the louder the volume, the larger the step size. This enhances the ability of ADPCM to accommodate sharp variations in waveform. The "feedback" method used to vary the step size depending on earlier samples can be optimized for the characteristic waveforms produced by the human voice. Probably the most widely used standard for ADPCM is the CCITT standard G.721, which stores 4-bit samples at a rate of 8000 samples per second (8KHz). As a rough guide, the highest frequency which can be represented is half the sampling rate, 8KHz thus accommodates the 4KHz maximum recommended for speech. The original Dialogic standard for ADPCM used 4-bit samples at a rate of just over 6KHz (6053Hz).

Continuously Variable Slope Differential Modulation, CVSD.

CVSD takes a rather different approach. A current gradient, or slope, is maintained by the algorithm as an approximation to the current rate of change of the waveform. As each new sample is measured, a new approximation to the slope is obtained, and the change in the slope is represented by a single bit. The steeper the slope, the larger the difference between reconstructed sample values. Thus, CVSD has a built-in adaptation to more rapidly changing values which could be compared to ACPCM. A typical rate required for voice transmission is 32,000 bits/second, as for ADPCM.

Vector Quantizing Code, VQC.

This approach is unique in that it deals strictly with groups of samples, rather than the current sample (plus, perhaps, the preceding sample). The waveform produced by the set of samples is compared with a table of templates, and the closest match is selected. VQC has the advantage that it can be optimized for specific tones and signals. For example, modem and fax signals vary quickly, but often can be covered by a relatively small number of templates, while speech varies more slowly and would have another characteristic set of templates. In fact, VQC even incorporates a limited amount of "error correction" for electronic modem signals, as long as the set of templates "knows" about the type of signal being transmitted, since even if the input signal to the coder is slightly distorted, there is still a good chance that the correct template will be matched.

High Capacity Voice, HCV.

To improve voice quality significantly without raising the standard data rate of 8Kbps requires a more sophisticated approach. HCV takes VQC to a whole new level by modeling the human vocal tract with characteristic sounds produced by the lips, vocal cords and so on. This demands number crunching beyond the limits of standard electronics, and requires *DSP* (*Digital Signal Processing*) technology utilizing custom-designed chips for high-speech calculations on digitized waveforms. A similar type of software methodology is used by text-to-speech systems to generate synthetic speech.

Dialogic Standard ADPCM

The digitization method originally used by Dialogic is a 4-bit ADPCM variant at a rate of 6053 samples/second . The "4-bit" designation means that each sample is represented by a 4-bit value. Older Dialogic literature sometimes referred to ADPCM files as *VOX* files (vox is the Latin for "voice"), and used the DOS file extension .VOX for files stored in this format. Dialogic derived their standard from the pre-existing *Oki* ADPCM standard, so you may sometimes see this format referred to as Oki or Oki ADPCM.

Telephony Sound Files

The first of the four bits is the *sign*, in other words whether the current sample is greater or less than the previous sample. If the current sample is greater, the sign is zero; if the current sample is less, the sign is one. The remaining three bits represent a value between zero and seven which represent the approximate magnitude of the difference between the samples. The representation of the magnitude is *non-linear*, which means that doubling the value in the magnitude does not necessarily double the amplitude of the encoded sound.

The ADPCM encoding algorithm works with three input values: two signed twelve-bit amplitude samples S_n, the current sample and S_{n-1}, the previous sample; and the current step size SS. The procedure is as follows:

1. Set these values to zero: B0, B1, B2, B3

2. Calculate the difference, $D_n = S_n - S_{n-1}$. If D_n is less than zero, set B3 to 1.

3. Set E to be the absolute value of D_n, i.e. if D_n is greater than zero, set E to D_n; if D_n is less than zero, set E to $-D_n$.

4. If E >= SS, set B2 to 1 and subtract SS from E.

5. If E >= SS/2, set B1 to 1, and subtract SS/2 from E.

6. If E >= SS/4, set B0 to 1.

After following these steps, the 4-bit ADPCM sample is the four bits:

B3 B2 B1 B0

The step size is re-calculated each time using the previous step size and the current 3-bit ADPCM magnitude M_n.

The step size SS can take one of 49 different values from the following table:

Nr	SS	Nr	SS	Nr	SS	Nr	SS
1	16	13	50	25	157	37	494
2	17	14	55	26	173	38	544
3	19	15	60	27	190	39	598
4	21	16	66	28	209	40	658
5	23	17	73	29	230	41	724
6	25	18	80	30	253	42	796
7	28	19	88	31	279	43	876
8	31	20	97	32	307	44	963
9	34	21	107	33	337	45	1060
10	37	22	118	34	371	46	1166
11	41	23	130	35	408	47	1282
12	45	24	143	36	449	48	1408
						49	1552

Note that a signed twelve bit sample ranges in value from -2048 to 2047, so the maximum step size of 1552 can take a sample from minimum to maximum in three steps. To calculate the step size, use the current 3-bit magnitude Mn from the current ADPCM sample and find X from the following table:

Mn	X
000	-1
001	-1
010	-1
011	-1
100	2
101	4
110	6
111	8

Use the value to adjust the current position of the step size in the table. For example, if the current step size is number 24, with step size 143, and Mn is 100, giving a change of 2, the new step size will be number 26, i.e. 173. If this would change the step size to less than number 16, use 16; if this would change the step size to greater than number 49, use 49.

Telephony Sound Files

This procedure may seem strange, but it was developed by extensive analysis of speech: the algorithm is effective for storing the human voice.

To decode an ADPCM value, the above procedure is reversed. The step size is adjusted from sample to sample in exactly the same way as for encoding. To decode, start by calculating the amplitude of the difference M_n:

$$M_n = B_2 * SS + B_1 * (SS/2) + B_0 * (SS/4) + SS/8$$

If B_3 is 0, set $D_n = M_n$

If B_3 is 1, set $D_n = -M_n$

Then the new output 12-bit linear sample S_n is calculated from the previous:

$$S_n = S_{n-1} + D_n$$

To initialize, the sample before the first twelve-bit sample is considered to be zero (at the middle of the scale), and the step size is set to the minimum value of 16 (number 1 in the table).

An important feature, and sometimes drawback, of ADPCM encoding methods is that they are *context-dependent* — in other words, the interpretation of a given set of samples depends on the samples which precedes that set. This means that you cannot simply "cut and paste" fragments of an ADPCM file without adjusting samples to accommodate the current context.

The 4-bit ADPCM algorithm can be "reset" to its initial state by a sequence of 48 samples of plus and minus zero (0000, 1000) in either order.

VOX Files

Dialogic has established a family of sound file formats for call processing hardware known as VOX (from the Latin word "vox", meaning "voice"). VOX is not compatible with the Windows .Wave format, however Parity Software offers a product called VOX Studio which can convert between VOX and Wave files and also record and play VOX files on Wave devices.

A digitized audio recording is characterized by a number of parameters.

> The number of bits used to store a sample. A sample is a binary number specifying the amplitude (loudness) of the sound at the instant a measurement was taken. Sample sizes will generally be 4, 8 or 16bit.
>
> The number of samples recorded per second, the sampling rate, which is usually measured in thousands of samples per second, KHz. Windows Wave files typically use 11 KHz or 22 KHz, VOX files use 6 KHz or 8 KHz.
>
> The compression scheme used, if any. There is a family of schemes known as ADPCM (Adaptive Differential Pulse Code Modulation) which is commonly used in voice processing hardware. ADPCM is designed especially for encoding the human voice. Files recorded with PCM (Pulse Code Modulation) are not compressed.
>
> Companding scale. Companding is a technique to squeeze a bigger range out of a given number of bits. The human ear is more sensitive to small changes in amplitude when the volume is low, companding exploits this effect by using a nonlinear mapping of sample values (the binary number used to store the sample) to sample sizes (the absolute value of the amplitude specified by the sample). A file format is referred to as linear if no companding is used. There are two common companding scales in use in computer telephony: the mu-law scale which is used internally by the digital telephone network in North America, and the A-law scale, which is used in most other countries.

Different manufacturers have different flavors of ADPCM, so be aware that a file recorded using equipment from one vendor may not work on another vendor's board.

Dialogic boards offer the following VOX file format options:

4bit ADPCM or 8bit PCM sampling,

6 KHz or 8 KHz sampling rate,

A-law or mu-law companding.

This gives a total of $2 \times 2 \times 2 = 8$ formats. At the time of writing, all Dialogic boards currently shipping except the D/21E, D/41E and D/81A offer either A-law or mu-law companding, but not both. The D/x1E boards can use A-law or mu-law as required on a per-file, the format defaults to mu-law. The D/81A supports both, but the selection is once and for all when firmware is loaded and cannot be changed on a per-file basis. The following table shows the companding types currently supported by each Dialogic board type.

Board Model	Companding supported
D/21D	mu-law only
D/21E	A-law and mu-law. Can be selected on a per-file basis.
D/41D	mu-law only
D/41E	A-law and mu-law. Can be selected on a per-file basis.
D/42xx (PBX integration boards)	mu-law only
D/21Dxx and D/41Dxx (international boards)	mu-law only
D/81A	A-law or mu-law. Selected at firmware load time by fontend=analog (mu-law) or frontend=digital (A-law) in configuration file, cannot be set on per-file basis.
D/4xA, D/4xB and D/2xB (older model boards)	mu-law only
D/121B	mu-law only
D/160SC	mu-law only
D/160SCLS	mu-law only

Board Model	Companding supported
D/240SC	mu-law only
D/240SCT1	mu-law only
D/300SCE1	A-law only
D/320SC	A-law only

Most boards therefore offer four different options: 4bit or 8bit samples at 6 KHz or 8 KHz. These four different options are as shown in the following table.

Bits/sample	Type	Rate	Kbits/sec	Kb/min	Mb/hour
4	ADPCM	6 KHz	24 Kbps	180 Kb	10.8 Mb
4	ADPCM	8 KHz	32 Kbps	240 Kb	14.4 Mb
8	PCM	6 KHz	48 Kbps	360 Kb	21.6 Mb
8	PCM	8 KHz	64 Kbps	480 Kb	28.8 Mb

(Note: the above table uses *K*b and *M*b in the old-fashioned metric sense of Kilo and Mega, i.e. 1,000 and 1,000,000 respectively. Disk storage is often measured in Kb=1024 bytes and Mb=1,048,576 bytes. The difference is a little more than 2% and usually not important. Note also that Kbps means Kilobits per second, not Kilobytes per second).

If you are using Dialogic international products, you should be aware that upgrading your hardware may involve recreating your voice prompts. For example, if you are now running an 8line analog system using the LSI/80 and D/81A, and want to upgrade to a digital system on an E1 trunk using the D/300SCE1, you will have to convert your VOX files from mu-law to A-law. Our recommendation is to create all your system messages as high-quality, 22 KHz Wave files and use VOX Studio to do the conversion. If you follow this procedure, it will be relatively easy to do your upgrade.

The public telephone network uses 8 KHz 8bit PCM encoding identical to the 64 Kbps VOX format, so this can give quality as good as theoretically possible in digital transmission using the network. There is little point in providing a higher quality format today, although future applications using the SC bus will be able to combine several PCM channels into single, high-fidelity channels which could be used,

Telephony Sound Files

for example, to carry CD-quality sound or video. This type of application will use hardware which has not progressed beyond the early development stages at the time of writing.

The default format is the most compressed, and therefore the lowest quality: 24 Kbps, that is 4bit ADPCM sampled at 6 KHz.

Note that VOX file compression acts by reducing to the number of bits required to store a sample value by a fixed amount. All VOX file formats have a fixed transfer rate, i.e. number of bits per second, as shown in the above table. If you know how many bytes of VOX data you have, you know exactly how long it will take to play that data.

There is a tradeoff between sound quality and the resources required to run your application. The best quality VOX format requires almost three times as much disk space and three times as much data throughput from disk storage as does the default format. Using the best quality format may limit the number of lines which can be run successfully, particularly if the VOX files are located on a network file server and must therefore be transported across a LAN.

A good compromise may be to record your system prompts in the best format (64 Kbps) and store them in a RAM drive so that they can be delivered very efficiently to the voice card. New messages from callers could be recorded in the most compressed format (24 Kbps) to save disk space and disk/network load.

There is a major weakness of VOX files. The only data in a VOX file is raw sample data, there is no file header to allow application software to determine the exact flavor of VOX file, or indeed to allow an application to distinguish between a VOX file and any other type of file. This is in contrast to Wave files, which include a header which specifies the sample size and sampling rate of the file.

Dialogic boards are not designed for high-fidelity recording. However, you can use a multimedia PC with a sound card, microphone and Wave file editor to create high-quality Wave files, then convert them to VOX for use on a Dialogic. This can produce results which are noticeably better than recordings made directly on a Dialogic card. Parity

Software's VOX Studio product offers all the software needed to do this plus a CDROM with more than 300 sound effects in 8bit and 16bit Wave format which can be converted to VOX. This will allow you to produce professional prompts for your system using desktop PC technology rather than an expensive digital recording studio.

Cutting And Pasting ADPCM Data

You will probably use voice editor software if you need to manipulate VOX or Wave files. If for some reason you need to do your own cutting and pasting of VOX files you should be aware that there is a pitfall for the unwary. PCM data can be cut and pasted without problems, but ADPCM files are different because of the compression scheme. To interpret ADPCM data you must begin at the start of the file and keep track of a scale factor used to determine the sample value. If a section is cut from the middle of an ADPCM file and added to another file, the data must be adjusted for the current scale factor at the point where the data is added. VOX file editors, of course, know how to do this automatically for you.

Sometimes users will try to combine two or more ADPCM files using a binary copying procedure such as the MSDOS COPY /B command. If you try this, you will likely hear a clicking sound or other glitch at the boundary between the two files.

Indexed Prompt Files (IPFs)

An Indexed Prompt File (IPF) is a collection of VOX files combined into a single file. Each VOX file stored within an IPF is called a prompt. An IPF begins with a header which includes an index listing all of the included prompts. Unfortunately the IPF file format was designed at a time when Dialogic only supported 4bit ADPCM encoding, and the only specification given of the VOX file type is the sampling rate. IPFs are sometimes called VBASE 40 (there is an old voice file editor of that name). The index entry for each IPF prompt specifies the position and size of each prompt in the file and also a comment field, which can be an arbitrary text string. Most applications will use this text string to store the words spoken in the prompt.

No additional file handles are needed to play files or phrases which only use IPFs. One or more file handles may be needed for each channel if VOX files are played. On systems with many phone lines, this may approach the maximum number of files which the operating system can have open at one time.

There are fewer files to administer, for example when distributing your application, if several VOX files are combined into an IPF.

There are also disadvantages of IPFs.

You can generally only play from, not record to, an IPF prompt. (This is because the application must either waste space, a new version of a prompt must be recorded at the end of the IPF, leaving an unused gap where the old version is stored; or alternatively must use significant overhead re-organizing the IPF when the recording is complete). IPFs are therefore suitable only for prompts which rarely or never change in your system since you would need to stop your application and copy a new IPF to change a prompt.

IPFs may become very large. This may be a problem in distributing your application since the IPF may not fit on a single diskette.

You have an additional step of building the IPF when you are developing the application. When your application is changing rapidly under development, it may be easier to deal with individual VOX files.

IPF File Format

For C programmers who wish to support IPFs in their application, we will give the format of the IPF header and index entry here:

```
typedef struct        // IPF Header
  {
  long MaxPrompts;    // Max prompts allowed
  long DigRate;       // Sampling rate
  long UsedPrompts;   // Nr active prompts
  long reserved1;     // Reserved (should be 0)
  long UsedBytes;     // Bytes used in file
  long reserved2;     // Reserved (should be 0)
  } IPFHDR;
```

```
typedef struct       // IPF Directory Entry
{
    long Pos;            // Address of prompt data
    long Bytes;          // Length in bytes
    long PosComment;     // File offset of ASCIIZ
                         // comment
} IPFDIRENTRY;
```

Data Compression

Files in ADPCM format are already highly compressed, and in most cases you will get very little extra compression by using utilities like Stacker, DoubleSpace or PKZIP. If you are mostly using an ADPCM format, Parity Software recommends that you do not use any disk compression software since it will add very little space (most of your hard drive will probably contain voice file data) and there is therefore little motivation for the added complexity and chance for problems caused by compression software.

PCM data, on the other hand, may compress relatively well, perhaps by up to 50%. However, the compression factor achieved will be very dependent on the types of compression algorithms supported by the software, some compression algorithms will fail to compress PCM by much at all.

Also, when playing 11 KHz Wave, the Dialogic firmware will have to convert on the fly from 11 KHz to 8 KHz sampling and also convert the companding to telephone network standards. This means that better audio quality will be obtained by converting Wave to 8 KHz PCM off-line with a dedicated conversion tool such as Parity Software's VOX Studio or Bitworks' Audio Work Station. This software will generally convert slower than real-time but will be able to perform higher-quality conversions. For this reason, even users with requirements for the highest possible audio quality will still have good reason to look seriously at using VOX format files for their application.

Should I Use Wave Files?

Microsoft Windows applications use an audio file format called Wave as a standard. This is not compatible with most sound files used on call processing boards. Parity Software's VOX Studio product can be used to convert between the two formats. With VOX Studio you can use a

Telephony Sound Files

high-quality microphone and Wave editor to create your system prompts, then when your editing is complete mix down to Dialogic-compatible formats. This will generally give much better sound quality than an editor such as VFEdit which uses the Dialogic board digitizer to create the sound files.

With the advent of TAPI, and some support for Wave files in the proprietary Dialogic API, you may be faced with a decision on using VOX or Wave files in your application. The pros and cons of Wave will be considered in detail in later chapters.

Chapter 24

Programming For Dialogic

Dialogic Drivers And APIs

An application program uses Dialogic boards by making function calls which invoke a device driver. The set of functions provided by each driver is called an *Application Programming Interface* (*API*). Usually an

API is provided as a library of functions which can be called from C or C++, although other techniques may be used.

Dialogic system software may be divided into two parts: *drivers* and *firmware*. Firmware is code which is downloaded to a board and executed, typically by a DSP processor, on that board. (A *Digital Signal Processor*, or *DSP*, is a special type of CPU chip which is designed for real-time signal analysis). The firmware does most of the "real work" of the board. Drivers are software components which are executed on the host computer (usually an Intel chip in a PC). To a first approximation, the job of a driver is to provide a layer which communicates between the application program and firmware. A typical API function inside the driver formats input parameters as required, passes the request on to the firmware, waits for a response from the firmware (often reported via a hardware interrupt) and then passes the response back to the application either as a return value from a function call or as an "event" or "message".

Operating Systems And Driver Types

Dialogic provides a number of different APIs on several different operating systems.

The following gives an overview of the different driver types available at the time of writing.

MS-DOS

Drivers are provided as TSRs. The main TSR is called D40DRV.EXE, this includes a driver for voice cards (D/xx, Proline/2V etc.) and also an interface which is used internally by other drivers, such as FAXDRV.EXE for fax cards. All API calls go through D40DRV. The programming interface to D40DRV is similar to that of MS-DOS itself: API calls are made by setting values in CPU registers and making a software interrupt. Dialogic provides C source code libraries for all the published API functions which can be compiled by most DOS C and C++ compilers. No support is provided for "DOS-extended" applications which load and run some of their code in extended memory (memory above the 1 MB MS-DOS limit). This is unfortunate, because there is often a severe shortage of memory in the lower 1 MB range in larger telephony systems because several different Dialogic driver types may be loaded, which together with network and other drivers, may take several hundred kilobytes. It is possible for DOS-extended applications such as Parity Software's 32-bit VOS for DOS engine to use the DPMI (DOS Protected Mode Interface) API to switch to real mode and call Dialogic functions, but this is a significant programming effort.

OS/2

Native drivers are provided for OS/2. The application program, which may be 16- or 32-bit, makes API calls via DLLs.

UNIX

Dialogic drivers are provided for several UNIX flavors. The range of boards supported may be quite different for each type, for updated information on which UNIX flavors are supported and which features are available in each driver type, the reader should check with their Dialogic hardware vendor. At the time of writing, the supported UNIX flavors are:

- SCO System V, release 3.2 (all versions).

- Interactive System V, release 3.2, v2.0.2 or later.

- UnixWare System V, release 4.2, v1.1 or later.

- AIX v3.2 or later.

- Solaris X86 v2.4 or later.

Several UNIX flavors are binary compatible. For example, an executable designed for and built on SCO System V can usually be copied unmodified to Solaris X86. Unfortunately, the Dialogic driver design breaks this compatibility so Dialogic applications must be rebuilt and retested on each target platform. The significantly increases the time and cost of developing Dialogic applications for multiple UNIX platforms.

16-bit Windows

Dialogic does not provide 16-bit drivers for Windows. As with DOS-extended programs, 16-bit applications for Windows 3.x or Windows 95 can use the MS-DOS drivers (which must loaded before starting Windows) via the DPMI interface which is built into Windows, but this is a lot of work (similar to creating DOS-extended applications), and you get no help from Dialogic software. DOS applications for Dialogic can be run in a DOS box under Windows 3.1 or 95, the Dialogic drivers can be loaded either before Windows (which takes memory away from all DOS boxes and from all Windows programs) or in a single DOS box, as required.

Neither DOS applications nor 16-bit Windows applications for Dialogic can be run under Windows NT.

32-bit Windows

Dialogic provides several different types of driver for 32-bit applications under Windows 95 and NT. This is because NT prevents direct access to hardware except through native Windows NT drivers, unlike Windows 3.1 and 95.

Under Windows 95, native drivers are provided together with a set of DLLs. Currently, only a few popular, low-density boards are supported under Windows 95. Windows NT's superior multi-tasking and robustness makes NT the platform of choice over 95 for professional telephony systems.

Under Windows NT, native drivers are scheduled for release in the summer of 1997 and so should be available by the time this edition of *PC Telephony* is published. Currently, Dialogic's NT drivers are based on a port of older UNIX drivers and require the NT Streams service to be installed in order for the drivers to operate. The native drivers are expected to offer significant performance benefits over the Streams-based drivers, and fortunately most applications written to the Streams-based drivers should run unchanged on the native drivers.

As on Windows 95, DLLs are provided for linking the driver to the NT application. Unfortunately, even though the function call names and parameters on 95 and NT are very similar (identical in most cases), an

Programming For Dialogic 363

application built for one will not run on the other if the default linkage method to the DLLs is used. This is because set of functions exported by the DLLs is different between the two operating systems. This is a basic design error in the Dialogic drivers – most other 32-bit Windows APIs are immediately portable between NT and 95 and this creates a significant problem for the application developer who wants to target both operating systems. Application binary executable files which are designed to run both on 95 and NT, such as Parity Software's VOS engine and CallSuite ActiveX controls, must circumvent the Windows loader and include their own code to load the Dialogic DLLs into their address space and import the required API functions. Alternatively you can build two separate executables, one for 95 and one for NT, which doubles the work in creating installers, testing, maintenance, version control etc.

Dialogic does provide a "compatibility library" with their Win32 drivers which allows you to build a Win32-compatible EXE. It is important to note this since the default libraries provided do not include this feature.

Drivers are available for the Intel x86 and DEC Alpha versions of Windows NT. These drivers are source-code compatible with each other, but a re-build of your application is necessary because of the different CPU architecture.

Dialogic also provides TAPI drivers (TAPI Service Providers, or TSPs) for several low-end voice boards. The Dialogic TSP is itself implemented using calls to the Dialogic proprietary API. TAPI therefore gives no new functionality, and in fact lacks several features which are available in the proprietary API due to limitations in the TAPI specification and in Dialogic's TSP implementation. TSPs are available both for Windows 95 and Intel x86 Windows NT. The main advantage of TAPI is that it allows an application quite easily to support low-end, single-channel voice modems in addition to Dialogic cards (Dialogic does not make a one-channel board).

A Sneak Peak

Before getting into detailed descriptions of the APIs, we will give a very quick look at a simple example to give a brief impression of writing code for a Dialogic application. This will introduce and illustrate some of the issues which will be discussed in more detail later.

We'll take a basic voice card such as a Proline/2V, Dialog/4 or D/41ESC and write code for one phone line (channel 1). We'll wait for a call and then answer the call by going off-hook. We'll overlook error handling to keep the code short and simple, a real application of course can't ignore the issue of checking for and processing errors.

MS-DOS API

DOS applications must start by locating the D40DRV TSR, since the software interrupt vector used can be changed by the user. This is done by the *getvctr* function. Applications must call *startsys* before calling other API functions, and must call *stopsys* before exiting.

```
int NrChannels; // Number of voice channels
                // installed

getvctr();
startsys(0, SM_EVENT, 0, 0, &NrChannels);
```

We can now wait for an incoming call, which is reported to the application via an *event*. Events are messages from the driver to the application. Events are obtained by calling the *gtevtblk* ("get event block") function, which takes a pointer to an *EVTBLK* structure as an argument. The return value from *gtevtblk* is −1 to indicate an event, or 0 to indicate that there is no event. If there is an event, the fields in *EVTBLK* are filled in with the channel number (*devchan*) and an integer identifying the event (*evtcode*). For example, event code *T_RING* (numerically 21) indicates an incoming call. Constants such as *T_RING* are defined in the *d40.h* header file provided with the Dialogic C source code libraries. To wait for an incoming call, the application polls the *gtevtblk* function until a *T_RING* event is returned.

```
EVTBLK EventBlock;

code = gtevtblk(&EventBlock);
if (code == -1 && EventBlock.evtcode == T_RING &&
    EventBlock.devchan == 1)
        // have an incoming call on line 1
```

To answer the call, the channel must be taken off-hook. This is done by the *sethook* function, which takes two arguments: the channel number and the hook state, *H_ONH* for on-hook or *H_OFFH* for off-hook.

```
sethook(1, H_OFFH);   // answer call on line 1
```

It takes half a second to set the hook switch. The *sethook* function returns immediately with a code to indicate that the operation was successfully started (zero) or that an error occurred (value greater than zero). When the board completes the operation, it posts a *T_OFFH* event to the application. The application polls *gtevtblk* until this event is reported.

In the DOS API, all time-consuming operations such as changing the hook state, playing and recording, waiting for digits etc., work in a similar fashion to *sethook*. The function which starts the operation returns immediately, an event (called a *terminating event*) is later returned to report the completion of the operation. This architecture allows an application to control two or more channels simultaneously.

Windows And UNIX Proprietary API

While there are a few differences in detail, the 32-bit Windows API for Windows NT and Windows 95 (*Win32*) and UNIX APIs are very similar. We will refer to the common set of functions as the Win32/UNIX API. Carefully-written telephony code can be ported between the two platforms with little or no modification. This happy circumstance derives from the origin of the NT API as a port from source code for the UNIX drivers. (Other Dialogic APIs have been designed and implemented by separate teams with no little or no view to consistency or portability).

Both Win32 and UNIX are multi-tasking operating systems. This means that several executing programs can be running simultaneously. Win32 and some UNIX flavors also support multi-threading, where a single program can have multiple execution paths executing in parallel, all with access to the same code and data. Multi-threading can give significant performance benefits compared with running multiple executables or multiple instances of a single executable. Writing code for a multi-threaded application is very similar to writing code for multiple executables. In most cases, the code will operate on a single phone line and doesn't need to take any account of what happens on other lines. This makes writing code much easier: time-consuming API functions such as setting the hook state, playing, recording and getting digits simply "block" or are "put to sleep" until the operation completes. The flow of the application can therefore be reflected in the flow of control of the program code, unlike on DOS where a state machine architecture must be used to control multiple lines. Actually, state machine and other programming models can also be used in the Win32 and UNIX APIs. For example, VOS for SCO UNIX uses this technique to gain the maximum possible performance since SCO does not support multi-threading, and running many binaries in parallel can be very slow under UNIX. Since most applications use the "blocking" (*synchronous*) model, we'll illustrate that here.

To use a channel, an application first opens the channel using *dx_open*, which returns a *device handle*, which is a small integer similar to a file handle. Other API calls require a device handle to identify the channel. The first voice channel is always named *dxxxB1C1*.

```
hdevLine1 = dx_open("dxxxB1C1", 0);
```

The function *dx_wtring* is a convenient method (other ways can be used) of waiting for a ring:

```
dx_wtring(hdevLine1, 1, DX_OFFHOOK, -1);
```

The arguments following the device handle specify to wait for one ring, to go off-hook when the ring is detected, and to wait indefinitely for a ring event with no time-out. So much easier! With these two function calls, we are done.

OS/2 API

The OS/2 API is quite similar to the Win32/UNIX API, but unfortunately is different enough in the details that writing portable code is a non-trivial effort. The function names are almost all different, and the arguments are often different. For example, a channel is opened using *dl_open* (versus *dx_open* on Win32 and UNIX), and you wait for rings using *dl_wtrng* versus *dx_wtring*. A significant drawback of the

Programming For Dialogic

OS/2 driver implementation is that it does not allow asynchronous mode API calls, which means that you can't create very high performance applications by writing state machines as you can on Window or UNIX.

Since the issues with programming the OS/2 API are very similar to (though a subset of) the Win32/UNIX API, and in light of the waning popularity of OS/2 as a telephony platform, we will not be discussing OS/2 programming any further.

TAPI

Getting a small application started with TAPI is a fair amount of work. We'll outline the steps here. The most important startup function is *lineInitialize*:

```
DWORD NrLineDevices;
LONG hTAPI;
void CALLBACK CallBack(DWORD hDevice,
    DWORD dwMessage, DWORD dwInstance,
    DWORD dwParam1, DWORD dwParam2, DWORD dwParam3);

lineInitialize(&hTAPI, hInstance, CallBack,
    "MyApp", &NrLineDevices);
```

This establishes a connection between your application and Windows. The *hTAPI* variable is a handle which identifies your application to the Windows TAPI DLL. As always, *hInstance* is the instance handle of your application. *CallBack* is a call-back function, similar to a window procedure, which is called by Windows when it has a message for your application. It is up to you to write a *CallBack* function to process these messages.

Each Dialogic channel is a *line device* in TAPI terminology. Devices are numbered 0, 1 … (*NrLineDevices* -1) where *NrLineDevices* is a parameter returned by *lineInitialize* to report the number of installed devices. This number is called the *device identifier*, we'll use a variable called *DeviceID* for this value. If the Dialogic board is the only TAPI device installed, then the first phone line will have *DeviceID* = 0.

The next step is version negotiation. This is a process where the application and the TSP agree on a version of TAPI. This is to allow old applications to run on new drivers, and vice versa. See the TAPI documentation for full details. I recommend this call as working on most current TSPs:

```
lineNegotiateAPIVersion(hLineApp, DeviceID,
   0x10000,0x100000, &TAPIVersion,
   LineExtensionID);
```

Individual line devices are opened using *lineOpen*:

```
LineErr = lineOpen(hLineApp, DeviceID, &hLine,
    TAPIVersion, 0, 0,
    LINECALLPRIVILEGE_MONITOR+
     LINECALLPRIVILEGE_OWNER,
    LINEMEDIAMODE_AUTOMATEDVOICE, 0);
```

The *TAPIVersion* variable is returned by the *lineNegotiateAPIVersion* call (so you can't skip the negotiation step. The *hLine* variable is a device handle returned by *lineOpen* which is used in subsequent API calls referring to this channel. As you can see, *lineOpen* is a function with many options, refer to the TAPI documentation for more details.

Once *lineOpen* has been called successfully, TAPI will start sending messages to your *CallBack* function. To signal an incoming call, *CallBack* will be called by the TAPI DLL with the following parameters:

```
dwMessage = LINE_CALLSTATE
dwDevice = (DWORD) hCall;
dwCallbackInstance = (DWORD) hCallback;
dwParam1 = LINECALLSTATE_OFFERING;
dwParam2 = (DWORD) CallStateDetail;
dwParam3 = (DWORD) CallPrivilege;
```

The *hCall* parameter is a *call handle*, an identifier used to track a call through different TAPI functions. The *hCallback* parameter is a value originally passed to *lineOpen*, allowing a single device to be open on multiple handles. (In our example we passed zero to *lineOpen* for this parameter, so *hCallback* will be zero).

Programming For Dialogic

Your *CallBack* function will typically look like an old-fashioned, "Petzold-style" window procedure, with one or more big switch statements for the message types and sub-parameters.

When an incoming call is detected, you answer the call using the *lineAnswer* function. The only argument you usually need to worry about is the call handle, *hCall*, which was passed to you in the *dwDevice* parameter to *CallBack*:

```
RequestID = lineAnswer(hCall, NULL, 0);
```

The return value from *lineAnswer* is a *request identifier*. TAPI does not provide synchronous functions like the Dialogic Win32/UNIX API, it is more like the MS-DOS API in that all API functions return immediately without waiting for the completion of an operation. Functions like *lineAnswer* that start an operation which takes time to complete return a numerical code called a request identifier. When the operation completes, your *CallBack* function is called from the TAPI DLL with these parameters:

```
dwMessage = LINE_REPLY
dwDevice = (DWORD) 0;
dwCallbackInstance = (DWORD) hCallback;
dwParam1 = (DWORD) RequestID;
dwParam2 = (DWORD) Status;
dwParam3 = (DWORD) 0;
```

The *Status* value passed in *dwParam2* is zero to indicate success, or otherwise is an error code. The request identifier passed in *dwParam1* matches the value returned by the function which started the operation. This mechanism allows more than one operation to be in progress at one time on a given device, although this is usually not legal in the current versions of TAPI.

Dialogic API Pros And Cons

As a software engineer, I know what it's like to meet a new, large, complex API for the first time. It takes a while to develop a "feel" for how the API was designed and how best to use it. You continually ask yourself, "How do I get this to work", "Is it really this way, did I miss something?", and "Surely they didn't make it that way!". I'm therefore going to make some general comments on the API designs and documentation, and also step out on a diplomatic limb point out some of the difficulties you may encounter when dealing with the Dialogic APIs.

Please note that the author has been working intensively with the Dialogic API for many years, and knows where many pitfalls lie. No API is perfect, and the reader should not interpret any warnings about these pitfalls as a caution about using Dialogic products, which clearly lead the industry. I would be very surprised if other vendors' SDKs didn't have their own problems, but I am not familiar with them and can't comment on that. I see my job as complementing the Dialogic marketing material and documentation, which naturally tends to emphasize the features and benefits over possible problems.

Since Windows is probably the most popular platform for new development, I will devote a separate chapter comparing TAPI and the Dialogic Win32/UNIX API in detail.

The following paragraphs describe my own point of view (not entirely endorsed, I'm sure, by Dialogic) on the design of the Dialogic APIs and some of the problem areas you may encounter.

Dialogic APIs Are Designed For OEMs.
When Dialogic designs and implements an API, the target developer is an OEM. (Not unreasonable, since most of Dialogic's business is with OEMs). The OEM takes a computer, adds some Dialogic boards and system software and writes an application, integrating all the pieces into a complete "black box" system. The OEM sells the complete system, not just a software application. The API requirements of an OEM are, however, not as extensive as those of developers who want to create "shrink-wrap" software which can be run on a variety of different board configurations, or for a tools vendor such as Parity Software who wants to create tools which can run on *any* Dialogic system. A vendor who sells software to a customer who has an unknown combination of boards and drivers needs to write software which can query to find the installed hardware configuration and its capabilities, and adapt "on-the-fly" to different types of hardware and different driver versions. It is typically in these areas that the Dialogic API can come up short.

Binary Compatibility And Portability

The IBM PC standard and MS-DOS, later with Windows, created a hardware and software platform where an application could be written once and would then run on any computer. Today, Windows NT, Windows 95 and Windows 3.1 can all run most MS-DOS and 16-bit Windows applications. The application adapts to different video capabilities, for example, by querying Windows to find out the size and color depth supported by the adapter and monitor. A software vendor only needs to build a single binary executable version of the application and test on a reasonable range of computer types. Unfortunately, this type of binary compatibility has often been significantly impaired by simple, avoidable design decisions in the Dialogic drivers. Dialogic's philosophy has been that an application should be built for a specific hardware configuration with a specific driver version loaded, and should be linked with the API library for that particular driver release. Attempting to run a new executable on an old driver, or vice versa, has caused problems in several different cases. Dialogic has undertaken to improve this in future releases, but problems of this type continue to show up. For example, the Win32 API is not portable between Windows 95 and Windows NT – unlike all other Windows APIs which are available on both operating systems. It is possible to write an application which is portable, but this requires significant programming effort which should have been taken care of by the drivers. A similar example is the unnecessary driver incompatibilities between different UNIX flavors, where non-Dialogic binaries can simply be copied over without changes, different builds of Dialogic applications must be made for each flavor.

Consistency

It helps to learn and use an API if there is a consistent use of related concepts, consistent naming of functions and structures, and so on. Unfortunately the Dialogic APIs have not been designed consistently within a given platform or across platforms. To give a typical example, on MS-DOS voice device channels are addressed by a channel number 1, 2, ... (number of channels installed), as in the example *sethook(chan, H_OFFH)*. T-1 interface devices, however, are addressed by board number and time-slot number, an example would be *chgidle(board, slot, set)* which enables or disables the transmission of an "idle" pattern on a channel. PRI ISDN interface devices use yet another scheme, where the *is_Open* function is used to obtain a device handle, the device handle is then used in other API calls. Unlike *dx_open* in the Win32/UNIX API which returns a device handle, *is_Open* returns an error code, the first argument is a pointer to a variable where the device handle is returned. As our earlier sneak peak examples showed, API function calls performing identical operations have quite different names and arguments on different operating systems. Data structures and parameter names are also very different across operating systems.

Version Dependency

API capabilities are continually being improved. For an application to adapt itself to different driver versions, the API needs to provide version information so that the application knows what capabilities are present. While some version information is available via the Dialogic APIs, it is not complete. Also, the Dialogic documentation does not specify which versions of the drivers support a given feature, so it can be a very considerable challenge to create software which behaves gracefully on multiple driver releases.

There are also dependencies on the hardware or firmware used. For example, some boards default to A-law companding when playing and recording VOX files, some default to mu-law. Some boards are capable of both companding types, some are capable of only one. The API does not provide a feature to determine the capabilities (A-law only, mu-law only, or both), or the default companding for a given voice channel.

Another example of a dependency issue is NFAS signaling, which is a variant of ISDN where multiple trunks share a single D-channel. You might have three trunks connected to a single PC, two of them would have 24 B channels and one would have 23 B + 1 D. An application needs to know the number of B channels on each board in order to open the channel devices and wait for a call. The API gives no way for an application to check if NFAS is present, and if so which board carries the D channel. Attempting to open a D channel as if it were a B channel has unpredictable results, including causing an application to lock up.

As a result of the lack of a complete configuration API and the many version dependency issues, an application or tool which is designed to

Programming For Dialogic

support a range of different Dialogic hardware and firmware types will probably need a configuration database of some kind which must be set up and maintained independent of the Dialogic drivers. Providing a user interface for reliably setting up and maintaining this database on each new computer, in particular making sure that it is consistent with the installed boards and drivers, is another significant challenge facing the developer.

In the following sections we will discuss a few specific API issues of particular importance.

File Handles And errno In Win32

There are a number of issues related to file handles in the Dialogic Win32 API which are not correctly or fully described in the Dialogic manuals and which can cause several different types of problems in application programs.

Functions which internally need to do file input/output, including *dx_play* (play file), *dx_rec* (record file), *fx_sendfax* (transmit file as fax) and *fx_rvcfax* (receive file as fax) require handles to files. Since the Win32 drivers are ports from UNIX, the Dialogic driver source code accesses these file handles internally using the *open, read, write* and *lseek* functions (collectively called *io.h* functions from the standard UNIX header file name). On UNIX, these are direct calls to the kernel, but on Windows they are implemented as run-time library functions. This difference is very important, because the file handles returned by *open* <u>cannot</u> be shared between DLLs or between an EXE and a DLL which the EXE has loaded. The reason is this: the *io.h* functions on Windows have an internal table of open files. Like all other data, this is stored in the data segment of the binary file (DLL or EXE) which is built by the linker. Each binary file will have its own copy of the run-time library, its own copy of the *io.h* functions, and therefore its own separate file handle table. This means that file handle 13 in A.DLL is s completely different handle than file handle 13 in B.DLL – even if the handle refers to the same physical file (which it probably doesn't), the two copies of the runtime library will maintain independent attributes of the handle such as the current position in the file.

For this reason, Dialogic provides functions named *dx_file*xxx for each *io.h* function. These are simply "wrappers" for *io.h* functions. In the Dialogic driver source code, a function such as *dx_fileseek* does little more than call *lseek*, if you could see the source code it would probably look something like this:

```
long __export dx_fileseek(int hFile, long Pos,
    int Mode)
{
return lseek(hFile, Pos, Mode);
}
```

These *dx_file*xxx functions in effect give the calling process access to the "private" file handle table which is linked into the Dialogic driver DLLs. You should be sure to use *dx_fileopen* rather than *open*.

The Dialogic documentation currently states that the *dx_file*xxx functions, and some others, set the *errno* variable in case of error. This is not correct. Just as with the file handle table, the *errno* variable is stored in the runtime library data area, so a process using the Dialogic driver will have its own *errno* which is different from the Dialogic DLL's variable of the same name. The *dx_file*xxx functions set the Dialogic DLL *errno* variable, and the only way to access it is through the (undocumented!) function *dx_fileerrno*, which returns an *int* and takes no arguments. It is used as follows:

```
hFile = dx_fileopen(filename, O_READ);
if (hFile == -1)
    {
    DialogicErrno = dx_fileerrno();
    // .. handle the error...
```

If you see a claim in the Dialogic manuals that the *errno* variable is set, for example under *vr_open*, don't believe it! The *errno* variable that is set is a copy in one of the Dialogic DLLs, not the one you can access directly. Fortunately, the various Dialogic DLLs LIBDXXMT.DLL, LIBVRXMT.DLL, LIBFAXMT.DLL etc. share a single copy of the Microsoft Visual C++ runtime library by using the *dx_file*xxx functions internally. This means that you can always retrieve the correct *errno* value through *dx_fileerrno*.

User I/O In Win32 And UNIX

Another area which can cause unexpected problems is *user i/o*. This is the term given by Dialogic to a feature which allows the programmer to provide custom data input/output functions for playing and recording. This is used for playing and recording data which is stored in some medium other than a traditional flat file. For example, you might have a software-based text-to-speech engine which works by providing you with a buffer of speech data every couple of seconds, or you might want to play voice data saved as a *binary large object,* or *BLOB,* in a relational database. To do this, you would use is *dx_setuio,* which takes a *DX_UIO* structure as an argument. The *DX_UIO* structure has three fields *u_read, u_write* and *u_seek* which are pointers to functions with the same semantics as the UNIX *read, write* and *lseek* functions. Win32 defaults could be set as follows:

```
DX_UIO uio;
uio.u_read = dx_fileread;
uio.u_write = dx_filewrite;
uio.u_seek = dx_fileseek;
dx_setuio(&uio);
```

If user i/o is requested by setting the *IO_UIO* flag in the *DX_IOTT* structure when playing or recording, then the Dialogic driver will call these functions in order to read, write and position in the file which is being played. The "file handle" is passed through to the user's functions, which can then use it as an internal index for any purpose. It might specify, for example, which text-to-speech synthesizer channel is providing the data for a given voice channel device. Remember that there may be several simultaneous plays and/or records in progress at one time, calls to the user i/o functions may therefore be interleaved in an unpredictable fashion.

The problem with this scheme is that a call to *dx_setuio* globally affects all Dialogic APIs in all attached binaries (in Win32, the parent EXE and any loaded DLLs) and all threads in those binaries.

To see how this can create problems suppose, for example, that you are a text-to-speech software vendor. You want to create a Win32 driver for your product which you will offer to Dialogic developers. The natural way to do this is for you to write a DLL which provides functions to speak on a given Dialogic channel. You might have functions like *TTS_SpeakString* and *TTS_SpeakFile*, the first argument would be the Dialogic voice channel number:

```
TTS_SpeakString(hDevVoiceChan, pszString);
TTS_SpeakFile(hDevVoiceChan, pszPathName);
```

How would you create these functions? To "speak", you would have your speech synthesis engine generate buffers of voice data which you would store in memory. You would set user i/o to re-direct *read* input to an function which you provide. You would invoke the Dialogic *dx_play* function with the *DX_UIO* bit set. Then, whenever the Dialogic driver wants a new buffer to play, it would call your *read* function and you would feed it a new chunk of voice data.

The problem with this scheme is that there can only be one component in the application which utilizes user i/o. If the caller's application also employs user i/o, the combination won't work. Whoever called *dx_setuio* most recently wins, overwriting the previous caller's pointers. Let's say your customer's application calls *dx_setuio* before your text-to-speech DLL does so, then your custom *read* function will receive read requests which were designed to go to the customer's main application.

The solution to these issues is the new *dx_setdevuio* function in the Win32 API, which allows an application to set user i/o functions on a per-device handle basis. Be sure to use *dx_setdevuio* and not *dx_setuio* in your applications.

Playing And Recording VOX Files

This issue applies to all Dialogic platforms. One of the most basic and often-used feature of Dialogic voice boards is the ability to play and record sound files in Dialogic's proprietary VOX format. The developer should be aware of the following issues which are completely or partially missing from the Dialogic documentation:

- Several boards support only some of the VOX file formats.

- The default play-back rate is changed when a play-back rate is explicitly set, this will affect all subsequent plays and records unless they also explicitly set a rate.

- The API does not allow an application to query the VOX file capabilities or current play/record-related defaults (companding, rate) on a channel.

The VOX file format is discussed in detail elsewhere, but to review quickly there are six different VOX file formats:

- 4-bit, 6 kHz ADPCM
- 4-bit, 8 kHz ADPCM
- 8-bit, 6 kHz PCM with mu-law companding
- 8-bit, 8 kHz PCM with mu-law companding
- 8-bit, 6 kHz PCM with A-law companding
- 8-bit, 8 kHz PCM with A-law companding

A VOX file contains audio data only in one of these six formats, there is no file header and therefore no way for an application program or the Dialogic driver to determine the format automatically. Also, many Dialogic boards support only a sub-set of these six formats. It is therefore the application's responsibility first to know the type of VOX file that is being played or recorded, and that the voice channel device supports this type of VOX file.

While the API functions used to play and record files differ considerably in detail on each platform, they all use a *mode* argument which is a bit-map of options. To take the Win32/UNIX function *dx_play* as an example, the mode argument includes a bit-wise "or" of bits which may include the following:

PM_ALAW	Use A-law companding
PM_SR6	Use 6 kHz rate
PM_SR8	Use 8 kHz rate
PM_ADPCM	Use ADPCM encoding
PM_PCM	Use PCM encoding

The Dialogic documentation does not mention the following important details:

Specifying an explicit rate bit of *PM_SR8* changes the default rate to 8 kHz, so that subsequent calls using a *mode* argument of zero, or a *mode* argument without a rate bit, will not play at the expected default of 6 kHz.

Some boards default to using A-law companding, some boards default to using mu-law companding. Some boards which default to mu-law companding are able to play back A-law files if the *PM_ALAW* bit is set, some are not. However, there is no way in the API to query the board to find out which type of companding is the default, there is no way in the API to explicitly request mu-law companding, and there is no way in the API to find out if a mu-law board is able to play A-law files.

The Dialogic documentation gives only sketchy information about which board types have which capabilities in terms of playing different types of VOX and Wave files and how they react to different combinations of *mode* bits in different configurations. When researching this chapter, I had hoped to give a complete table showing the capabilities and behavior of all Dialogic boards, but I was unable to obtain complete, reliable information. (And considering the range and complexity of the Dialogic product line, it is impractical for me, or for most software developers, to set up complete tests to determine this type of basic information).

Programming For Dialogic   379

These issues might sound obscure, so here are some scenarios where they can bite you.

Suppose you have lots of application source code which uses *PM_NORM* (zero) for the play and record mode. That's not unusual, you've never needed to worry about the mode bits because all your files are 6 kHz ADPCM. If you add a single API call anywhere in your application which plays an 8 kHz file, you must now go through all your source code and make sure that you change the mode on all your other calls to include at least a *PM_SR6* bit. If you don't do that, then following your 8 kHz message all your files recorded at 6 kHz will play back at 8 kHz and will sound like Mickey Mouse. And, remember, the Dialogic API doesn't give you a way to check the board model of a voice channel or the VOX file capabilities, so you will have to implement this check through your own configuration database.

For another example, suppose you have a system which includes a board which can only do A-law encoding, such as the D/300SC-E1, together with a board which can be configured for either A-law or mu-law, such as the D/41ESC. Suppose further that your source code was developed in the US so you've never worried about setting companding. First, you will have to re-record any PCM files which you use as system prompts because the D/300SC-E1 can't play them (they will have mu-law companding). Second, you will have to check your source code carefully to make sure that you specify A-law encoding in the appropriate places in the source code which play files on the D/41ESC (this may require a test of the channel number), otherwise the D/41ESC channel will default to mu-law and your play-back will be distorted. If your product may be targeting, say, both T-1 and E-1 systems, you need to take these issues into account when you first design and build your software, this can save you considerable grief when you later decide to port to E-1 and have to confront the A-law/mu-law issue.

Fast-Forward, Pause and Rewind In Wave Files

The Dialogic API provides for transferring data from parts of a VOX or Wave file as opposed to the whole file. By specifying a start position in bytes and a number of bytes to play, you can implement features like *VCR controls* (fast-forward, pause, rewind). Unfortunately, current implementations of the driver (June 97) don't implement these features when playing or recording with a Wave file (this is not mentioned in the documentation). If you set up the data structures to position partway through the data, the play or record will ignore the setting and start at byte offset zero. Unfortunately, this makes it very difficult to implement VCR controls for Wave files. The only possible implementation is to use user i/o. The problem is that you have to write your own code to parse the Wave file format, which is quite challenging. This is discussed in more detail in the chapter comparing TAPI/Wave with the Dialogic API.

Creating EXEs For Both Windows NT And 95

The Windows 95 and Windows NT operating systems provide very similar APIs to application developers, the common set of functions is known as the Win32 API. Most applications written for Win32 (Microsoft Word...) will run on either 95 or NT. As we have mentioned, applications which use standard import libraries for the Dialogic Win32 DLLs are not immediately portable between 95 and NT because of the DLLs export different sets of functions on 95 and NT.

This section explains how to work around the issue by loading the DLLs "manually", this is the technique used by the "compatibility library" provided by Dialogic, this would be the recommended method for developers who want to create applications which will run both on 95 and NT.

To work around the differences between the DLLs on each O/S, you must do the job of the Windows loader and "patch" your application to call the correct addresses, and you must be sure to do this by referencing function names rather than ordinals.

Programming For Dialogic

The Windows API functions you need are *LoadLibrary*, which is used to load a DLL into memory, and *GetProcAddress*, which gets a pointer to a function inside a DLL. Suppose you want to call *dx_stopch* in LIBDXXMT.DLL. First, you must load the DLL:

```
HANDLE hLIBDXXMT;

hLIBDXXMT = LoadLibrary("LIBDXXML.DLL");
if (hLIBDXXMT == NULL)
    // .. error
```

Then you need the address of the *dx_stopch* function. For this, we will need a pointer variable, so we need to define the type. The *dx_stopch* function takes one integer argument and returns an integer:

```
typedef int (*type_fn_dx_stopch)(int hDev);
```

You can now use the *GetProcAddress* function to get the address:

```
type_fn_dx_stopch *PtrTo_dx_stopch;

PtrTo_dx_stopch = LoadLibrary(hLIBDXXMT,
"dx_stopch");
if (PtrTo_dx_stopch == NULL)
    // .. error
```

From now on, when you need to call the *dx_stopch* function, you use the pointer:

```
// Abort an action on channel:
(*PtrTo_dx_stopch)(hDev);
```

By repeating this for all required Dialogic API functions, you will create an application which will adapt to both 95 and NT versions of the Dialogic driver.

Chapter 25

Dialogic Devices

Introduction

Almost every Dialogic API call refers to a *device*. We will spend some time here explaining devices in detail since this topic not well covered in the Dialogic documentation.

With a couple of minor exceptions, there are two types of device: *board devices* and *channel devices*.

A *channel device* processes a single stream of audio and/or signaling data, generally the data for a single telephone call.

A *board device* is a special type of device to which a set of channel devices belongs. Sometimes a board device corresponds to a physical Dialogic board, but in many cases a physical board will have two or more board devices. For this reason, board devices are sometimes referred to as *virtual boards*.

To give a simple example, consider a basic 4-channel voice board like a Dialog/4 or D/41ESC. From the API's point of view, this type of board contains these devices:

- Four voice channel devices.
- One voice board device.

(Actually, there are four hidden analog interface devices as well, but in most cases you don't need to know that; we will return to this issue later).

The voice channel device is the one that is most used. This is the device which plays messages, records messages, detects digits and so on.

An application rarely needs to use a voice board device. In fact, the MS-DOS API doesn't provide a single function which addresses a voice board device. In the Win32/UNIX API there are a few functions such as *ATDX_FWVER()* to get the firmware version running on the board, and *ATDX_PHYADDR()* which gets the shared memory address of the board.

For another example, consider a D/240SC-T1. This board contains these devices:

Dialogic Devices

- 24 voice devices.
- 6 voice board devices.
- 24 digital interface (DTI) channel devices.
- One DTI board device.

The voice channel devices on the D/240SC-T1 are very similar to those on the analog boards. The main difference is the lack of an analog trunk interface, so functions like *sethook* (DOS) and *dx_sethook* (Win32/UNIX) generate errors. Unfortunately, there is currently no completely reliable way for an application to determine through the API whether an analog interface is attached to a voice channel device, so this is another item which must either be hard-coded into an application or entered into a custom configuration database.

Note that there are six voice board devices, not one as you might expect. On MS-DOS, this doesn't matter because you can't talk to a voice device anyway. On the Win32/UNIX API, this is very important because voice channel device names are based on board and channel number. The first four voice channel devices on a D/240SC-T1 are named *dxxxB1C1* to *dxxxB1C4*, the next four are *dxxxB2C1* to *dxxxB2C4*, and the last voice channel device is *dxxxB6C4*. This scheme originated with Dialogic's first higher-density board firmware for the old D/12x PEB boards; the D/12x firmware emulated three D/4x boards so that only minimal changes would be needed for the D/4x drivers to support the D/12x.

Since voice board devices are so rarely used, we will generally abbreviate "voice channel device" to "voice device". "Voice board device" will always be spelled out in full.

The digital board device, unlike the analog board device, is important for applications. Digital board devices always correspond to a single trunk. For example, there are trunk-level events such as a report of loss of synchronization, and trunk-level API calls such as setting alarm handling mode, these are handled through digital board devices.

Device Types

A complete list of Dialogic device types is given in the following table.

Device Type	Channels per device	Device name (Win32/UNIX)	Description
Voice channel (VOX)	2, 3 or 4	dxxxBnCm	Voice processing (play, record, get-digits, dial, analyze call progress, etc.), n is the virtual voice board number, m is the channel number 1, 2.. in that virtual board.
Voice board	N/A	dxxxBn	
Analog trunk interface channel (LSI).	2 or 4	dxxxBnCm	Interface to standard telephone lines ("loop start"). LSI channels are addressed through a voice channel.
Digital trunk interface (DTI) channel	23, 24 or 30	dtiBnTm	Digital interface to T-1, E-1 or PRI ISDN trunk, n is DTI virtual board number, m is time-slot number of the channel.
DTI board	N/A	dtiBn	
Fax (FAX)	4	dxxxBnCm	Fax channels on FAX/40 or VFX/40. Fax channels are part of the voice channel.
Modular Station Interface (MSI) station	8, 16 or 24	msiBnCm	Station (headset or phone) interface on MSI/SC board, n is board number, m is station number.

Dialogic Devices

Device Type	Channels per device	Device name *(Win32/UNIX)*	Description
MSI board	N/A	msiBn	
DCB processor	N/A	dcbBnDm	Conferencing processor on DCB ("Spider") board. The processor number m is 1, 2 or 3 as there are up to 3 processors on a DCB board.
DCB board	N/A	dcbBn	
Voice recognition channel.	4	vrxBnCm	Voice recognition resource channel on a VR/xxp PEB-based voice recognition board, n is the virtual board number, m is the recognizer number 1, 2... on that board.
Voice recognition board.	N/A	vrxBn	

The following table lists the devices on some of the more common Dialogic boards.

Board	Devices
Proline/2V, D/21D, D/21H	1 voice board. 2 voice channels.
D/41D, D/41E, D/41H, D/41ESC, Dialog/4	1 voice board. 4 voice channels.
D/240SC-T1, DTI/241SC running T-1 firmware	6 voice boards 24 voice channels.

Board	Devices
	1 DTI board.
	24 DTI channels.
D/240SC-T1, DTI/241SC running ISDN firmware	6 voice boards
	23 voice channels.
	1 DTI board.
	23 DTI channels.
D/300SC-T1, DTI/301SC running E-1 or ISDN firmware	8 voice boards.
	30 voice channels.
	1 DTI board.
	30 DTI channels.
D/240SC	6 voice boards.
	24 voice channels.
D/320SC	8 voice boards.
	32 voice channels.
DTI/240SC	1 DTI board.
	23 (T-1) or 24 (ISDN) DTI channels
DTI/300SC	1 DTI board.
	30 voice channels.
D/240SC-2T1	Exactly equivalent to D/240SC-T1 + 2 x DTI/240SC. For example, there will be 2 virtual DTI boards.
D/300SC-2E1	Exactly equivalent to D/300SC-T1 + 2 x DTI/300SC.
D/480SC-2T1	Exactly equivalent to 2 x D/240SC-T1.
D/600SC-2E1	Exactly equivalent to 2 x D/300SC-E1.

Dialogic Devices

Board	Devices
VFX/40, VFX/40E or VFX/40ESC	1 voice board. 4 voice channels. 4 fax channels.

Transmit And Listen/Receive Channel Devices

All channel devices have two halves, one called *transmit* and the other called *listen* or *receive*.

The listen half of a channel takes an existing stream of audio and/or signaling data and analyses or uses this data. For example, the listen part of a voice channel is responsible for recording messages and detecting digits. The listen part of a T-1 DTI channel is responsible for detecting when a caller hung up by reporting an A signaling bit value change from 1 to 0.

The transmit half of a channel generates a stream of audio and/or signaling data. For example, the transmit part of a voice channel is responsible for playing messages and dialing digits. The transmit part of a T-1 DTI channel is responsible for setting the A and B signaling bit values which are sent out on the trunk.

Most API functions address only one of the two halves of a device channel. For example, the DOS function *xplayf* and the Win32/UNIX function *dx_play* (each of which play audio from a file) address only the transmit half of a voice channel. The DOS function *recfile* and the Win32/UNIX function *dx_rec* (which record audio to a file) address only the record half of a voice channel. It is possible to route the two parts of a device independently. The usual routing has a full-duplex connection between a voice channel and a trunk interface channel (LSI or DTI). However, on an SC Bus or PEB board, the listen half of a voice channel may be routed to a different interface channel from the transmit half.

Voice Channel Devices

There are several different flavors of voice device. Most capabilities and API functions apply to all flavors.

The main capabilities of voice channel devices are:

- Playing audio files.
- Recording audio files.
- Dialing DTMF, MF and pulse digits
- Detecting DTMF, MF and (with optional firmware) pulse digits.
- Generating custom tones.
- Detecting custom tones.
- Analyzing call progress.

Most voice boards, including all those named D/xx and "basic" voice boards such as the Proline/2V and the Dialog/4, have fully-capable voice channels. A few boards, named LSI/xx1SC or DTI/xx1SC, have limited voice channels which can perform all functions except playing and recording audio.

Voice devices are tied in non-obvious ways to analog interface (LSI) devices and to fax devices. Dialogic boards are currently designed so that analog interfaces and fax channels are always paired with voice channels. There is no "pure LSI" board with no voice devices corresponding to a DTI/240SC board which only has DTI devices, and similarly there is no "pure fax" board which just has fax resource devices. The API handles this in a rather confusing way, which will be discussed in more detail in the LSI and fax paragraphs that follow.

Analog Interface (LSI) Devices

The main capabilities of LSI devices are:

- Controlling the hook state of the trunk connector on the board.
- Detecting incoming call (ring) and caller hang-up (disconnect).

In the underlying hardware, the analog interface to each phone line is a separate electronic device. However, the Dialogic API does not treat LSI as a separate device type with a separate set of API functions, the

Dialogic Devices 391

LSI API is all implemented via functions which take a voice device channel number or device handle.

The exception is LSI boards with SC bus interfaces. The *ag_getxmitslot*, *ag_listen* and *ag_unlisten* functions which control SC bus routing (these functions have the same name on all operating systems) truly address an LSI device which is separate from the paired voice device. It is therefore possible to have an LSI device routed differently from the corresponding voice device. You might use this feature to share resources in a configuration such as D/41ESC + VFX/40. However, the application must be careful to treat this situation correctly. Suppose, for an artificial example, that you have a single D/41ESC. Normally voice channel 1 will be in a full-duplex routing with LSI channel 1, and so on for the remaining channels. Suppose we change this default and make a full-duplex routing of voice channel 1 with LSI channel 2, and also of voice channel 2 with LSI channel 1. This is perfectly legal and can be made to work robustly. Now, however, the API works in a way which may be unexpected. Let's define phone line 1 as the line which is attached to LSI interface 1, and phone line 2 as the line attached to LSI 2. To go off-hook on line 1 you must call *sethook* (DOS) on voice channel number 1 or *dx_sethook* (Win32/UNIX) on device *dxxxB1C1*, but you will play a greeting message on channel number 2 or *dxxxB1C2*. Similarly, a hang-up event from line 1 will be reported on channel number 1 or *dxxxB1C1*.

The situation with PEB-based LSI devices is quite different. In PEB boards, the LSI has no programming interface. The LSI interacts with voice devices via a signaling bit which is carried on each PEB time-slot. For example, when a voice channel goes off-hook (*sethook* or *dx_sethook*), the transmit half of the voice channel begins inserting a 1 value for the signaling bit into the PEB time-slot. The LSI channel which is listening to this time-slot uses the value of this signaling bit to determine its hook state, so it reacts to the bit value change by going off-hook. Conversely, the LSI converts the presence or absence of loop current on the analog line into the signaling bit which the LSI in its turn inserts into the other half of the PEB time-slot. (PEB time-slots, unlike SC bus time-slots, have two halves called transmit and receive). The voice channel listens to this time-slot, when it detects a change in the signaling bit from 1 to 0 it reports a hang-up event. Therefore, in PEB-based systems, both resource (play, record..) functions and signaling (hook state, hangup event...) functions for a given phone line always refer to the same voice channel number even if the device routings have been changed.

DTI Channel Devices

The main capabilities of DTI channel devices are:

- Detecting incoming call and caller hang-up via digital signaling.
- Making out-going calls (on ISDN only, tone dialing on T-1 or E-1 requires a voice device as well to do the dialing)
- Disconnecting calls.
- Inserting a silence pattern for the audio (for situations where there is no voice channel routed).

PEB-based DTI devices with T-1 firmware support a mode called *transparent signaling* which allows all the trunk signaling to be done via the voice device channel, making applications immediately portable from analog to T-1. With SC bus systems, the bus hardware does not support signaling, and the API does not support emulation for a transparent mode. It is therefore up to the application to make DTI calls on digital interfaces and LSI calls on analog interfaces.

Fax Channel Devices

The main capabilities of fax channels are:

- Receiving fax transmissions and saving as document files in "raw" or TIFF/F file format.
- Transmitting fax documents in "raw", TIFF/F or ASCII text format (multiple files in mixed formats may be combined in one transmission).

Like LSI devices, fax devices are always paired with voice devices. However, unlike with LSI, there is a separate fax API which has separate channel numbers (DOS) or device handles (Win32/UNIX). On Win32/UNIX, there is a curious hybrid: there are *fx_* functions to address the fax devices, but *fx_open* takes the voice device string as an

Dialogic Devices 393

argument. For example, if you have a VFX/40SC, the first voice channel is opened by *dx_open("dxxxB1C1", 0)*, and the first fax channel is opened by *fx_open("dxxxB1C1", 0)*. This call uses the same device name string, but will return a different device handle which must be used when calling other *fx_* functions.

The SC bus functions *fx_getxmitslot*, *fx_listen* and *fx_unlisten* affect the routing of the paired voice channels as well as the fax devices. In other words, you cannot have a fax channel routed differently from its paired voice channel. Note how this differs from LSI devices, which can be routed differently.

On DOS the fax channel number may be different from the paired voice channel. For example, a system with a D/41ESC with board ID 0 and a VFX/40SC as board ID 1 will have eight voice channels numbered 1 .. 8 and four fax channels numbered 1 .. 4. However, fax channel 1 is paired with voice channel 5 (the first voice channel on the VFX). In this configuration, it would be illegal to call *fx_open* on *dxxxB1C1*, the first fax channel device would be opened by *fx_open* on *dxxxB2C1*.

MSI Station Devices

The Dialogic MSI boards allow a PEB or SC bus system to connect to telephone-like devices such as operator headsets. Each connection is similar to an LSI device, except that it is like the switch end, not the customer end, of a telephone line: the MSI provides battery voltage to the line.

The main capabilities of MSI devices are:

- Providing battery voltage to an analog line (called a *station*).
- Providing ring voltage to ring a station on demand (MSI/SC-R only).
- Connecting a bus time-slot to a station.
- Detecting hang-up by a station.

Voice Recognition Resource Devices

The Dialogic VR/xxp PEB-based voice recognition board family has an API through the *vr_* series of functions. SC bus based systems use software-only voice recognition or the Antares board. Each software-only product and each Antares firmware vendor uses its own proprietary API, creating a big mess for software developers who don't want to be tied to one system.

Device Names And Numbers

In a system with a single board, it is pretty easy to figure out which device names and numbers refer to which devices. With multiple boards, it requires careful analysis to figure out how to address each device. This is one of the biggest problem areas of the Dialogic API – it does not allow the application to query the installed physical configuration and determine how to address each installed device through the API.

The following sections walk you through some of the important information needed so that you can understand how the Dialogic drivers determine device names and numbers.

Board Numbers

The first and most important step is to determine the order which Dialogic assigns to installed boards.

Many Dialogic boards have a board ID number which is set by a rotator switch on the top of the board. This type is known as a *BLT* (for *Board Locator Technology*) board.

If all your boards have ID numbers, then the boards are ordered following increasing ID.

Some boards do not have an ID switch. To determine the board numbering when these types of boards are installed, you need to know the shared memory address or I/O port address assigned to each board. The shared memory address is assigned through a configuration program and/or by setting DIP switches on the board, depending on the board model. Boards are ordered by increasing shared memory address, boards sharing one address are ordered by ID. Boards with ID switches will all share the same memory address. Some boards do not have a shared memory address, but do have an I/O port address range.

Dialogic Devices 395

For these boards (DTI/1xx, DTI/2xx, DMX and MSI/C), boards are numbered in order of increasing I/O port address.

To illustrate these rules, here are some examples.

If you have a D/41D at address D000 and a Proline/2V at DB00, then the D/41D will be board 1 and the Proline/2V will be board 2. If you have a mixture of boards with and without ID locators, then the shared memory address is used to determine the ordering. If you have a Proline/2V at D000 and one or more boards with IDs at DA00, then the Proline/2V will be board 1, the remaining boards will be numbered in order of increasing ID.

Here's another example to illustrate the board numbering for voice channels.

Physical Board	ID	Address	Device Name (Win32/UNIX)	Channel Nrs. (DOS)
D/41ESC	0	D000	$dxxxB1$	1 to 4
D/240SC-T1	1	D000	$dxxxB2..dxxxB7$	5 to 28
Proline/2V	N/A	D800	$dxxxB8$	29 and 30
Dialog/4	N/A	DB00	$dxxxB9$	31 to 34

If the BLT boards were re-configured to use shared memory address DB00, and the Dialog/4 were changed to use D000, then the assignments would change to the following:

Physical Board	ID	Address	Device Name (Win32/UNIX)	Channel Nrs. (DOS)
Dialog/4	N/A	D000	$dxxxB1$	1 to 4
Proline/2V	N/A	D800	$dxxxB2$	5 and 6
D/41ESC	0	DB00	$dxxxB3$	7 to 10
D/240SC-T1	1	DB00	$dxxxB4..dxxxB9$	11 to 34

These rules are always followed when numbering boards:

- Boards in a family are numbered 1, 2 ... (number of installed boards of that type). There are no gaps in the numbering.

- Virtual boards on a single physical board are numbered consecutively.

"Family" refers to all boards of a given type: voice, DTI etc.

On DOS, DTI, MSI and DMX boards belong to a single family, but on the Win32/UNIX API belong to three different families. DTI, MSI and DMX boards are collectively known as *CCM* boards, and are based on similar hardware architectures. On DOS, CCM boards have APIs with a common subset of functions which applies to all three board types. A system with two DTI/201s, one MSI and one DMX might be assigned board numbers for DOS as follows:

Physical Board	I/O Port	Board Nr *(DOS)*
DTI/211	310	1
DTI/211	320	2
MSI/C	330	3
DMX	340	4

The board number is used as the first argument to DTILIB, MSILIB and DMXLIB functions like *gtbdattr*.

In the Win32/UNIX API the DTI, MSI and DMX boards belong to separate families. Device names would be assigned as follows:

Physical Board	I/O Port	Device Name *(Win32/UNIX)*
DTI/211	310	*dtiB1*
DTI/211	320	*dtiB2*
MSI/C	330	*msiB1*
DMX	340	*dmxB1*

In a mixed CCM system with BLT and non-BLT boards, eg. a DTI/211 and a D/240SC-T1, the BLT boards will all be numbered before the CCM boards. For example in a system with two of each board this could be the result (showing DTI board devices only):

Dialogic Devices

Physical Board	ID	I/O Port	Device Name *(Win32/UNIX)*	Board Nr. *(DOS)*
D/240SC-T1	0	N/A	*dtiB1*	1
D/240SC-T1	1	N/A	*dtiB2*	2
DTI/211	N/A	310	*dtiB3*	3
DTI/211	N/A	320	*dtiB4*	4

Voice Channel Numbers

With the Win32/UNIX API, you start by listing the virtual voice board devices $dxxxB1$, $dxxxB2$... and their associated physical boards, which will be numbered following the order of the underlying physical boards. For each virtual board, you can work out how many voice channels there are:

- 2 channels on a two-line card such as a Proline/2V, D/21D or D/21H, or on the last virtual voice board for a given an E-1 or E-1 style ISDN type interface (which has 30 channels, giving seven 4-channel virtual boards and one 2-channel virtual board),

- 3 channels on the last virtual board for each T-1 style ISDN interface which has a D channel (remember that with NFAS there will only be one interface with 23 channels, all others will have 24 channels).

- 4 channels on all other virtual boards.

On DOS, voice channels are numbered from 1 up to the total number installed, with no gaps in the numbering. Channels are ordered by increasing board number.

The only exception to this rule is with T-1 type ISDN firmware, which has 23 usable voice channels for each virtual DTI board with a D channel. In this case, the 24th channel (corresponding to the D channel) is skipped. For example, if you have a D/240SC-T1 running ISDN firmware as the first board and a Proline/2V as the second board, then the D/240SC-T1 has voice channels 1 to 23 and the Proline/2V has voice channels 25 and 26. It will be illegal to address voice channel number 24 in this configuration. If there is more than one ISDN board and NFAS firmware is loaded, then there will be 24 available voice channels on all the virtual DTI boards except for the one with a D channel. Unfortunately, the API does not permit an application to determine if NFAS is loaded, nor to find out which board has the D channel in the case where NFAS is loaded.

Fax Channel Numbers

In the Win32/UNIX API, fax channels are opened by calling *fx_open* on the paired voice device. For example, if you have a D/41ESC as board ID 0 and a VFX/40SC as board ID 1, then you open the first fax channel by calling *fx_open* on device *dxxxB2C1* because *dxxxB2* is the VFX/40SC board. The return value from *fx_open* is the device handle, which is used in all other *fx_* API calls.

On MS-DOS, fax channels are numbered independently of voice channels, so fax channels are numbered 1, 2 ... (number of fax channels installed). The fax channel number is used in all *fx_* API calls. The application must take into account that the fax channel number may not match the paired voice channel number. In the case of a D/41ESC with ID 0 and a VFX/40SC with ID 1, fax channel 1 will be paired with voice channel 5.

The following table summarizes this example:

Board	Voice Device (Win32/UNIX)	Voice chan (DOS)	Fax Device (Win32/UNIX)	Fax chan (DOS)
D/41ESC	*dxxxB1C1*	1		
D/41ESC	*dxxxB1C2*	2		
D/41ESC	*dxxxB1C3*	3		
D/41ESC	*dxxxB1C4*	4		
VFX/40SC	*dxxxB2C1*	5	*dxxxB2C1*	1
VFX/40SC	*dxxxB2C1*	6	*dxxxB2C2*	2
VFX/40SC	*dxxxB2C3*	7	*dxxxB2C3*	3
VFX/40SC	*dxxxB2C4*	8	*dxxxB2C4*	4

DTI Channel Numbers

In the Win32/UNIX API, DTI channel devices are named *dtiBnTm*, where *n* is the virtual DTI board number, and *m* is the time-slot number of the channel in the trunk which is attached to that DTI

Dialogic Devices

board. Each virtual DTI board is connected to one and only trunk (T-1, E-1 or ISDN).

In the DOS API, DTILIB, MSILIB and DMXLIB functions take two arguments, a CCM board number and a time-slot number. The Dialogic manuals refer to the board number as a "device handle", but this is misleading since there is no function to open and return the handle, and the value is simply a board number,

Remember that the DOS board number may be different from the board number used in the Win32/UNIX device name because on DOS all CCM boards are numbered in the same sequence, but on Win32/UNIX DTI, MSI and DMX have separate sequences.

Consider a configuration with an MSI/SC as board ID 0 and D/240SC-T1 as board ID 1.

Physical Board	ID	Device Name (Win32/UNIX)	Board Number (DOS)
MSI/SC	0	msiB1	1
D/240SC-T1	1	dtiB1	2

In this configuration, suppose we wanted to clear (set to zero) the transmitted A signaling bit on time-slot 1 of the T-1 trunk. On DOS, this would be done using the *clrsignal* function on CCM board 2:

```
clrsignal(2, 1, A_BIT);
```

Using Win32/UNIX, this would be done using the *dt_setsig* function on the device handle for *dtiB1T1*:

```
hDevSlot1 = dt_open("dtiB1T1", 0);
// ...
dt_segsig(hDevSlot1, DTB_ABIT, DTA_SUBMSK);
```

Chapter 26

Programming Models

What Is A Programming Model?

Dialogic programming presents two significant challenges to the software designer which are not found in most other types of applications.

- Managing multiple telephone lines within a single application program.

- Dealing with unsolicited events (especially caller hangup) while processing, for example while playing a menu prompt or updating database records.

Dialogic offers several different *programming models* for designing applications. The term "programming model" refers to the techniques used in the program source code to manage multiple lines and control the execution flow of the program.

The simplest model is called the *synchronous model*. Time-consuming functions such as playing, recording and getting digits block until completion. A separate program is run for each phone line, relying on the operating system to provide multi-tasking.

At the opposite end of the spectrum is the *state machine programming model*. All function calls return immediately in so-called *asynchronous mode*. The main control loop of the application polls the driver for events which report the completion of time-consuming operations. The position of each phone line within the application flow is recorded in a table called a *state table*. Each time an event is received, the application consults the state table to determine the next operation to perform for that line. On MS-DOS, where the operating system does not provide multi-tasking, the state machine model is the only option which can control multiple lines.

State machine programming is also often the model which gives the best performance and lowest overhead. Because of this special importance of state machine programming, and its exceptional difficulty, we will devote a separate chapter to discussing the state machine model in detail.

In the rest of this chapter, we will discuss the other programming models in more detail.

Multi-Tasking, Multi-Threading and Fibers

All Dialogic APIs run on operating systems which provide support for multi-tasking, with the exception of MS-DOS (which is still important in telephony because of its very low memory and CPU overhead). Since software is very expensive to write, it makes sense to consider using the operating system's features for multi-tasking to simplify the software engineering effort.

There are three main types of multi-tasking support provided by Windows, OS/2 and UNIX: *basic multi-tasking*, *threads*, and *fibers*.

With basic multi-tasking, the operating system shares time between different applications which are running in parallel. With UNIX, OS/2 and 32-bit Windows applications (but not 16-bit!), the operating system performs *pre-emptive multi-tasking*. Several times per second (typically 20 to 100 times) a hardware interrupt is generated by a "clock tick". When the clock ticks, the operating system takes control and saves the state of the currently executing application. This involves saving CPU registers and other information into an internal database maintained by the operating system scheduler. The scheduler than decides which application will next receive some execution time (a *time-slice*), restores the CPU registers and other details of the machine state saved when that application was last stopped (*pre-empted*), and starts execution in that application at the point where it was interrupted.

A *thread* is an execution path through an executing application program. Most programs execute just a single thread. However, 32-bit Windows and some UNIX flavors support *multi-threading* where a single application binary can start two or more threads of execution. Each thread of execution has access to the same data and code, but has its own stack. The operating system performs pre-emptive multi-tasking on all the threads within a single application in a very similar way to basic multi-tasking. However, the scheduler can often switch between different threads in a single application much more quickly than it can switch tasks because most of the machine state (the *context*) remains the same. Multi-threading also typically has much less memory overhead because only a single copy of the application code and data need be loaded. Some UNIX flavors do not support multi-threading.

A *fiber* is a new Windows NT feature which is even more "light-weight" than a thread. A fiber is like a thread, an execution path through an application. The main difference is that fibers are not pre-emptively multi-tasked. Like 16-bit Windows applications, fibers use *cooperative multi-tasking*. In this scheme, a fiber explicitly calls a Windows NT (*SwitchToFiber*) function to yield control to another fiber in the same application. Fibers have even better performance than threads because there is less context-switching overhead and no pre-emption. However, more skill is needed when programming with fibers because the application assumes responsibility for scheduling.

Synchronous And Asynchronous Modes

In the Win32/UNIX API, functions such as *dx_play* which start time-consuming operations can be run in two different modes.

In *synchronous mode*, the function "blocks" (puts the calling task or thread to sleep) until the operation completes. Synchronous mode is normally used only when there is pre-emptive multi-tasking by the operating system. For example, you wouldn't normally call a function in synchronous mode from a fiber because fibers are not pre-emptively multi-tasked, and the result would be that not only the calling fibers but all other fibers would be suspended until the function completed.

In *asynchronous mode*, the function returns immediately with an indication of whether the operation was successfully started. The application later receives an event which reports the completion of the operation. There are several ways to receive this notification, we will discuss these in the next section.

For most time-consuming Win32/UNIX functions, the mode is specified by the last argument to the API function. For example, the *dx_sethook* function takes half a second to complete. To go off-hook in synchronous mode, use:

```
dx_sethook(hDev, DX_OFFHOOK, EV_SYNC);
```

To use asynchronous mode, change from *EV_SYNC* to *EV_ASYNC*:

```
dx_sethook(hDev, DX_OFFHOOK, EV_ASYNC);
```

Event Notification

In Dialogic terminology, an *event* is a report from the driver to the application. Events are very similar to Windows messages.

There are two types of event:

- *Terminating events*, and
- *Transition events* (sometimes designated *call status transition* events).

A terminating event is a report that an operation completed. Note that even if a function is executed in synchronous mode, a terminating event is still generated and stored internally by the Dialogic driver. It can then be retrieved by the application to determine the reason that the synchronous function stopped.

A transition event is an unsolicited event that reports a change in state of the trunk. The most common and most important transition events report caller hangup, which would be a drop in loop current on an analog trunk, a change in signaling bits on a T-1 or E-1 trunk, or a disconnect message on an ISDN trunk.

Events can be retrieved by an application using one or more of the following methods:

- Waiting,
- Polling,
- Callback, and
- Window message.

Event Notification By Waiting

The *sr_waitevt* function in the Win32/UNIX API waits for the next event to be reported. There is one argument, which is the maximum length of time in milliseconds to wait:

```
Result = sr_waitevt(msec);
```

The return value *Result* is one of the following:

Result	Meaning
-1	Function timed out without an event.
>= 0	An event is reported, the return value is then the remaining time in milliseconds before timing out.

If *msec* is set to −1, then *sr_waitevt* will wait indefinitely for an event.

The *sr_waitevtEx* function waits for an event on a given device handle, or on a given set of handles, whereas *sr_waitevt* waits for an event from any device.

You can use this method of event notification to create a programming model which is very similar to the asynchronous model, with some greater flexibility. For example, you might want to play a message saying "Please hold while we check your account information" and at the same time be accessing a SQL database to check the account. You can achieve this within a single thread by starting the play in asynchronous mode, accessing the database and then waiting for the event:

```
hDevLine1 = dx_open("dxxxB1C1", 0);
// ...
dx_playwav(hDevLine1, "wait.wav", &TPT, EV_ASYNCH);
AccessSQL();
Result = sr_waitevtEx(&hDevLine1, 1, -1, &hEvent);
```

Not to worry if your *AccessSQL* function takes longer than the message play, the event will be saved in the queue and the *sr_waitevtEx* function will return immediately. Events are queued by the SRL functions and returned in the order that they were received.

Event Notification By Polling

With polling, events are retrieved by calling the *sr_waitevt* function (or *sr_waitevtEx*) in the Win32/UNIX API, specifying zero or a very short time-out, or *gtevtblk* on DOS. These functions return an indication of whether an event is pending in the queue, or that the event is empty.

If an event is present, the details of the event are obtained by these functions in the Win32/UNIX API:

Win32/UNIX Function	Returns
sr_waitevt	Indication of whether an event is present.
sr_getevtdev	Device handle of device which generated the event.
sr_getevttype	Numerical code (the "event code") which indicates the type of event.
sr_getevtdatap, sr_getevtlen	A pointer to, and length of, a block of data associated with the event.

On DOS, the *gtevtblk* function fills out an *EVTBLK* structure when an event is reported. *EVTBLK* has the following fields:

EVTBLK field	Value
devtype	Type of device which generated event: DV_D40 for voice devices, DV_DTI for DTI devices etc.
devchan	Channel number, zero indicates a board device event.
evtcode	Numerical code (the "event code") which indicates the type of event.
evtdata	16-bit value containing additional data which may be associated with the event.

Event Notification By Callback

For callback notification, the application gives the Dialogic drivers the address of a function. This function is called when an event is triggered. This is similar to a window procedure, well-known to traditional Windows programmers. When you create a window, you give the address of a function to Windows which is called a window procedure or "winproc". (Visual C++ programmers using MFC: this process is usually hidden from you inside the *CWnd* class). When there is a message for your window, the winproc is called with arguments indicating the type of the message and any additional parameters related to the message, such as the coordinates of the mouse.

Callback functions are established by calling *sr_enbhdlr*:

    sr_enbhdlr(hDev, iEventType, lpfnCallBack);

The *hDev* argument specifies the device handle, which will previously have been returned by *dx_open*, *dt_open* or another "open device" function. If you use *EV_ANYDEV*, events from all devices will be sent to this handler. The *iEventType* argument specifies the event code, which corresponds to a possible return value from *sr_getevttype*. If you use *EV_ANYEVT*, then all events will be sent to this handler.

Dialogic refers to handlers defined by a single device handler as *device-specific*, handlers which are defined by *EV_ANYDEV* are called *device non-specific*. Handlers which are defined by a single event type are called *event-specific*, those defined by *EV_ANYEVT* are called *event non-specific*.

The *lpfnCallBack* argument is a pointer to the callback function, which has a single argument, the event handle. If the callback function needs details of the event, it calls *sr_getevtdev* and *sr_getevttype* which take the event handle as an argument.

The Dialogic API allows more than one handler for a given event type. If there is more than one handler, then handlers are called in the following order:

1. Device-specific, event-specific handlers.

2. Device-specific, event non-specific handlers.

3. Device non-specific, event non-specific handlers.

If there is more than one handler in a category then the Dialogic API does not define the order in which the handlers are called and the application should not rely on the handlers being called in any given order.

There are some subtle issues to be wary of in callback handler code.

- You cannot make Dialogic API calls in synchronous mode from within a handler.

- The callback handler mechanism in the Dialogic SRL is not "thread-aware". This means that handlers can be enabled or disabled in any thread.

- On operating systems which support multi-threading, the callback handler will be called by default in the context of a private thread which is created by the Dialogic library. This is very important – it means that the callback function may be executing in parallel with the rest of your application with pre-emptive multi-tasking swapping between them. This can create so-called "race conditions" where the callback handler is trying to update data which is also being accessed by the rest of your code, which can lead to subtle bugs if you have not designed your code correctly. You can change this so that all execution is performed in your primary thread by calling:

    `sr_setparm(SRL_DEVICE, SR_MODELTYPE, SR_STASYNC);`

    In this case, the callback handler will be called only from within *sr_waitevt* or *sr_waitevtEx*. The call to *sr_waitevt* or *sr_waitevtEx* will not return until all event handlers have returned. This fully "serializes" your application so that at any one time you can be running only either a single event handler or the main thread.

In a "pure callback" model program, the main thread will enable a series of callback functions, and will then end with this call:

`sr_waitevt(-1);`

If a callback handler returns a non-zero value, this call to *sr_waitevt* will return, otherwise the program will continue indefinitely receiving calls to event handlers whenever an event is triggered. This model is rarely used because it will typically require a state table implementation, which is hard to program.

Event Notification By Window Message

In the Win32 API you can also receive event notification through the standard Windows message queue (this does not work, for obvious reasons, on UNIX). This is done through the *sr_NotifyEvent* function:

```
sr_NotifyEvent(hWnd, uMsg, SR_NOTIFY_ON);
```

The *hWnd* argument is a standard window handle. The *uMsg* argument is the message number to be sent to the window. This should be obtained through the *RegisterWindowMessage* Windows API function. All events will now be sent to that window, there is no mechanism to dispatch events by device and/or my event code as there is with callback functions.

The *wParam* and *lParam* parameters to the message are not set by Dialogic. When the registered message is received by the window, the application should call *sr_waitevt(0)* followed by calls to the usual *sr_get*xxx functions in order to retrieve details of the event.

This technique may appeal to programmers who are comfortable with developing traditional Windows applications. However, as with TAPI, your application flow will be much more complex than with an application designed around synchronous mode API calls.

Hybrid Programming Models

It is very common for applications to use more than one programming model. For example, you may use the easy-to-program synchronous mode for all time-consuming functions but choose to get caller hang-up notification via a call-back function. This technique is especially useful when programming digital interfaces. Analog interfaces may be programmed to interrupt plays, records etc. when a hang-up is detected. Digital interfaces do not have this feature, so it is up to you to call *dx_stopch* when a hang-up is detected in order to abort any operation in progress. A convenient way to do this is to call *dx_stopch* inside an callback event handler.

Programming Models

When programming in a mixed model, it is helpful to understand how the Dialogic SRL dispatches an event. When an event is received, the SRL searches a list of criteria for possible methods of dispatch. When an active method is found, this method is used and the search is quit. The search proceeds as follows:

1. If an synchronous function is waiting for the event, the calling thread is un-blocked. (It is not possible for more than one thread to be in a blocked state for the same device, the call to such a function from a second thread would have resulted in a "device busy" error return from the driver).

2. If any threads are waiting in a call to *sr_waitevtEx*, the first thread found is un-blocked.

3. If any threads are waiting in *sr_waitevt*, the first thread found is un-blocked.

Chapter 27

Dialogic Programming For MS-DOS

Introduction

In this chapter we will begin developing C programs for MS-DOS which work with Dialogic voice cards. While some of the preliminary material should be understandable for non-programmers, we will quickly descend into hard-core coding. This discussion is not meant to replace the documentation your voice card vendor provides — rather, it is intended to provide complementary information, by giving working examples and pointing out some of the pitfalls.

Driver Functions

MS-DOS Driver functions fall into two main categories, Dialogic terms them *multi-tasking* and *non-multi-tasking*. Don't be misled by this terminology — the Dialogic MS-DOS driver does not provide a multi-tasking environment in the usual sense, however it does provide a mechanism where a single C program can control multiple, independent interactions with several telephone lines.

Non-multi-tasking functions take a negligible amount of time to execute, a typical example would set parameters on the board determining how call progress analysis should be attempted on a given channel.

Multi-tasking functions are those which take a significant amount of time to complete: playing and recording speech files, waiting for touch tone digits, and so on.

Dialogic distributes driver functions in the following source files:

D40LIB.C   C source code for driver functions.

D40FCNS.C   C source code for low-level utility functions required by D40LIB, including calling the Dialogic driver through a software interrupt, manipulating DOS file handles, and extracting segment and offset addresses from a *far* pointer.

D40.H   Header file defining constants and data structures used by the driver functions.

D40LIB.H   Function prototypes for D40LIB functions.

VFCNS.H   Function prototypes and constants for DOS file handle functions.

A typical C source code file will include the two important header files:

Dialogic Programming For MS-DOS

```
#include "d40.h"
#include "d40lib.h"
```

The DOS file handle functions (named with the *vh* prefix for "voice handle") are not recommended for several reasons to be described later, however if you wish to use them you should add:

```
#include "vfcns.h"
```

Your program will, of course, be linked with object code for *D40LIB.C* and *D40FCNS.C*.

The following is a list of the most-used D40LIB multi-tasking functions:

`callp`	Dial digits and do call progress analysis.
`dial`	Dial digits without call progress analysis.
`getdtmfs`	Wait for one or more digits.
`playbuf`	Play audio from memory buffer.
`playfile`	Play audio from DOS file handle (obsolete, use *xplayf* instead).
`playuser`	Play audio from EMS memory buffer.
`recbuf`	Record audio to memory buffer.
`recfile`	Record audio to DOS file handle.
`recuser`	Record audio to EMS memory buffer.
`ring`	Start ring sequence on AMX port.
`sethook`	Take channel on- or off-hook.
`setiparm`	Set channel parameter.
`wink`	Start wink protocol.
`xplayf`	Play audio from DOS file handle.

A typical example of a multi-tasking function is *xplayf*, which requests the driver to begin playing an audio file on a given speech card *channel* (phone line). The function returns a value zero if the play was successfully started, or a positive error code otherwise, a pattern followed by all the multi-tasking functions.

Some time after a successful invocation of *xplayf* the play operation will end, and the driver reports this to the application program by adding an *event* to the *event queue*. The event queue is the key to the mechanism which allows multiple phone lines to be controlled by one program.

There are two types of events: *terminating events*, which indicate that a multi-tasking function has completed, and *call status transition events*, which report that the board has detected a change in the signaling state on a channel, such as a drop in loop current indicating that a caller has hung up.

The *gtevtblk* function is used by a program to retrieve the next event in the queue, if any:

```
int gtevtblk(EVTBLK *event);
```

The argument is a structure of type *EVTBLK*, defined, as are all D/4x-related structures and constants, in the header file *D40.H* provided by Dialogic. The value returned by *gtevtblk* is one of:

```
 0    No pending events
-1    Event pending: details filled in structure referenced by the
      event argument.
```

As for most Dialogic functions, a returned value > 0 would indicate an error condition, for example if the Dialogic driver has not been installed, or if an invalid argument was passed to the function.

The structure members of the *EVTBLK* type which are filled in when an event is retrieved are:

```
devtype
```
Device type. This will take one of the values defined as a constant named with the *DV_* prefix in a Dialogic header file. For example, *DV_D40* indicates and event generated by a D/4x-type board (including a D/2x, D/8x or D/12x), or *DV_FAX* for a FAX/120 event.

```
evtcode
```
Event code. This is the code indicating which type of event occurred. Note that the interpretation of this value depends on the device type, typically each type of device has event codes numbered 1, 2 .. and so on.

Dialogic Programming For MS-DOS

    `devchan`   Device channel number. This is the "channel number" or "line number" responsible for generating the event. For example, if the event was a report that a file play had been completed, this would be the channel number where the play had been performed. For some events which relate to a specific board rather than a specific channel, this contains no useful value.

    `evtdata`   Event data. For some events, this contains a value with more information relating to the event. As an example, for a *T_CATERM* (call analysis complete) event, this value indicates the result of the analysis (Connect, Busy, etc.).

    `mempntr`   Memory pointer. For some events, points to a memory location containing data relevant to the event.

    `board`   Board number. Indicates the board number which generated the event.

Event codes for D/4x board events are defined as constants named with the *T_* prefix in *D40.H*. Most types are terminating events, reporting the completion of a multi-tasking function. The following are the call status transition events for the D/4x:

    T_LC   Loop current off transition — in other words, caller hang-up. Be warned that if a multi-tasking function is in progress when a loop current off transition is detected, and if loop current interruption is enabled, then the *T_LCTERM* (loop current termination event) will be reported instead.

    T_RING   Rings detected, a report of an incoming call requesting service.

    T_SILOFF   Silence off, in other words a transition from silence to sound in the received audio.

    T_SILON   Silence on, transition from sound to silence.

    T_LCON   Loop current on.

    T_WKRECV   Wink received.

A First Program

With these preliminaries out of the way, we can begin to write a working program. The program will perform the following operations:

1. Wait for an incoming call.
2. Answer the call (go off-hook).
3. Play an initial greeting.
4. Record a message from the caller.
5. Play the recorded message back to the caller.
6. Hang up (go on-hook).
7. Return to step 1.

If at any time the caller hangs up, the program returns to step 1. If at any time the [Esc] key is pressed from the keyboard, the program terminates.

The *startsys* program starts the operation of the Dialogic driver, a typical call would look like this:

```
int irq = 5;
int code, channels;
code = startsys(irq,SM_EVENT,0,0,&channels);
```

The *irq* argument gives the hardware interrupt level of the D/4x board, the *SM_EVENT* argument specifies that event queue rather than polling operation is requested, and *channels* is set by a successful call to the number of installed D/4x channels. As usual, a return value of zero indicates success. As a precaution, it is advisable to start with a call to *stopsys*, which stops the driver in case it was left in an undefined state by a previous program. The initial sequence in our test program will therefore be:

```
stopsys();
code = startsys(irq,SM_EVENT,0,0,&channels);
if (code == 0)
    printf("%d channels\n", channels);
else
    {
    printf("ERROR: startsys=%d);
    exit(1);
    }
```

Dialogic Programming For MS-DOS

Our simple program will operate on channel one only (note that channel numbers are counted with one as the first channel, not zero as C programmers might expect). Each time a multi-tasking function is set in motion, it will need to poll the *gtevtblk* function until a terminating event is detected. For single-channel example programs and prototypes, we can simply wait until a terminating event is reported, or until the caller hangs up. The following does the job:

```
#include <conio.h>
#define "d40.h"
#define Esc 0x1b
EVTBLK event;
/*
block: Wait for event. Return 0 when
terminating event,1 for ring, 2 for hang-up.
*/
    int block()
        {
        int code;
        for (;;)
            {
            /*  Check for [Esc] keystroke  */
            while (kbhit())
                if (getch() == Esc)
                    {
                    printf("[Esc] typed\n");
                    stopsys();
                    exit(0);
                    }
            /*  Check for event  */
            code = gtevtblk(&event);
            if (code == 0)   /*  No event  */
                continue;
            else if (code != -1)   /*  Error  */
                {
                printf("ERROR: gtevtblk=%d\n",
                    code);
                stopsys();
                exit(1);
                }
            else   /*  Have event  */
                switch (event.evtcode)
                    {
                    /*  If ring, return 1  */
                    case T_RING:
                        return 1;
                    /*  If caller hung up, return 2  */
```

```
            case T_LC:
            case T_LCTERM:
                return 2;
            /*  If other transition, ignore  */
            case T_SILON:
            case T_SILOFF:
            case T_LCOFF:
                continue;
            /*  Have a terminating event  */
            default:
                return 0;
            }
    }
```

As a bonus, this function checks whether a keystroke has been entered from the keyboard, and quits the program if the [Esc] key was pressed. If a program is terminated without the *stopsys* call being made to shut down the speech card operations, the driver will start clicking sounds on the PC loudspeaker.

Using *block*, we can develop a macro which waits for a terminating event, and which also jumps back to the beginning of the program if a caller hang-up is detected:

```
#define Wait() {if (block()==2) goto Start;}
```

The *xplayf* function takes three arguments:

```
int xplayf(unsigned channel, unsigned mode,
    RWB *rwb);
```

As usual, *channel* is the D/4x channel number, *mode* defines the playback mode — we will use the default value of zero for now, and *rwb* is a pointer to a structure known as the Read Write Block (RWB), defined as type *RWB* in *D40.H*. The RWB specifies a whole series of options which determines how a play, record or get-digits operation may terminate. The *clrrwb* function is a "helper" provided in *D40LIB* which sets a RWB to zero, giving default values:

```
void clrrwb(RWB *rwb);
```

RWB fields required by *xplayf* and *recfile* for our example program include:

Dialogic Programming For MS-DOS

filehndl	DOS file handle.
maxsec	Maximum seconds to allow for completion.
termdtmf	Set to '@' to allow any DTMF tone to interrupt.
loopsig	Set to 1 to allow caller hang-up to interrupt.
rwbflags	For *recfile*, set to 0x02 to generate tone when record operation starts.

The following function *play* provides us with the basic functionality of playing a speech file:

```
/*
play.c: Simple function to initiate file play
*/
#include "d40.h"
#include "d40lib.h"
#include <stdio.h>
#include <stdlib.h>
RWB rwb;

void play(int chan, int handle)
    {
    int code;
    clrrwb(&rwb);
    rwb.filehndl = handle;
    rwb.loopsig = 1;
/* Allow any tone to interrupt: */
    rwb.termdtmf = '@';
    code = xplayf(chan, 0, &rwb);
    if (code != 0)
        {
        printf("ERROR: xplayf=%d\n", code);
        stopsys();
        exit(1);
        }
    return;
    }
```

An outstanding question is getting a file handle. C compilers generally provide UNIX-like functions *open*, *close* and so on to manipulate file handles. The drawback of these functions is that they tend to add large amounts of code and to be rather inefficient since they must allow for different options such as translating between UNIX and DOS files (CR/LF or just LF at the end of a line). Dialogic provides somewhat leaner and meaner functions through the *dos...* and *vh...* series of functions. Beware, however, that the most important function, *vhopen*, is *brain damaged!* Valid DOS file handles are numbered from zero as the first handle, however a zero return value from *vhopen* is used to indicate an error. File handles can be in short supply in voice processing applications — DOS initially provides a limit of only 20 handles, and PC-based systems can easily accommodate more than 20 phone lines, each of which could be playing or recording to a different speech file handle. DOS starts each program with five predefined file handles (three for keyboard/screen input and output, one for the printer and one for the serial port), and it would be desirable to start each voice processing application by closing each of these pre-defined handles to make them available for speech files. Screen output and keyboard input can still be done through lower-level BIOS calls, for example, even when the DOS handles have been closed. Using *vhopen*, you'll get into trouble if you try to re-use handle zero. My recommendation is to edit the source files which Dialogic provides so that zero no longer represents an error return from *vhopen*, or better still write some small assembler routines that don't use those clunky old *int86* or *int86x* calls.

Beware, also, that the Dialogic driver doesn't take kindly to being passed invalid file handles, or to file handles which are closed while a multi-tasking operation using the handle is still in progress. All kinds of unpleasant things can happen, including the driver trashing parts of your program. This seems a little strange, since the driver goes to considerable lengths to check other function call arguments, but you have been warned.

The following function *record* can be used to initiate a record-to-file operation:

```
/*
record.c: Function to start recording to file
*/
#include "d40.h"
#include "d40lib.h"
#include <stdio.h>
#include <stdlib.h>
RWB rwb;

void record(int chan, int handle, int secs)
    {
    int code;
/* Clear digit buffer (otherwise a pending */
/* digit will terminate record immediately) */
```

Dialogic Programming For MS-DOS

```
   code = clrdtmf(chan);
   if (code != 0)
      {
      printf("ERROR: clrdtmf=%d\n", code);
      stopsys();
      exit(1);
      }
   clrrwb(&rwb);
   rwb.filehndl = handle;
   rwb.loopsig = 1;
   rwb.maxsec = secs;
   rwb.termdtmf = '@';    /* Allow DTMF term. */
   rwb.rwbflags = 2;      /* Tone at start */
   rwb.rwbdata1 = 2;      /* Tone duration */
   code = recfile(chan, &rwb, 0);
   if (code != 0)
      {
      printf("ERROR: recfile=%d\n", code);
      stopsys();
      exit(1);
      }
   return;
   }
```

A point to note here is the use of the *clrdtmf* function before starting the record. The Dialogic functions have cleverly been designed so that touch-tone interruption applies *even if the digit arrived before the function is started.* This may seem perverse, but in fact greatly simplifies application design providing that the caller is allowed to "type ahead" through menus and prompts. If any touch-tone digits are pending in the driver's input buffer when the *recfile* operation is started, and if *termdtmf* has been specified as '@' in the RWB, then the operation will be terminated immediately with a *T_TERMDT* event ("terminating DTMF digit received"). We obviously don't want that — we want to allow the caller to signal the end of the recording using a touch-tone, hence we start by clearing the digit buffer. The *secs* argument gives the maximum length permitted for the recording.

Another common operation is getting digits from the caller. The following function *getdigs* provides the basic functionality:

```c
/*
getdigs.c: Simple function to get DTMF digits
*/
#include <stdio.h>
#include <stdlib.h>
#include "d40.h"
#include "d40lib.h"
#include "vfcns.h"
#define MAX_CHANNELS 4
#define MAX_DIGITS   20
static char
   digit_buffer[MAX_CHANNELS*MAX_DIGITS];
RWB rwb;

/*
digbuf: Addr. of digit buffer for channel
*/
char *digbuf(int chan)
   {
   return digit_buffer + (chan  1)*MAX_DIGITS;
   }
void getdigs(int chan, int ndig, int secs)
   {
   int code;
   char *ptr;
   clrrwb(&rwb);
   rwb.maxdtmf = (byte) ndig;
   rwb.maxsec = secs;
   rwb.loopsig = 1;
   ptr = digbuf(chan);
   rwb.xferoff = d4getoff(ptr);
   rwb.xferseg = d4getseg(ptr);
   code = getdtmfs(chan, &rwb);
   if (code != 0)
      {
      printf("ERROR: getdtmfs=%d\n", code);
      stopsys();
      exit(1);
      }
   return;
   }
```

This function introduces two new fields in the RWB: *xferseg* and *xferoff,* which specify the segment and offset respectively of the buffer where the digits received should be stored. We defined a new function *digbuf* so that external functions can gain access to the digits which have been received.

Dialogic Programming For MS-DOS

Due to the way the Dialogic driver handles touch-tones, you don't have to worry about whether the caller typed ahead, interrupting earlier prompts, when you start your get-digits operation. If enough digits are pending in the buffer, the operation will terminate immediately with a *T_MAXDT* event, your program can treat this in exactly the same way as with an inexperienced caller who waits until all prompts have finished playing until entering touch-tones. In other words, your programs will rarely need to distinguish playing prompts that complete with a *T_EOF* (end-of-file on play) event from those which terminate with a *T_TERMDT* (terminating DTMF digit received).

With these building blocks in place, we can present the source code for the complete application:

```
/*
demo.c: Single line example application
*/

#include <stdio.h>
#include <stdlib.h>
#include "d40.h"
#include "d40lib.h"
#include "vfcns.h"

/* Functions defined in other source files */

int block(void);

void play(int, int);
void record(int, int, int, int);

#define Wait() {if (block()==2) goto Start;}

int irq = 3;         /* Dialogic h/w int. */
int chan = 1;        /* Channel number */
int code;            /* Return code   */
int channels;        /* Nr installed channels */
int handle = 1;      /* DOS file handle */

void main(int argc, char **argv)
    {
/* If given, command line arg is */
/* h/w interrupt    */
    if (argc > 1)
        irq = atoi(argv[1]);
    stopsys();
    code = startsys(irq,SM_EVENT,0,0,&channels);
```

```c
            if (code == 0)
               printf("%d channels\n", channels);
            else
               {
               printf("ERROR: startsys=%d\n", code);
               exit(1);
               }
            code = setcst(chan,
             C_LC|C_RING|C_ONH|C_OFFH, 1);
            if (code != 0)
               {
               printf("ERROR: setcst=%d\n", code);
               exit(1);
               }
Start:
/*   Close handle   */
            if (handle > 0)
               vhclose(handle);
            handle = 1;
/*   Go onhook   */
            code = sethook(chan, H_ONH);
            if (code != 0)
               {
               CSB csb;
               getcstat(chan, &csb);
               printf("sethook failed, code=%d\n",
                code);
               printf("cstat=%d\n", csb.status);
               stopsys();
               exit(1);
               }

            Wait();
            printf("\nReady for call...\n");
/*   Wait for ring   */
            while (block() != 1)
               ;
/*   Answer call   */
            code = sethook(chan, H_OFFH);
            if (code != 0)
               {
               printf("sethook failed, code=%d\n",
                 code);
               stopsys();
               exit(1);
               }
            Wait();
```

Dialogic Programming For MS-DOS

```
    /*  Create file for recorded message   */
    handle = vhopen("MESSAGE.VOX", CREATE);
    if (handle <= 0)
        {
        printf("ERROR: vhopen(MESSAGE.VOX)=%d\n",
            handle);
        stopsys();
        exit(1);
        }
    /*  Record message   */
    record(chan, handle, 60, 3);
    Wait();
    /*  Rewind message to start of file   */
    vhseek(handle, 0L, 0);
    /*  Play message   */
    play(chan, handle);
    Wait();
    /*  Prepare for next call   */
    goto Start;
    exit(0);
    }
```

We have used one new function: *setcst*, which determines which call status transitions the driver is to report as events, and how many incoming rings should be detected before reporting a *T_RING* (rings received) event. In our program, we set the number of rings to one, other applications may wait for more rings — for example, to give a live person a chance to answer the call, or for features such as toll-savers which wait for a set number of rings before answering to indicate that no new messages have been recorded.

This style of programming for a single channel, using *block* or a similar technique to simply "put the program to sleep" until an event occurs, is ideal for learning to use driver functions, building prototypes, and other development work. It has the major advantage that the flow-of-control constructions of the C language — *if..else, switch,* loops, function calls — can be used directly to mirror the flow of the application. Programming for multiple channels, unfortunately, is not so easy. This will be the subject of a later chapter.

Chapter 28

State Machines

Introduction

To control multiple telephone lines in a single thread, a whole new program structure is needed. Dialogic calls this technique *state machine* programming, Computer Science graduates may have learned to call state machines *Deterministic Finite State Automata*. Programmers have learned to call them *difficult*.

This chapter is intended primarily for experienced C programmers. While the basics of state machines can pretty much be grasped by anyone who understands a flow chart, the implementation is messy, and can tax the skills of the most experienced C veteran. Proceed at your own risk.

Take a simple example: the Greeter, an application which answers the phone, plays an initial greeting, accepts a touch tone digit 1, 2 or 3, then plays a message depending on the touch tone entered, and finally hangs up and waits for a new call. The flow chart is:

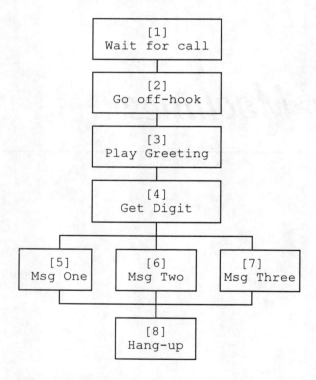

Greeter Application Flow Chart

To get things started, we can imagine the program resting in box [1] of the flow chart — we say that it is in state 1. It waits there until an incoming ring is detected, this causes the program to go to box [2]. In state machine language, we would say that the event generated by the

State Machines

incoming ring caused a *state transition* from state 1 to state 2. In state 2, the application waits until the off-hook operation has completed, which will be signaled by a terminating event. This terminating event will cause the program to begin playing the greeting, going to state 3.

A *state* may be thought of as a box in a flow-chart, and is generally an operation which takes an appreciable time to complete. The completion of the operation performed in a state is indicated by an event, which causes a *transition* to the next state. Several different events may be generated in any given state, these may cause transitions to the same or to different subsequent states.

The Greeter flow chart could be described as a list of states, for each state we specify what must be done (*actions*) when any given event occurs, and the corresponding transition to a new state to be made for that event:

    State 1, Waiting For Call.
        Event=Ring Detected:
            Start play of greeting file
            Transition to State 2

    State 3, Play Greeting
        Event=Play Completed:
            Start waiting for touch tone

    State 4, Get Digit
        Event=Digit 1 received:
            Start play of message 1
            Transition to State 5
        Event=Digit 2 received:
            Start play of message 2
            Transition to State 6
        Event=Digit 3 received:
            Start play of message 3
            Transition to State 7

State 5, Msg One
    Event=Play Completed:
        Start on-hook operation
        Transition to State 8

State 6, Msg Two
    Event=Play Completed:
        Start on-hook operation
        Transition to State 8

State 7, Msg Three
    Event=Play Completed:
        Start on-hook operation
        Transition to State 8

State 8, Hang-up
    Event=On-hook completed:
        Transition to State 1 (re-start)

This type of list can easily be adapted to a table format, a *state table*. This table can be represented as C program control structures, such as *switch*, or as data within the program. The current state of each active channel will be stored in an array, each event that occurs may result in the state for the corresponding channel being updated.

The State Machine Engine

The heart of a state machine program is a loop which, in outline, is something like this:

1. Check for an event.

2. If there is no event, go to step 1.

3. If there is an event, go to step 4.

4. Refer to state table, identify transition resulting from current state with this event type.

5. Perform operations (if any) associated with this transition.

State Machines

6. Update current state for channel associated with event.

7. Go to step 1.

Using this architecture allows multiple channels to traverse the same flow chart, represented as a state table.

This simple description glosses over many of the issues which must be addressed in a practical implementation of a state machine application. The state table sketched above ignores all exceptional conditions — what happens when an unexpected event happens, such as a time-out when waiting for a digit, or a caller hang-up at a point in the middle of the flow chart.

An important architectural decision is how to transfer control at each transition to the actions to be performed. Several models, of increasing usefulness and also increasing complexity in implementation, will be presented.

At a minimum, an array will be needed storing the current state of each channel:

```
#define MAX_CHANNELS   4   /* Nr ports */
int c_state[MAX_CHANNELS];
```

It would be nice to use the channel number as a subscript to the array, but C arrays count from zero. There is a neat trick which allows you to use the channel number as a subscript, but which avoids an unused array element for channel "zero" as found in some Dialogic demonstration programs:

```
#define state   (c_state-1)
```

This works because *state[chan]* will become *(c_state-1)[chan]*, which, according to the usual rules for interpreting a C expression, is the same as *c_state[chan-1]*. Fortunately, the expression *(c_state-1)* will be evaluated at compile/link time, so there is no run-time overhead associated with subtracting one from the channel number. This same trick can be used for any per-channel arrays.

The main loop could be structured something like this:

```
for (;;)
  {
  code = gtevtblk(&event);
  if (code == 0)
      continue;
  else if (code != -1)
      /* .. error .. */
  else if (event.devtype != DV_D40)
      /* Ignore non-D/4x event */
      continue;

  /* Process new event */
  switch (state[event.devchan])
      {
      case 1: /* Waiting for call */
          /* Process event in state 1 */
          /* ... */
      case 2: /* Going off-hook */
      /* .. etc for each state    */
      } /* end switch */
  } /* end for */
```

For example, in state 1 we are waiting for an incoming call, i.e. for ring voltage to be detected, signaled by an event of type *T_RING*. When the ring arrives, the on-hook operation should be started, so an outline of the case for state 1 would be:

```
case 1: /* Wait for call */
    if (event.evtcode == T_RING)
        {
        code = sethook(event.devchan, H_ONH);
        if (code != 0)
            /* .. error .. */
        state[event.devchan] = 2;
        }
    break;
```

The transition to state 2 is accomplished simply by updating the *state* array.

A Complete State Machine Program

The following is a complete listing of the Greeter application as a state machine program for MS-DOS.

```c
#include <stdio.h>
#include <stdlib.h>
#include <conio.h>
#include "d40.h"
#include "d40lib.h"
#include "vfcns.h"

#define Esc 0x1b      /* ASCII [Esc] key */

/* Functions defined in other source files */
void play(int, int);
void record(int, int, int, int);
void getdigs(int, int, int);
char *digbuf(int);

#define MAX_CHANNELS 4 /* Nr voice ports */

int c_state[MAX_CHANNELS];   /* Chan. state */
#define   state   (c_state1)
int c_handle[MAX_CHANNELS]; /* File handle */
#define   handle    (c_handle1)

EVTBLK event;       /* Event block */
int irq = 3;        /* Dialogic h/w interrupt level
*/
int chan = 1;       /* Channel number */
int code;              /* Return code */
int evtcode;        /* Event code */
int channels;       /* Nr installed channels */
int chan;              /* D/4x channel number */
char digit;         /* DTMF digit */

void main(int argc, char **argv)
   {
/*  If given, command line arg is   */
/*  h/w interrupt  */
   if (argc > 1)
      irq = atoi(argv[1]);
   stopsys();
   code = startsys(irq,SM_EVENT,0,0,&channels);
   if (code == 0)
      printf("%d channels\n", channels);
   else
      {
      printf("ERROR: startsys=%d\n", code);
      exit(1);
      }

/*  Initialize all channels to state 1   */
```

```c
        for (chan = 1; chan <= MAX_CHANNELS; chan++)
            {
            code = setcst(chan,
                C_LC|C_RING|C_ONH|C_OFFH, 1);
            if (code != 0)
                {
                printf("ERROR: setcst=%d\n", code);
                exit(1);
                }
            state[chan] = 1;
            handle[chan] = 1;
            }

    /*  Main loop: State Machine engine   */
        for (;;)
            {
            /* Check for Esc keystroke  */
            while (kbhit())
                if (getch() == Esc)
                    {
                    printf("[Esc] typed\n");
                    stopsys();
                    exit(0);
                    }
            code = gtevtblk(&event);
            if (code == 0)
                continue;
            else if (code != 1)
                {
                printf("ERROR: gtevtblk=%d\n", code);
                stopsys();
                exit(1);
                }
            else if (event.devtype != DV_D40)
                /* Ignore nonD/4x event */
                continue;
            chan = event.devchan;
            if (chan < 1 || chan > MAX_CHANNELS)
                {
                printf("WARNING: channel number %d\n",
                    chan);
                continue;
                }

            evtcode = event.evtcode;

    /*  Process onhook, doesn't matter  */
    /*  which state we were in  */
            if (evtcode == T_LC ||
                evtcode == T_LCTERM)
                goto Hang_up;
```

State Machines

```
    /* Process new D/4x event */
        printf(
"Channel %d  State %d  Event Code %d\n",
            chan, state[chan], evtcode);
        switch (state[chan])
            {
        case 1:  /* Waiting for call */
            if (evtcode != T_RING)
                continue;
            code = sethook(chan, H_OFFH);
            if (code != 0)
                {
                printf(
"ERROR: sethook(OFFH)=%d\n", code);
                stopsys();
                exit(1);
                }
            state[chan] = 2;
            continue;

        case 2:  /* Going offhook */
            if (evtcode == T_OFFH)
                {
                handle[chan] =
                    vhopen("GREET.VOX", READ);
                printf("handle = %d\n",
                    handle[chan]);
                if (handle[chan] <= 0)
                    {
                    printf(
"ERROR: Cannot open GREET.VOX\n");
                    stopsys();
                    exit(1);
                    }
                play(chan, handle[chan]);
                state[chan] = 3;
                }
            continue;

        case 3:  /* Playing greeting */
    /* Check for possible play term. events */
            if (evtcode == T_TERMDT
                || evtcode == T_DOSERR
                || evtcode == T_EOF
                || evtcode == T_TIME)
                    {
                    vhclose(handle[chan]);
                    handle[chan] = 1;
                    getdigs(chan, 1, 3);
                    state[chan] = 4;
```

```c
            }
          continue;

      case 4: /* Getting digit */
          if (evtcode == T_MAXDT)
             {
    /* Got one digit  retrieve from buffer */
             digit = *digbuf(chan);
             code = clrdtmf(chan);
             if (code != 0)
                {
                printf("ERROR: clrdtmf=%d\n",
                   code);
                stopsys();
                exit(1);
                }
             printf("digit = %c\n", digit);
             switch (digit)
                {
                case '1':   /* pressed [1] */
                   handle[chan] =
                     vhopen("MSG1.VOX", READ);
                   state[chan] = 5;
                   break;
                case '2':   /* pressed [2] */
                   handle[chan] =
                     vhopen("MSG2.VOX", READ);
                   state[chan] = 6;
                   break;
                case '3':   /* pressed [3] */
                   handle[chan] =
                     vhopen("MSG3.VOX", READ);
                   state[chan] = 7;
                   break;
                default: /* Invalid digit */
                /* Invalid digit  give up */
                   goto Hang_up;
                }
             if (handle[chan] <= 0)
                {
                printf(
   "ERROR: Can't open MSG?.VOX\n");
                stopsys();
                exit(1);
                }
             play(chan, handle[chan]);
             continue;
             }
          else if (evtcode == T_TIME)
          /* No digit entered  give up */
             goto Hang_up;
```

```
        case 5: /* Playing message 1 */
        case 6: /* Playing message 2 */
        case 7: /* Playing message 3 */
/* Check for possible play term. events */
            if (evtcode == T_TERMDT
             || evtcode == T_DOSERR
             || evtcode == T_EOF
             || evtcode == T_TIME)
                {
                vhclose(handle[chan]);
                handle[chan] = 1;
                code = sethook(chan, H_ONH);
                if (code != 0)
                    {
                    printf(
"ERROR: sethook(ONH)=%d\n", code);
                    stopsys();
                    exit(1);
                    }
                state[chan] = 8;
                }
            continue;

        case 8: /* Hanging up */
            if (evtcode == T_ONH)
                state[chan] = 1;
            continue;
            }
    Hang_up:
    /* Close file if one is open */
        if (handle[chan] >= 0)
            {
            vhclose(handle[chan]);
            handle[chan] = 1;
            }
        code = sethook(chan, H_ONH);
        if (code != 0)
            {
            printf(
"ERROR: sethook(ONH)=%d\n", code);
            stopsys();
            exit(1);
            }
        state[chan] = 8;
        continue;
        }
    }
```

Real Life

The Greeter application is fine as an educational tool, but does not represent a very realistic model for a full-blown application. Real-life programs will have large numbers of states, and the *switch* construction may not be an appropriate design. The C compiler may not be capable of supporting large numbers of cases within a *switch*, or may produce less efficient code. Even if the C compiler is up to the task, it is hard for a programmer to deal with a single, large source file containing the entire program flow.

It would be nice to make more of the application flow table-driven. One method is to use the ability of C to call a function whose address is stored as an array element. State transitions, and the action to be taken when the transition is made, can then be stored as a structure, a skeleton for this structure could be:

```
struct state_transition
    {
    int event_code;     /* Event causing trasn. */
    int (*action)();    /* Function to call     */
    int new_state;      /* State after transn.  */
    } trans[NTRANS];
```

Here, *NTRANS* is a constant defined to be the total number of state transitions. Each state needs a variable number of these structures listing the possible transitions, this can be accommodated by a table:

```
struct state_table
        {
        int ntrans; /* Nr of elements in tr[] */
        int *trn;
        } table[NSTATES];
```

The constant *NSTATES* is the total number of states. The integer array *trn[]* for each state would reference the *trans* table.

Using this organization, the central event loop would work like this:

```
        for (;;)
            {
            int evtcode, devchan, current_state;
            int i;
            struct state_transition t;
```

State Machines

```
for (;;)
   {
   int evtcode, devchan, current_state;
   int i;
   struct state_transition t;

   evtcode = /* event code */
   devchan = /* channel */
   current_state = state[devchan];
   t = table[current_state];
   for (i = 0; i < t.ntrans; i++)
      if (evtcode == t->event_code)
         {
         state[devchan] = t->new_state;
         (*(t->action))();
         break;
         }
      else
         t++;
   }
```

A significant drawback of this approach is that a function address is used as part of the table, which does not lend itself well to the development of "application generator" or "computer-aided software engineering" style tools, where software tools are built to aid in the construction of tables.

States And Super States

If you were to sit down and develop a flow-chart for a voice processing application, you would be unlikely to include as much detail as the Greeter flow-chart which opened this chapter. Instead, you would probably have boxes with titles like "Main Menu" or "Get PIN Number". Many of the boxes you drew would probably involve a series of steps something like this:

1. Play one or more speech files

2. Get one or more touch-tones from the caller

3. Jump to another flow-chart box, depending on the touch-tones entered by the caller.

To implement this idea, we can broaden the idea of a "state" to be something you might call a *super-state* — a series of operations which match a "template". There would be a series of templates to cover different types of flow-chart boxes which could arise in different voice processing applications. An application can then be described as a series of "filled-in" templates.

The template would include some "blanks" which would be filled in to specify each flow-chart box. A typical template might look something like this, with XXXX representing blanks to be filled in:

1. Play message XXXX

2. Get XXXX touch-tones from the caller

3. If the caller entered 0, jump to template XXXX
   If the caller entered 1, jump to template XXXX
   If the caller entered 2, jump to template XXXX
   ... etc.

This approach has been taken by many companies for in-house development tools, and lies at the core of some of the commercially available DOS-based *Application Generator* products, which provide menu-driven input screens for defining applications built up out of super-states.

To implement the idea in a program, the central idea is to define a C structure which encompasses all possible variants of a super-state (template types). The run-time engine will interpret the resulting table, an array of super-state structures. A super-state structure which can cover all possible variations will probably include one or more *unions* of different sub-structures, each of which describes the parameters of a different variant.

Without attempting to give a complete prescription, we can get a flavor of a super-state structure by looking at a few basic variants. Using the three-step sequence described above as a model, we could start by specifying a speech file to play as an integer number. The integer would somehow be converted to a DOS path name or to a prompt number

State Machines

within an indexed file by an algorithm defined elsewhere in the program. A reserved value such as -1 would be used to indicate that no speech file should be played in this super-state:

```
int message_file;
```

The next field could give the number of touch-tone digits to get in response:

```
int nr_digits;
```

The would be a number of additional parameters, for example:

```
int seconds;
  /* Max time for get-digit operation */
int silence;
  /* Inter-digit time-out */
unsigned bitmap;
```

The *bitmap* field could specify discrete Yes/No decisions such as whether touch-tone interrupt of the message play should be permitted.

The action to be performed would be specified by another integer field:

```
int action_type;
```

One example of an action to be taken could be a jump based on a single touch-tone digit received, this can be specified as an array:

```
int tone_goto[16];
```

where the 16 elements are for the digits 0 - 9, *, #, and a - d. The integer value in the *tone_goto* array would be the number of the super-state to execute if the given touch-tone had been received. This array might be extended with jumps to execute if a time-out or inter-digit time-out were detected, or other exceptional conditions were encountered.

Another example of an action which could be taken would be to issue a speech card command, such as to go on-hook. This command would not require additional parameters. Other commands might require parameters, such as closing an AMX matrix switch-point, which would require x and y coordinates:

```
struct AMX_coords
    {
    int x;
    int y;
    };
```

With the variants we have listed so far, the super-state structure looks like this:

```
struct super_state
    {
    int message_file;
    int nr_digits;
    int seconds;
    int silence;
    unsigned options;
    int action_type;
    union action_parms ap;
    };
```

where *action_parms* is a union of the possible parameters for the given action, the value in the *action_type* field will determine which of the union members should be taken:

```
union action_parms
    {
    int tone_goto[16];
    struct AMX_coords amx_xy;
    };
```

The entire program can be stored in an array of super-states:

```
struct super_state program[NSTATES];
```

which, since all information is now in numerical form with no function or other address pointers, can be stored in a disk file.

The run-time engine interpreting the program will need to be enhanced considerably to accommodate super-states. The original *state* array which kept track of the current state of a channel will be extended to contain at least two separate values: the current super-state number, and an indication of the step number within the super-state which is currently being executed.

The super-state idea has the advantage that it is relatively easy to design and implement for a competent C programmer, and also lends itself reasonably well to building software development tools where super-states are described by "scripting" commands in a text file. For those who are less programming-oriented, the fixed lists of possible action sequences lend themselves to a menu-driven approach where a data entry screen is presented to a user who can select any of the possible options by selecting from menus. Many service bureaus and other companies producing custom audiotex software have developed tools based on these ideas.

However, there are drawbacks: super-state constructions can be rather rigid, and can waste a good deal of memory in table storage since many structure elements will typically be zero or inactive in any given super-state. Sophisticated applications require more traditional programming-oriented features such as variables, loops, subroutines and other decision-making constructions, which can be awkward to implement well in this type of environment.

There's More To Life Than Voice

So far, our discussion of state machines has concentrated on handling multiple speech card channels. Many voice processing applications will need other tasks performed in parallel with the caller interaction, such as monitoring a serial port for communication with a host computer, providing a screen interface for a keyboard user to update options in the system such as mailboxes. Integrating these types of tasks provides yet another layer of complexity for the hapless state machine programmer.

For example, a screen/keyboard interface must perform the "blocking" operation of waiting for a keystroke. This fits into the state machine model: a typical state will wait for a keystroke, the event caused by typing a key will cause a transition to a new state, the corresponding action will probably update the screen. However, commercial library packages for performing interface functions do not fit this model, and even if source code is available it may be a difficult task to re-structure them appropriately. The typical problem is that the pre-written code will wait for keyboard input in some deeply nested function call. In state machine code the point where the program begins to wait for a keystroke is indicated by setting the current state and then continuing with the main loop, which would mean exiting from all levels of nested function calls.

As another example, suppose that it is necessary to copy a speech file at some point in an application. A speech file may be many hundreds of kilobytes, and take too long to copy in a single operation. It is therefore necessary to break the copying into pieces, one buffer will be copied at a time, control will be returned to the main event loop once completed.

Chapter 29

PEB Programming

Introduction

The PEB carries 24 channels, known as *time-slots*. Each time-slot has two independent channels: one is the receive channel, which is the incoming data from the telephone trunk, and the transmit channel, which is the data sent to the telephone trunk.

Sometimes newcomers to the PEB are confused about the terminology "transmit" and "receive". It may help to remember that these names always are from the point of view of the network module which attaches the PEB to the telephone trunks. "Transmit" data is data which is transmitted to the telephone network, typically a message played to a caller or a digit dialed. "Receive" data is data received from the telephone network, such as the caller's voice or a touch-tone dialed in response to a menu.

Receive may be abbreviated to Rcv, transmit to Tx.

Each PEB time-slot thus has four components:

1. Received audio.
This is the digitized audio signal put onto the PEB by the network module (eg. LSI, DTI or MSI).

2. Received signaling bit.
Each PEB time-slot carries one signaling bit which has value 0 (usually indicating no connection) or 1 (connection active). If the network module is receiving a digital signal, the PEB receive bit value will probably be exactly the value of the A bit received from the T-1 network trunk. An analog module such as the LSI or MSI will translate the presence or absence of loop current on the line to the signaling bit value on the receive time-slot.

3. Transmitted audio.
Digitized audio signal sent to the network module for transmission out to the network line.

4. Transmitted signaling bit.
Values of the bit determining the signaling state of the transmitted signal. On a digital network module, A and B bits will probably be transmitted as copied from the PEB signaling bit. An analog device such as the LSI or MSI board will translate the transmit signaling bit state to a hook-switch state: 0=on-hook, 1=off-hook.

PEB Programming

Signaling bits are carried on the PEB using "robbed-bit" signaling, where least the significant bit of a sample is sometimes replaced by the signaling bit.

Every PEB time-slot originates (receive) or terminates (transmit) at a network module:

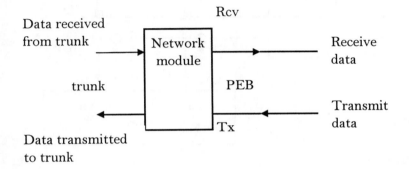

A voice processing card, such as the D/12x, may be connected to the receive and/or transmit data, depending on the data routing which can be turned on and off under program control. Any D/12x channel can be connected to any PEB time-slot on the PEB to which it is connected. The receive and transmit data may be routed independently. The D/12x is a typical example of a *resource module*: a board which can be connected via the PEB to process the received and/or transmitted data.

We will try to develop a diagram method of discussing PEB routing. Hopefully this will assist the reader in understanding the capabilities and programming functions which are used by PEB devices.

We will start with one network module and one resource module. There may be several resource modules attached to the same PEB, but there will only be one network module for each time-slot.

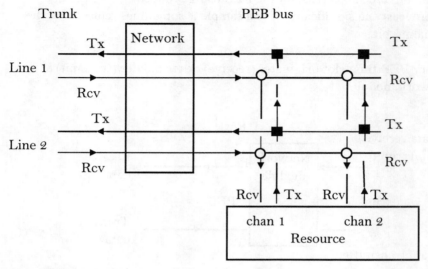

PEB architecture.
For simplicity, only two time-slots and two channels are shown. "Network" is the network interface board, such as an LSI or DTI. "Resource" is a resource module such as a D/12x. The lines with arrows represent data routings: data samples may be copied in the direction of the arrows to transmit audio to and from the trunk. A possible transmit routing is represented by a black square ■, a possible receive routing is represented by a white circle O.

The terminology we will use is that the PEB has time-slots, while resource modules have channels. A PEB time-slot may also be referred to as a channel, but we will avoid this usage.

Channels are also divided into receive and transmit halves. The distinction between transmit and receive channels is from the point of view of the resource board.

A typical transmit channel will play audio files or generate a text-to-speech voice, a typical receive channel will perform voice recognition or record an audio file. Notice that this naming convention means that transmit channels will connect to transmit time-slots and receive channels will connect to receive time-slots.

In our diagrams, time-slots will be shown as horizontal lines, channels as vertical lines. The flow of data is indicated by arrows.

PEB Programming 451

Time-slots may be connected in various ways to channels on resource boards. In the diagram, the connections (switch-points) marked ■ and ○ may be turned on and off.

There are important rules associated with these switch points:

1. A transmit channel on a resource board may transmit to any number of transmit time-slots.

2. Only one channel may be transmitting to any given transmit time-slot.

3. A receive time-slot can be routed to any number of receive channels.

4. A receive channel on a resource board may only receive from one PEB time-slot.

Using the diagram, these rules might be formulated visually as follows (here "line" refers to a line —— drawn in the diagram):

1. Vertically, any number of the switch-points ■ on a single line (i.e. for a given channel) may be "on".

2. Horizontally, at most one of the switch-points ■ on a single line (i.e. for a given time-slot) may be "on".

3. Horizontally, any number of the switch-points ○ on a single line (i.e. for a given time-slot) may be "on".

4. Vertically, at most one of the switch-points ○ on a single line (i.e. for a given channel) may be "on".

These rules may seem confusing, but they all derive from one intuitive principle from the underlying electronics:

All data routing is done by copying 8-bit samples.

To see how this is so, we will digress a little to explain more about the internals of the PEB bus.

The PEB carries data in a very similar way to a digital trunk using T-1 or E-1. The technique is called *Time Division Multiplexing*, or *TDM*. The transmit and receive half of the PEB are each a single stream of bits, carried at 1.544 bps (T-1 type PEB, 24 time-slots) or 2.048 bps (E-1 type PEB, 32 time-slots). The bit stream is divided into *frames*. Each frame is a fixed number of bits in length and carries one 8-bit PCM sample from each time-slot, followed by synchronization bits so that the end of each frame can be located:

```
01010011 10001001 11110010 ... 11111101    1..
Slot 1   Slot 2   Slot 3       Slot 24/32  Sync bits
```

Typical PEB frame consists of one 8-bit sample from each time-slot.

The PEB is built so that 8,000 8-bit samples per second for each time-slot are carried. Therefore, 8 KHz 8-bit PCM data can be carried for each conversation.

A transmit channel on a resource device will be generating 8-bit PCM samples at a rate of 8,000 8-bit samples per second. The PEB electronics allows each sample to be copied into the PEB bit stream at any position (time-slot) in the frame, and into more than one time-slot if desired.

A receive channel on a resource device requires 8,000 8-bit samples per second. The PEB electronics allows any one position (time-slot) in each frame to be copied to the receive channel.

The rules of PEB data routing can be understood from the principles:

> 1. A given 8-bit audio sample must have originated from only one source. The PEB cannot combine two samples, for example by adding the two amplitudes and dividing by two. This is the function of conferencing devices, it cannot be achieved with the electronics of the PEB.

PEB Programming

2. One sample can be copied to as many different positions (time-slots) in a PEB frame as required.

The following illustrations may help clarify these points.

PEB transmit data (one frame shown)

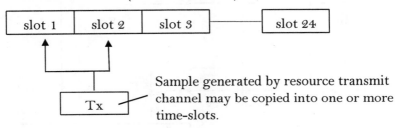

Sample generated by resource transmit channel may be copied into one or more time-slots.

Transmit routing. Routing is done by copying. In this (not typical) example, we illustrate a transmit channel routed to two transmit time-slots, numbers 1 and 2. Each rectangle represents an 8-bit sample. Data generated by the transmit channel is copied both to positions 1 and 2 in the PEB frame. Because slots 1 and 2 are now "filled in", no other resource device can be transmitting to these slots.

PEB receive data (one frame shown)

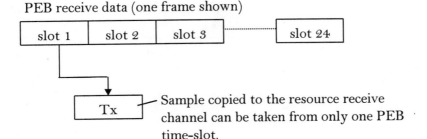

Sample copied to the resource receive channel can be taken from only one PEB time-slot.

Receive routing. In this example, we illustrate a receive channel routed from receive time-slots 1. Each rectangle represents an 8-bit sample. The PEB cannot combine data from more than one time-slot, so no other time-slot can be routed to this receive channel without un-routing slot 1.

In a typical situation, both the receive and transmit halves of each resource channel are connected to the corresponding PEB time-slot:

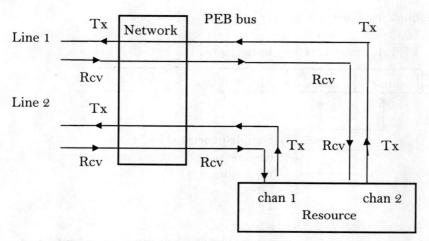

Typical routing. In the usual routing, time-slot 1 will be connected to channel 1, time-slot 2 to channel 2 and so on.

Another possible routing would have channel 1 transmitting audio to both time-slots 1 and 2:

PEB Programming

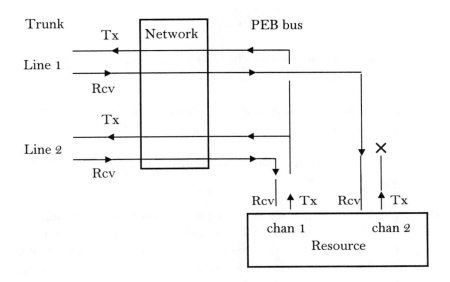

Unusual routing example where channel 1 broadcasts audio to both time-slots 1 and 2. Audio transmitted from channel 2 is not connected to any time-slot. The × symbol is used to indicate that the data on a particular time-slot or resource channel reaches a "dead end" because it is not connected to anything. Remember that a transmit channel can be connected to two or more PEB transmit time-slots, but a receive channel can only be connected to one PEB receive time-slot.

To summarize the general rules which apply to PEB connections from trunk to network interface to resource module:

1. Network interface boards "hard-wire" trunk connections to PEB time-slots: line 1 to slot 1, line 2 to slot 2 and so on. In other words, the connection between analog lines (LSI, DID) or time-slots on a digital trunk (DTI) and PEB time-slots are fixed and cannot be changed under software control. An exception is the MSI, which can connect analog stations to any PEB time-slot under program control.

2. Resource module channels can transmit to zero, one or many PEB transmit time-slots.

3. Resource module receive channels can only accept data from at most one PEB receive time-slot.

4. Resource module receive channels cannot receive data from PEB transmit time-slots.

5. Resource module transmit channels cannot transmit to PEB receive time-slots.

6. Only one transmit channel may transmit to any given PEB transmit time-slot.

Rules 4 and 5 may sound strange: why would you want to transmit to a receive time-slot or vice versa? This question arises in a configuration where there are two or more network modules. Consider, for example, a call center design where there is a T-1 trunk attached to a DTI board, a D/121B voice card and an MSI for connecting to operator headsets. The configuration could be diagrammed as follows:

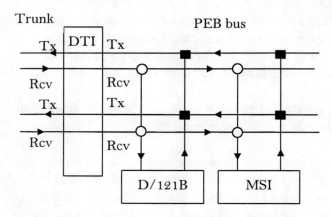

PEB configuration with two resource modules. An important limitation of the PEB architecture is illustrated by this configuration with a DTI and MSI on the same PEB. The D/121B cannot play a message to an operator headset connected to the MSI: this would require transmitting from the D/12x to a receive PEB time-slot.

As the above diagram shows, the D/121B in this configuration can play messages to a caller on the T-1 but not to an operator on a headset attached to the MSI. The reason for this is that the operator can hear audio from the MSI, which takes it from the receive half of the PEB time-slot because this is where the audio from the caller is carried, and (rule 5) the D/121B cannot transmit to a receive time-slot. If a PEB cross-over cable were used, the D/121B could send and receive through the MSI, but it would then be unable to communicate with a caller on the T-1. In order to interact with both callers and operators, two D/xx cards would be needed for each time-slot.

PEB Routing Functions In The MS-DOS API

The *sb_route* function is used to connect both receive and transmit data to a PEB D/xx device such as a D/121B or D/81A:

```
code = sb_route(channel, timeslot);
```

The *channel* argument refers to the D/xx channel number, *timeslot* is the PEB time-slot number. Using a time-slot number of -1 will disconnect the D/xx channel from the currently connected time-slot (if any).

Receive and transmit channels can be controlled independently by using the *sb_rtrcvxmt* function:

```
code = sb_rtrcvxmt(channel, rcv, tx);
```

where:

> channel is the D/xx channel number,
>
> rcv is the receive time-slot number, zero if no connection is to be made, or -1 to disable the connection.
>
> tx is the transmit time-slot number, zero if no connection is to be made, or -1 to disable the connection.

For example, it might be useful to be able to record part of a conversation or detect touch-tones dialed by the caller from the network while allowing transmit data to come from another source on the PEB (for example, a text-to-speech generator, or another caller routed through the DMX). To achieve this, a call to *sb_rtrcvxmt* with *tx* set to zero would route the received data to the D/xx for recording without changing the current routing of the transmit data.

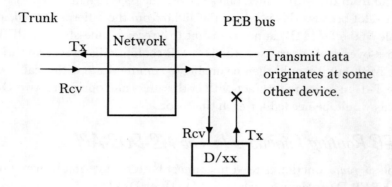

Receive time-slot only routed to D/xx channel. This situation allows the D/xx to detect an in-coming touch-tone from the network or to record the received audio while still allowing transmission of audio from another source on the PEB, such as a text-to-speech board, or the voice of another caller via a DMX. The independent routing of receive and transmit data is accomplished by the sb_rtrcvxmt function.

Routing Example: DMX

Consider the example configuration discussed earlier which allows a conversation between any two callers on an LSI/120. The three boards used are the LSI/120, a D/12x and a DMX.

The caller is initially routed to the D/12x, which plays menus and gets touch-tone responses from the caller. When a given option is chosen, the caller is placed in a one-on-one conversation with another caller.

The situation when the call arrives will be usual routing of the PEB time-slot to a D/12x channel set by *sb_route*.

PEB Programming

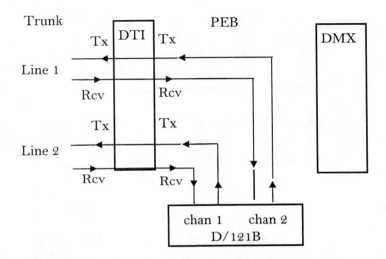

When the caller requests a live conversation, the transmit data must be taken from the DMX rather than from the D/12x so that the callers can hear each other, this requires the following call on both time-slots:

```
sb_rtrcvxmt(slot, slot, -1);
```

where *slot* is the time-slot of one of the callers, which will (in this configuration) be the same as the D/12x channel number. Prompts can no longer be played to the caller from the D/12x. However receive channel functions, such as DTMF detection, hang-up detection, or recording audio, can still be performed.

The connection between the two time-slots on the DMX board is made using the *dm_route* function:

```
code = dm_route(device, slot1, slot2, type);
```

where:

> *device* is the DMX device number (set by the DTI driver through a configuration file),
> *slot1* is the first time-slot to connect,
> *slot2* is the second time-slot to connect,
> *type* is the type of connection to make.

The connection types are defined by constants in the *DMX.H* header file, for our example we need to make a two-way (full-duplex) talk path between the two time-slots, this would be made by using the *DMX_MAKE* type:

```
dm_route(device, slot1, slot2, DMX_MAKE);
```

The DMX will "cross over" the receive and transmit channels so that the audio received on one channel will be transmitted on the other, and vice versa, as shown in the following diagram.

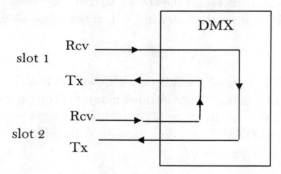

DMX connecting two time-slots with full duplex. This connection provides a complete talk-path between two time-slots.

To return to our example, when the *sb_rtrcvxmt* call has been made to disconnect the transmit time-slot from the D/12x, and when the *dm_route* call has been made to get transmit data from the other caller, the situation will look like this:

PEB Programming

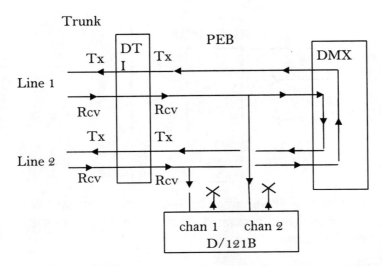

Routing with DMX talk path between two callers. The DMX is used to make a two-way talk path. The D/12x is still available for DTMF detection, recording, hang-up detection and other services provided by the receive channel.

The DMX can also be used to make a half-duplex connection, where the receive data from slot 1 is sent to the transmit channel of slot 2, but no connection is made in the reverse direction:

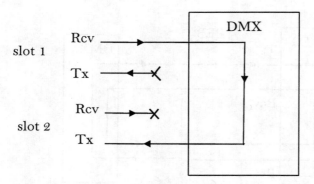

DMX connecting two time-slots with half duplex. With a half duplex connection, data is sent in only one direction, from the receive channel of one time-slot to the transmit channel of the other time-slot.

Time-Slot Routing For Drop And Insert

The DMX board provides for flexible drop and insert applications where any conversation on one network module can be connected to a conversation on any other network module.

It is also possible to create "hard-wired" drop and insert configurations where a cross-over PEB-0 cable makes the connection between in-bound and out-bound time-slots.

A diagram of hard-wired drop and insert is as follows.

PEB Programming

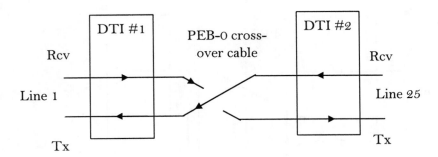

Hard-wired drop and insert configuration. This type of configuration was commonly used, before the DMX board was available, in order to connect two T-spans by hard-wiring two DTI/1xx boards with a PEB-0 cross-over cable.

The connection between the two DTI boards is via a PEB-0 cross-over cable. PEB-0 is a precursor of the current PEB bus found only on DTI boards. It is not pin-compatible with other PEB interfaces, and is only used for hard-wired drop-and-insert configurations.

The main disadvantage of hard-wired drop and insert configurations is that the mapping between in-bound and out-bound conversations is fixed: if a call arrives on time-slot 3 on one network module, it can only be connected to a call on time-slot 3 on the opposite module.

The following diagram shows a two-way drop and insert configuration with two network cards and two speech cards. The connection between the two time-slots could either be through a DMX board or a hard-wired cross-over cable.

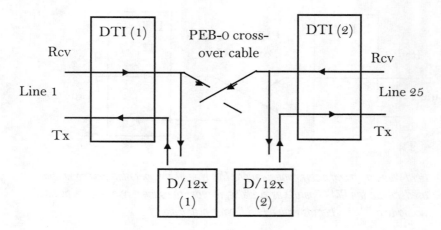

Two-way drop-and-insert, both sides dropped to D/12x. This diagram shows a two-way drop and insert configuration where time-slots on both sides have been dropped to D/12x channels. The two callers cannot hold a conversation since the audio transmitted to each caller originates from a D/12x channel rather than from the opposite caller. Note also that, as the diagram shows, D/12x(1) can only communicate with the caller on network module (1), and D/12x(2) can only communicate with network module (2).

This diagram illustrates an important point: a D/12x channel can only communicate with the time-slot its "own" network module, i.e. on the network module which is on the same side of the cross-over cable or DMX board as the D/12x. If D/12x (1) plays a speech file, it will only be heard by the caller on network module (1) even though the connection is made between the time-slots across the PEB. Similarly, if the caller on network module (2) dials a touch-tone, it will be detected by D/12x (2) only: there is no way to route the received data on network module (2) to the receive channel on D/12x (1).

Note, however, that most conversations start from analog devices like regular telephones. Analog lines just carry sound, there is no distinction between received and transmitted audio. This means that any sound sent on the transmit portion of a time-slot that is routed through an analog device will end up "looping through" to the receive channel of the same time-slot. In a drop and insert situation, for example, a touch-tone dialed on one side may loop through the phone at the opposite end of the conversation and be detected on the opposite time-slot.

Signaling

Signaling refers to the information conveyed by a telephone connection in addition to the sound. The most important signal is an indication of whether the connection is active. On an analog line, this signal is the presence or absence of loop current. On a digital connection, this will usually be signaled by the value of the A bit: $A=1$ means the connection is active, $A=0$ means that there is no connection.

To review, the important signals are:

Ring.
On an analog line, ring is signaled by an AC voltage applied to the line. This can be detected by a device even when on-hook. On a digital trunk, ring is signaled by the received A bit, which will be set to 1.

Seize.
Line seizure is an indication that a device wishes to initiate a call. On an analog line, a device seizes a line by going off-hook, causing loop current to flow. On a digital trunk, seizure is signaled by setting the transmitted A bit to 1.

Disconnect.
A device terminates a call by issuing a disconnect. On an analog line, this is done by putting the device on-hook, i.e. hanging up, stopping the flow of loop current. On a digital trunk, disconnect is signaled by setting the transmitted A bit to 0. The disconnect signal (on-hook or A bit 0 state) should last a minimum period, often one second, in order to be recognized as a disconnect rather than a flash-hook or line glitch.

Flash-hook.
Used to request a service from a switch without ending the call. An analog line issues a flash-hook by putting the line on-hook briefly, a digital connection by setting the A bit to zero briefly. In other words, a flash-hook is a brief hang-up. To distinguish a flash-hook from a line glitch or a true hang-up, the on-hook period should be within limits set by the switch servicing the call; generally the optimal period is about 0.5 seconds.

Wink.
Less common than the other signals, winks are used as part of a "handshake" protocol between a device and a switch. For example, a wink may be used to acknowledge receipt of ANI or DNIS digits from a switch and to indicate that the device is now ready to receive the call. On an analog line, a wink is transmitted by reversing the polarity (+ and −) of the loop current: this must be done using the external DID/xx device. On a digital trunk, a wink is signaled by setting the A bit briefly to 1.

Each PEB time-slot carries signaling information in addition to audio channels. Signaling bits may be carried through "robbed bits" from the audio channel, where the least significant PCM sampling bit is occasionally used to contain an A or B bit value, or in a separate channel devoted to signaling for all time-slots. Robbed bit signaling is the usual method.

Analog network modules such as the LSI and MSI must translate between analog protocol and the digital signaling on the PEB. For example, if the transmit A bit on the PEB is high, the LSI must ensure that the line is off-hook. If loop current stops flowing, the LSI must set the received A bit on the PEB to zero.

In order for the VRU software to monitor a call, a device which is capable of detecting a disconnect signal must always be connected to the received time-slot. Similarly, a device capable of setting signaling bits must always be connected to the transmit time slot so that a known signal state is transmitted to the network.

The devices capable of monitoring and setting signaling bits are:

D/xx.
The D/xx has complete facilities for detecting and generating signaling.

PEB Programming

MSI.
The MSI is able to detect but not generate signaling. The events which the MSI will detect are on-hook and off-hook on a station channel.

DTI/xx.
The DTI/xx boards are able to detect and generate the most important signaling states and transitions:

> Detect any received A or B bit transition
>> This allows detection of seize (received A bit changes from zero to one), and hang-up (received A bit changes from one to zero) without a D/12x being connected to the time-slot.
>
> Set transmitted A and B bits.
>> This allows the DTI/xx board to seize (set transmitted A bit to one) or to disconnect (set transmitted A bit to zero). The DTI/xx boards are able to set the transmitted signaling bits when in *inserted signaling* mode where the transmit signaling bits on the PEB are over-ridden by values selected in the DTI board. When in *transparent signaling* mode, the DTI simply transmits the bits as set on the PEB.

It would be nice if the DMX and MSI boards had similar functions, but (at least at the time of writing), this is not the case.

DTI Transmit Signaling

In the following diagrams, a thin line represents signaling bit values:

```
_____     A and B bits,
```

while a thick line represents audio:

```
▬▬▬▬▬▬▬▬▬▬▬▬     Audio data
```

The DTI/xx network module boards have the ability to set the transmitted A and B bit values. Usually, the DTI is set to *transparent signaling* mode, where A and B bit values are taken from the PEB. In order for the signaling bit values to be set by the DTI, the DTI time-slot must be set to *inserted signaling* mode.

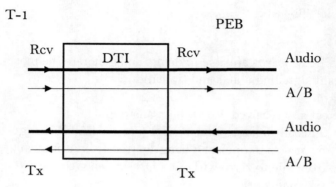

DTI in transparent signaling mode. In transparent signaling mode (the usual mode), the DTI simply passes signaling bit values from the transmit time-slot on the PEB. Received data is not affected by the transparent / inserted mode of the DTI.

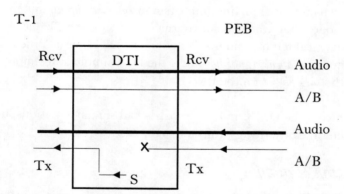

DTI in inserted signaling mode. In inserted signaling mode, the DTI "inserts" signaling bit values from a signal generator on the DTI board, shown as "S" in the diagram.

To set the signaling mode of the DTI board, the *xmittype* function is used:

```
code = xmittype(device, timeslot, type);
```

where:

PEB Programming

device is the device number of the DTI board as determined by the configuration file for the DTI driver,

timeslot is the PEB time-slot number, the transmit channel of which will be affected,

type is one of the two values *SIGINS* (for inserted) or *TRANSP* (for transparent) defined in the *DTI.H* header file.

The bit values generated by the DTI when in inserted signaling mode are determined by the *setsignal* and *clrsignal* functions:

```
code = setsignal(device, timeslot, state);

code = clrsignal(device, timeslot, state);
```

The *setsignal* function changes the bits indicated by the *state* argument to one, the *clrsignal* function changes the bits indicated by the *state* argument to one. The *state* value will be one of:

A_BIT	A bit only,
B_BIT	B bit only, or
A_BIT | B_BIT	Both A and B bits.

It is important to know the state of the DTI signal generator when the board is switched from transparent to inserted signaling. If *xmittype* is called before *clrsignal* and *setsignal* are used to set the bits to a known state, this may introduce a "glitch" into the transmitted bit values, which may be interpreted by the T-1 switch as a hang-up, flash-hook or other signal.

The following procedure is therefore recommended for switching to inserted signaling mode.

1. Set the signal generator bits to zero.
This is done with the call:

```
clrsignal(device, timeslot, A_BIT|B_BIT);
```

2. If required, set A and/or B bit to one.
This would be done using *setsignal*.

3. Go into inserted signaling mode.
This is done with the call:

    xmittype(device, timeslot, SIGINS);

For the same reason, if a time-slot is to be dropped to a D/12x channel, which is then to take over transmitted signaling, the D/12x channel should be set to a known state before going to transparent signaling mode. This is done by the following procedure:

1. Ensure that D/12x channel is in known signaling ("hook") state.
This can be achieved with the *sethook* function. (When sent to a D/12x channel, *sethook* commands are used to set a signal generator on the board which applies signaling bits to the transmitted data. On an analog board such as a D/4x, the *sethook* command directly controls the hook switch on the board).

2. Go into transparent signaling mode.
This is done with the call:

    xmittype(device, timeslot, TRANSP);

The same consideration applies for any other source of signaling information. For example, if a time-slot is routed through the DMX board, be sure that the signal source is set to a known value before going to transparent mode to accept the signal information from the other time-slot.

Idling A DTI Time-Slot

The DTI board has one more capability to process transmitted time-slot data: transmitting silence, otherwise known as "idling" a time-slot.

When a time-slot is idled, the bit pattern hex 7F, binary 01111111, or hex FF, binary 11111111, is transmitted as the PCM data for the time-slot. This is a constant value of + or - the maximum amplitude, and will therefore be heard as silence. (The more obvious value of hex 00, binary 00000000 cannot be used because of consistency checks on the T-1 data stream which expect a certain number of 1 bit values). When using

PEB Programming

robbed-bit signaling, as will be usual on a T-1 connection, this value will always result in a 1 in the least significant bit, and will therefore overwrite the transmitted A and B bits with 1. It is therefore recommended that a time-slot never be idled unless it is already in inserted signaling mode, which will ensure that the desired transmit signaling bit values are preserved.

Idling is turned on and off by the *chgidle* function:

```
code = chgidle(device, timeslot, state);
```

where *state* is one of the following two values defined in *DTI.H*:

ENABLE Enable silence transmission,

DISABLE Disable silence transmission.

The main use for time-slot idling is in a hard-wired drop and insert configuration, where it may be desired to suppress the audio coming over the PEB from the other time-slot until it is time to "patch" the conversation through.

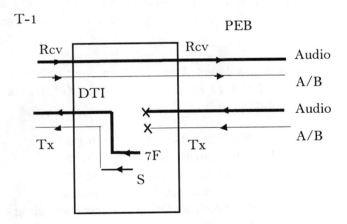

DTI in inserted signaling mode with idle enabled. When idle is enabled for a time-slot, an "all-ones" pattern ("7F" in the diagram) is transmitted on the audio channel, which is heard as silence. To ensure that this pattern does not interfere with the correct value

of the transmitted signaling bits, this mode should always be used in conjunction with inserted signaling mode, which transmits A and B bits from the signal generator S.

DTI Signaling Detection

The DTI board has the ability to detect transitions in the received signaling bits. The important transitions are:

Seize.
Received A bit changes from zero to one.

Disconnect.
Received A bit changes from one to zero, indicating that the caller hung up.

Wink.
Received A bit changes briefly from zero to one and back to one.

The DTI board reports transitions in a similar way to the D/4x and D/12x: an event is posted to the event queue. Events are retrieved by calling the *gtevtblk* function. When a DTI event is reported, the *devtype* field in the returned *EVTBLK* structure is set to the *DV_DTI* constant, defined in *DTI.H*.

When a signaling transition is detected by a DTI, an event will be posted with *devtype* set to *DV_DTI*, *evtcode* set to *T_DTSIGNAL*, and *evtdata* set to one of the following:

DTMM_AON	A bit transition from zero to one
DTMM_AOFF	A bit transition from one to zero
DTMM_BON	B bit transition from zero to one
DTMM_BOFF	B bit transition from one to zero
DTMM_WINK	Brief A/B bit ON period (corresponding ON/OFF transitions will not be posted as separate events).

The reporting of transitions is enabled and disabled using the *setmsgmsk* and *clrmsgmsk* functions, which together are analogous to the *setcst* function for voice devices. They are called as follows:

PEB Programming

```
code = setmsgmsk(device, timeslot, events);

code = clrmsgmsk(device, timeslot, events);
```

where the *events* argument is a bit-wise "OR" of the events which are to be reported, using the same constants defined in *DTI.H*:

DTMM_AON, DTMM_AOFF, DTMM_BON, DTMM_BOFF, DTMM_WINK.

Events not mentioned in the *events* bit-mask are not changed. As you would expect, *setmsgmsk* enables detection of events mentioned in the bit-mask, *clrmsgmsk* disables detection.

For example, to detect hang-up and wink:

```
setmsgmsk(device, timeslot, DTMM_AOFF | DTMM_WINK);
```

DTI Hard-Wired Drop And Insert Example

In this sub-section we give a more complete example of an application flow.

The configuration we will consider will require the full capabilities of the DTI board to detect and generate signaling information: a two-way hard-wired drop and insert with one D/12x board for each DTI/101.

The outline of the application flow is as follows.

1. DTI board (1) monitors the in-bound time-slots for incoming calls.

2. When a call arrives, the program checks the status of the system to find a free channel on D/12x board (1). (If no channel is available, the call cannot be processed. This will hopefully be a rare problem).

3. The call is "dropped" to the free D/12x channel, which presents menus to the caller. The caller selects an option which

requires the call to be patched through to an out-going call. The caller is asked to wait while the call is completed.

4. A free channel on D/12x board (2) is identified (this is guaranteed to succeed because of the structure of the application).

5. D/12x board (2) goes off-hook, dials a number and completes the call.

6. The two D/12x channels are un-dropped, passing audio through and creating a talk path between the caller and the called party. The two D/12x channels are now free to process new calls, the DTI boards will monitor the conversation for a disconnect from one side or the other.

Since the D/12x channels only participate in a call for a short time in the initial menuing and call setup, this application may function acceptably without needing a second D/12x for each T-1.

In detail, the function calls required will be as follows.

1. To monitor for an incoming call.

We will design the application so that the DTI board is always responsible for disconnect monitoring, even when a conversation is dropped to a D/12x channel. This is achieved by using *setmsgmsk* with the A to zero transition:

```
setmsgmsk(dti1, slot1, A_OFF);
setmsgmsg(dti2, slot1, A_OFF);
```

(Note that the time-slot number will always be the same on DTI (1) and DTI(2) owing to the hard-wired connection).

The DTI board must be set to inserted signaling with transmitted zero A and B bits to indicate that it is ready to receive a call. If A and B were set to one, this would indicate a call in progress and result in a busy tone if a call were sent to that channel. To ensure that the A and B bits are zero, the call:

PEB Programming

```
clrsignal(dti1, slot1, A_BIT | B_BIT);
```

would be made, followed by a change to inserted signaling mode:

```
xmittype(dti1, slot1, SIGINS);
```

As noted earlier, the calls should be made in this order to avoid glitches in the A and B bits.

The *setmsgmsk* function will be called with a mask including *DTMM_AON*, which indicates an incoming ring:

```
setmsgmsk(dti1, slot1, DTMM_AON);
```

The application will then wait for a *T_DTSIGNAL* event for that channel with a data code of *DTMM_AON*.

2. To drop the call to the D/12x and answer the call.
The application should ensure that the D/12x channel is on-hook:

```
sethook(channel1, H_ONH);
```

It is then safe to go to transparent signaling so that the D/12x can control the call:

```
xmittype(dti1, slot1, TRANSP);
```

The *sb_route* function will be used to drop the call:

```
sb_route(channel1, slot1);
```

Here, *channel1* is a free channel on D/12x (1), *slot1* is the time-slot number where the call arrived.

Now the D/12x channel can answer the call:

```
sethook(channel1, H_OFFH);
```

Menus and responses will be processed in the usual way with calls to *xplayf* and *getdtmfs*.

3. To start the out-bound call.

The DTI (2) channel should start out as the DTI (1) channel: in inserted signaling mode with A and B bits set to zero. Silence generation will be initially be enabled to suppress any audio passed through the cross-over. This situation is achieved by the following calls:

```
clrsignal(dti2, slot1, A_BIT | B_BIT);
xmittype(dti2, slot1, SIGINS);
chgidle(dti2, slot1, ENABLE);
```

Note the order that the calls are made to prevent glitches.

When an out-bound call is to be started, the D/12x (2) channel should be forced on-hook with:

```
sethook(channel2, H_ONH);
```

The DTI (2) time-slot (which will have the same number as the DTI (1) time-slot because of the hard-wired cross-over) will be dropped to the D/12x (2) channel:

```
sb_route(channel2, slot1);
```

At this point, it will only be the transmit audio, not signaling, which is taken from the D/12x channel. Since the signaling state of the D/12x channel is known to be on-hook, it is safe to go to transparent signaling so that the D/12x (2) channel controls the out-going signaling:

```
xmittype(channel2, slot1, TRANSP);
```

and the call can then be started in the usual way by going off-hook with:

```
sethook(channel2, H_OFFH);
```

and *dial* or *callp* can be used to complete the call.

PEB Programming

4. To patch the call through.

When the second call has been completed, the original caller can be connected through to the new called party as follows.

The DTI boards will regain control of the transmitted signaling since the D/12x boards are to be disconnected. Stable calls are now in progress, so the signaling bits to be transmitted should be set to 1:

```
setsignal(dti1, slot1, A_BIT | B_BIT);
setsignal(dti2, slot1, A_BIT | B_BIT);
```

It is then safe to return to inserted signaling on both ends:

```
xmittype(dti1, slot1, SIGINS);
xmittype(dti2, slot1, SIGINS);
```

The silence generation should be disabled on DTI (2), which will allow the called party to hear the original caller:

```
chgidle(dti2, slot1, DISABLE);
```

and finally the two D/12x channels can be disconnected from the call, creating a talk path between the two callers:

```
sb_route(channel1, -1);
sb_route(channel2, -1);
```

The call will continue until one of the callers hangs up, which will be detected by one of the DTI boards as a received A bit transition to zero.

The DTI/124

The DTI/124 is similar to the DTI/211, except that D/4x connections are provided instead of a PEB for the D/12x. A PCM expansion connector (PEB-0) is still available for drop and insert applications. The D/4xs, or other devices with a compatible interface such as an AMX card, is connected to the DTI/124 via an AEB cable:

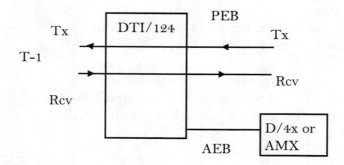

*DTI/124 provides both PEB-0 and AEB connectors.
A typical configuration will include a DTI/124 and up to 6 D/4x cards.*

Most of the functions provided by the DTI/211, in particular all the functions relating to detection and generation of signaling bits, are also supported by the DTI/124.

The most significant difference is in the routing of time-slots. The *enablech* and *disablech* functions control the routing of a time-slot to an AEB connector:

```
code = enablech(device, timeslot, devchan);
code = disablech(device, devchan);
```

The *device* and *timeslot* arguments are the usual, *devchan* is the "device channel", numbered 1 to 24 on each installed DTI/124, of the AEB connection. The *enablech* function is analogous to *sb_route* called with a given channel number, *disablech* to *sb_route* called with a channel number of -1.

When a time-slot is dropped to an AEB channel, transmit data is taken from the AEB and receive data is sent to both the AEB and the PEB. This is the same behavior as on a D/12x board after an appropriate *sb_route* call.

In fact, if the correct Dialogic driver and firmware versions are installed, *enablech* and *disablech* will function as *sb_route* if the designated board is a DTI/211. This is convenient for applications which mix

DTI/211s and DTI/124s in the same PC. It is recommended that *sb_route* be used if possible for D/12x applications.

Controlling The MSI Board

The MSI board is a lot like a DTI/124 from a programmer's point of view. The following functions perform in more or less the same way:

enablech
Routes a time-slot to an analog station channel number in the range 1 to 24 for each installed MSI.

disablech
Disables the time-slot routing.

setmsgmsk
clrmsgmsk
Enables and disables on- and off-hook detection on the analog station. The events supported are:

MSMM_OFFHOOK	Station went off-hook (seize).
MSMM_ONHOOK	Station went on-hook (disconnect).

The MSI performs one useful function inexplicably missing from the D/4x API: it can report the current hook state of a channel. The *ms_getcst* function (misleadingly named, since it has nothing to do with a call status transition) does the job:

```
code = ms_getcst(device, devchan);
```

This is a multi-tasking function which will post a *T_MSGETSG* event to the event queue with *evtdata* set to:

MSCH_ON	Channel is on-hook, or
MSCH_OFF	Channel is off-hook.

Chapter 30

SC Bus Programming

Introduction

If you have just struggled through the PEB Programming chapter and were mystified by some of the caveats and gotchas, you will be relieved to learn that the SC bus presents a much more logical and consistent set of capabilities to the software developer.

The SC bus carries a number of time-slots. The base number of time-slots starts at 1024, but may be more, depending on the clock speed. Unlike the PEB, time-slots are not divided into receive and transmit halves. You might consider all SC bus time-slots to be transmit time-slots.

How The SCbus Works

To help understand what SC bus functions really do, we'll give a simplified account of how the bus works. An electrical engineer might quibble about some of the details; the intention is to give a picture which is accurate enough to understand software commands fully without drowning in technicalities. We'll assume that there are 1024 time-slots, the bus may actually be configured for a different number depending on the clock speed.

The SC bus has a wire carrying voltage. If you attached a volt-meter to the wire, you'd see the voltage on the wire vary with time. This varying voltage represents bits. A fixed number of times per second (set by the clock speed of the bus), each SC bus chip on the wire measures the voltage and decides if the bit value carried by the bus at that time is a zero or a one.

If we wrote down all the bits, we'd have a long trail of zeroes and ones:

```
110000111110011111110111111100111 ...
time ->
```

There is another wire on the bus which carries a short voltage spike once every 8,192 bits. This is called a synchronization signal. When an SC bus chip sees the synchronization signal, it knows that the bit on the data line is the first bit in a frame of 8,192 bits. There are 8,000 frames per second, making 8,000 x 8,192 = 65,536,000 bits per second.

The SC bus chip interprets the frame as 1,024 8-bit values. Each value is called a time-slot. The first 8 bits are time-slot 0, the next 8 bits are time-slot 1, and so on.

SC Bus Programming

```
110000111110011111110111111100111 ...
└──────┘└──────┘└──────┘└──────┘
time-slot    0       1       2       3
```

Where do the bits come from? Attached to each SC bus chip are one or more devices. A device may be able to generate data. For example, a voice transmit device may be able to play a speech file or dial touch-tones. Generating data means producing 8-bit samples 8,000 times per second. We say that the device is generating 8 KHz 8-bit PCM data. Each 8-bit sample represents the loudness of the sound at that instant. Let's suppose that we want this voice channel to produce the bits for time-slot number zero. (Time-slots are numbered 0, 1, ... 1023). We'd send a command to the SC bus handler for that board telling it: "connect that voice channel to SC bus time-slot zero." Every time the SCbus chip sees the synchronization pulse, it knows to get an 8-bit sample from the voice channel and use it to set the voltages for the next 8 bits on the data wire. If the routing was to time slot 75, the chip would know to look for the synchronization pulse and then wait for 75*8 clock ticks and then apply the voltages. We call this process "transmitting data to an SC bus time-slot."

Clearly, only one device can be transmitting to a given time-slot. If two or more devices are trying to set the line voltage at the same time, you'll get unpredictable (bad) results.

Other devices take data from the bus instead of transmitting to the bus. For example, the receive part of a voice channel may want to record sound to a file or analyze the incoming audio to detect touch-tones. This channel needs to be provided with 8,000 8-bit samples every second. The SC bus chip is informed which time-slot number is to provide samples to the receive channel. 8,000 times per second, the chip waits the right amount of time after the synchronization signal (the amount of time is given by the time-slot number), and then copies 8 bits to the receive channel input register. Those 8 bits will of course be that time-slot from the current frame. 8,000 times per second, the receive channel reads data from the register and passes it to the firmware for processing. This process is called "listening to an SC bus time-slot." It is possible to *unlisten* a device, which means that it is not connected to any time-slot. The input register for the device will always contain 8 zero bits, and the device hears silence.

It is pretty obvious from the way this works that for each receive channel, there can only be one SC bus time-slot providing data to that channel. However, many different receive channels may be listening to any given time-slot.

A *routing* is the process of copying samples to the bus or from the bus.

A *transmit routing* means that data is copied to a time-slot from a transmit channel.

A *listen routing* means that data is copied to a receive device channel from a time-slot.

SC Bus Devices

Attached to the SC bus will be one or more *devices*. A device will in most cases be either:

- A network device which copies a conversation to and from an external trunk or phone line of some kind, or

- A resource device which processes conversation data, such as a voice or fax processing channel.

Each device will have a receive and transmit *channel*. The receive channel is often called the *listen* channel.

Note that the terms transmit and receive/listen are from the point of view of the bus, *not* from the point of view of the network device as in PEB programming. For resource devices, this terminology is the same as for the PEB, but for network devices the transmit and receive terms are reversed. This is confusing if you are used to thinking about the PEB, or just an obvious way to describe the situation if you have never done PEB programming.

For example:

SC Bus Programming

· On a D/xx voice processing resource device, the listen channel will record audio and detect touch-tones, the transmit channel will play audio and dial touch-tones.

· On a T-1 network device, the listen channel will take audio and signaling data from an SC bus time-slot and copy it to the transmit time-slot on the T-1 span which sends data to the central office, the transmit channel will take audio from the central office and copy it to the SC bus.

· On a voice recognition resource, the listen channel will take audio from an SC bus time-slot, the transmit channel will probably be used only to send a beep prompting the caller to begin speaking.

The routing of a device transmit channel to an SC bus time-slot is fixed at the time the system is started and cannot be changed by an application. As you will soon see, this does not place any functional limits on the application providing that you don't exceed 1024 device channels.

SC Bus Routing

To think about SC bus routing functions, I like to draw diagrams where SC bus time-slots are represented by horizontal lines, routing to devices by vertical lines.

To give a specific example, consider a D/240SC (24-channel voice processing resource board) and a DTI/240SC-T1 (24-channel network interface board) connected by an SC bus.

When the system is initialized, each D/xx voice channel will be routed to one SC bus time-slot, as shown in the following diagram.

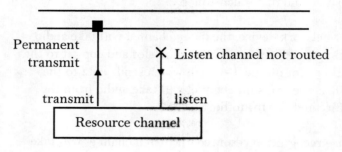

At start-up, the transmit channel of a resource device is permanently connected to an SC bus time-slot. This connection is never changed. The listen channel is not connected at start-up, it is the application's responsibility to make a connection.

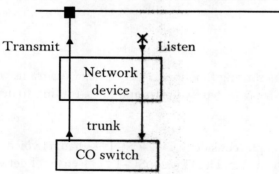

At start-up, the transmit channel of a network resource is permanently connected to an SC bus time-slot. This connection is never changed. The listen channel is not connected at start-up, it is the application's responsibility to make a connection. Note that the data _received_ from the central office is _transmitted_ to the SC bus by the transmit half of the network device channel, the terms "transmit" and "listen" are from the point of view of the S C bus, unlike PEB terminology where "transmit" and "receive" is from the point of view of the network device.

SC Bus Programming

The usual routing for an application will be that each voice channel is connected to one network channel. This required two routing commands:

(1) The voice device listen channel is connected to the SC bus time-slot which is getting data from the network transmit channel, and

(2) The network device listen channel is connected to the SC bus time-slot which is getting data from the voice transmit channel.

When these two routings are made, there is a data path both from the network to the voice listen channel and also from the voice transmit channel to the network.

The following diagram shows the routing when the two commands have been successfully completed.

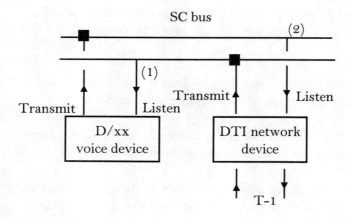

To make a normal voice channel routing, the two connections marked (1) and (2) must be made. These connections create audio paths from the network to the voice channel and vice versa.

SC Bus Routing API

The Dialogic API provides a logical set of SC Bus routing functions. Each function name has a prefix which identifies the type of device:

dl_	Voice processing resource device (D/xx).
ag_	Analog network interface device (-LS).
dt_	Digital network interface device (T1 or E1).
fx_	Fax resource device.
ms_	MSI/SC station interface device.

The function suffix denotes the operation to be performed:

getxmitslot	Returns the SC bus time-slot which is permanently routed to the transmit channel of a device.
listen	Connect the listen channel of the device to an SC bus time-slot.
unlisten	Disconnect the listen channel.

Devices are specified by a device argument, which is generally derived from a board number and a channel number as follows:

```
Device = (BoardNr << 8) | ChannelNr;
```

Time-slots are stored in a structure of type *SC_TSINFO*, which contains two members:

```
typedef struct
{
long sc_numts;
long *sc_tsarrayp;
} SC_TSINFO;
```

The *sc_numts* member specifies the number of time-slots, the *sc_tsarrayp* member points to an array of time-slot numbers. In our examples, we will always set *sc_numts* to 1 since there is always one and only one time-slot attached to a transmit device channel. (More than one listen channel may be receiving data from a time-slot, in this situation more time-slot numbers may be used, but this is not needed for the SC bus routing command API).

To make connection (1), we need first to find the transmit time-slot of the DTI network transmit channel:

```
SC_TSINFO.sc_numts = 1;
dt_getxmitslot(DTIdevice, &SC_TSINFO);
```

A value pointed to by *SC_TSINFO* structure is set by *dt_getxmitslot* to the required time-slot number. The application will never need to look at this value since the *SC_TSInFO* structure is simply handed off to the *dl_listen* function which does the routing:

```
dl_listen(DXXdevice, &SC_TSINFO);
```

Making connection (2) is just the reverse process:

```
dl_getxmitslot(DXXdevice, &SC_TSINFO);
dt_listen(DTIdevice, &SC_TSINFO);
```

In general, to make a full-duplex (both-way) connection between two devices *xx* and *yy*, the sequence is:

```
SC_TSINFO.sc_numts = 1;
xx_getxmitslot(XXdevice, &SC_TSINFO);
yy_listen(YYdevice, &SC_TSINFO);
yy_getxmitslot(YYdevice, &SC_TSINFO);
xx_listen(XXdevice, &SC_TSINFO);
```

This sequence is encapsulated in the *nr_scroute* function:

```
nr_scroute(DeviceX, DeviceTypeX,
  DeviceY, DeviceTypeY, Type);
```

The DeviceType arguments are set to *SC_VOX* (voice channel), *SC_LSI* (loop-start channel) etc. to indicate the device type, this is used by *nr_scroute* to determine which device type prefix (*dl_*, ag..) to use for the API functions.

If Type is set to *SC_FULLDUP*, then *nr_scroute* performs exactly the sequence above.

If Type is set to *SC_HALFDUP*, the *nr_scroute* performs a one-way connection only, as in the following:

```
xx_getxmitslot(XXdevice, &SC_TSINFO);
yy_listen(YYdevice, &SC_TSINFO);
```

This means that DeviceY listens to DeviceX, but not the other way around.

In some older Dialogic documentation for the *nr_scroute* function (SCbus Routing Guide 05-0289-001), the description has the half-duplex case backwards, the documentation says that DeviceX listens to DeviceY, but this is not correct.

In fact, the device argument could have been used internally by the device driver to identify the device type, thus greatly simplifying the API by providing a single set of device-independent functions which would not need the *xx_* prefix or the device type arguments to *sc_scroute*. Unfortunately, Dialogic did not take advantage of this opportunity and a switching layer in your application will need to link in libraries for each specific device type which you will need to support.

SC Bus Programming 491

The above examples have assumed that the devices are in a state where no data is currently being routed to listen channels. A listen channel can only take data from a single time-slot.

If it is desired to break an existing listen routing, the *xx_unlisten* call for the appropriate device can be used.

The *nr_scunroute* function will make the one or two unlisten calls required to break a full- or half-duplex connection:

```
nr_scunroute(DeviceX, DeviceTypeX, DeviceY,
    DeviceTypeY, Type);
```

Again, *Type* can be *SC_HALFDUP* or *SC_FULLDUP*. If *SC_HALFDUP* is used to break a half-duplex connection then *DeviceX* and *DeviceTypeX* are ignored, the only call which is made by the function is *yy_unlisten*. This is confusing (when I first saw the function description I couldn't understand why the *DeviceX* and *DeviceTypeX* arguments were needed): a better design for the "convenience" functions might have been to have separate routing functions for full- and half-duplex make and break operations.

Intra- And Inter- Board Connections

It was conceptually cleaner to think of a separate network and voice resource board in the above examples, but there is no difference between a D/240SC+DTI/240SC-T1 combination and the equivalent single-board D/240SC-T1 device. Devices attached to the SC bus can be routed to each other in exactly the same way whether they are located on the same board or on a different board.

Chapter 31

CTI And TSAPI

Introduction

In this chapter, we will examine *Computer Telephone Integration, CTI,* and Novell's TSAPI product. TSAPI is a particular implementation of a CTI protocol, which has gained momentum in the market and may be destined to become a significant standard Computer Telephone Integration (CTI)

The phrase computer telephone integration, CTI, has a specific meaning in the call processing industry: it refers to linking a phone system, which might be a PBX or ACD, to a network server, generally but not necessarily a file server. CTI is an example of a client-server architecture. Client PCs are connected to the file server via a conventional Local Area Network (LAN) and are able to send commands to and receive notifications from the telephone system via the LAN.

The term *client-server* is a general computing term used when resources used by a PC (the *client*) are provided by a separate computer (the *server*). The server's resources may be shared by two or more client PCs. The connection between the client and server is made by via a LAN. Most people are familiar with the idea of a file server: a central computer containing hard disk files which can be accessed from other PCs (clients) on the network. The client/server philosophy extends this idea from files to printers, databases and other services such as telephony.

The following diagram shows the typical configuration of a CTI system.

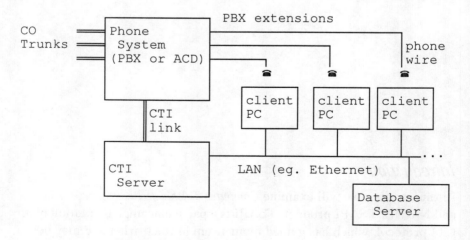

Typical call center design using CTI, eg. TSAPI-based. The CTI Server may include the network file server, CTI link and database server integrated into one computer, or these functions may be divided over two or three machines.

CTI And TSAPI

The CTI link is an RS-232 serial line, an Ethernet connection or any other type of link which allows the phone system to receive commands from and send notifications to the file server. Commands and notifications are together referred to as *messages*.

For example, an incoming call might be detected by the PBX and the ANI digits collected. A message would be sent to the CTI server including the trunk number where the call was received and the digits collected. The CTI server is likely to be a program running on the file server, a *Netware Loadable Module* (*NLM*) in the case of Netware or a Win32 program in the case of Windows NT, for example. The CTI server would send a message to the applications program containing the incoming call details. The application will then want to route the call either to a live agent or to a automated voice response unit. This will be done by sending a command to the CTI server, which would in turn send a command over the CTI link requesting a connection to be made between the incoming trunk and the appropriate extension.

TSAPI is Novell's CTI API. Its main characteristics are as follows.

· The CTI server is part of a file server running Netware. The file server will have a hardware connection to the phone system, such as an Ethernet card or COM port; this hardware connection will be dependent on the specific make and model of phone system used and will generally be provided by the phone system vendor.

· A PBX-independent NLM will be running to provide an interface to the network, this will use a PBX-dependent switch driver to communicate via the CTI link.

· Client PCs will be running Microsoft Windows. A client API is also provided for other Netware servers on the same network, but this will rarely be used in practice.

· From the point of view of the applications developer, TSAPI is hardware-independent: a TSAPI application can be written in such a way that it will run on any network with a TSAPI-based

CTI server. This is a major advantage compared with using a proprietary solution tying the developer to a single PBX vendor, which until the advent of TSAPI was the available option.

The conceptual model of the application is primarily a call center with a number of live agents who speak to customers, the agents have headsets connected to an ACD accepting in-bound calls or a predictive dialer making out-bound calls, although a broader range of systems can certainly be created with TSAPI.

Missing: VRU Media Control

Neither TAPI nor TSAPI covers the *media control* for an automated voice processing computer. Media control means how to play and record messages, get touch-tone digits with talk-off and play-off suppression, detect disconnect tones and so on. TAPI and TSAPI do not replace specific media APIs, such as Dialogic's proprietary API or the open ECTF S.100 API (the SCSA API) standard, or alternative media control interfaces such as Windows' multi-media API which is used to play and record Wave sound files.

CSTA Interface

The *Computer Supported Telecommunications Application* (*CSTA*) standard is a definition for CTI links based on message-passing between the CTI server and PBX/ACD. CSTA was produced by the *European Computer Manufacturers Association* (*ECMA*). The CTI messages in the CSTA specification are used as the basis for TSAPI function calls. A copy of the specification is available from ECMA, address:

114 Rue du Rhône
CH-1204 Geneva
Switzerland

Telephone: +41 (22) 735 36 24.

TSAPI Components

The following are the components of a TSAPI-based system.

CTI Link

This is a hardware-specific connection between the Netware server and the PBX/ACD. Typical examples will be RS-232 serial links or Ethernet cables.

CTI Link Hardware

This is the hardware-specific hardware mounted in the Netware server required for the CTI link. It may be as simple as a COM port or standard Ethernet card, or could be a proprietary card supplied by the PBX/ACD manufacturer.

Switch Driver

The switch driver is a set of programs (NLMs) running under Netware which communicate between the hardware-independent components of TAPI, supplied by Novell, and the hardware-specific components, provided by the phone system vendor. The switch driver might translate a proprietary CTI protocol to TSAPI, for example, and provide any supported Operation, Administration and Maintenance (*OA&M*) capabilities offered by the switch.

Switch Driver Interface

This is a software layer running under Netware which passes messages between the application and the switch driver.

Telephony Services NLM

This NLM provides a layer converting between LAN message packets (IPX/SPX) and messages to and from the switch driver. This layer is responsible for security restrictions defined in a telephony services user database.

Telephony Server

This is a convenient shorthand for a Netware server containing all of the above components.

Telephony Services API

TSAPI itself is a set of function calls and messages which can be used on a Windows (or Netware) client PC on a network with a telephony server. The application developer will primarily be concerned with writing a program, most likely in C or C++, which uses TSAPI function calls and responds to TSAPI messages.

TSAPI Client Library

C and C++ applications running on Microsoft Windows will link with this library in order to use TSAPI. Microsoft would term this library a Software Development Kit (SDK). If you are an application developer, your primary concern is to understand and utilize the capabilities of the function calls in this library.

TSAPI Library

Application components written as NLMs may be run on the telephony server or on another Netware server on the same LAN, transparently to the application. The TSAPI Library contains the function calls available to an NLM.

Application Types

The designers of TSAPI anticipated a number of application categories. The following is an outline, distilled from the TSAPI specification v1.9, of the most important categories.

- Call management.

- Call screening.

- Call logging.

- Directory dialing from personal (client), workgroup (server) and corporate directories.

CTI And TSAPI **499**

· Dialing and integration with other applications.

· Integration of message waiting indicator for e-mail and other messaging applications.

· *Screen-pop* (display of account information on agent PC screen when customer call routed to his/her station) in in-bound call center (ACD).

· Predictive-dial out-bound call center.

How To Get TSAPI

The TSAPI Software Development Kit, including software to simulate a telephony server on a network, is available from Novell, Inc., Provo, UT phone (801) 429-7000.

Chapter 32

TAPI

Introduction

When I arrive at my office in the morning, I have a routine. I check my in-tray for faxes which may have arrived overnight. I check my voice mail for messages. I start up my desktop PC, log on to the network and check my e-mail. A good deal of my morning, and perhaps the rest of the day, can be spent replying to these messages and acting upon them.

Like a great many people who work in offices, I spend most of my day communicating: by voice (meetings or telephone calls to colleagues, customers and vendors), text (e-mail and fax) and graphic images (fax). Today, these different communications media are rarely coordinated and integrated, even though many of them pass through my desktop PC. To check my voice mail, I reach for my telephone, dial the voice mail extension and dial touch-tones to indicate my mail box and password. Despite the fact that the computer with the voice mail machine is connected to the same LAN as my desktop PC, I can't use my PC to check for messages. To make a phone call, I'll look up a telephone number in the company's database, pick up the telephone and manually dial the phone number which I can see on the screen. To make an outgoing fax, I can simply "print" to our fax server from any Windows application. If the reply needs information from a non-Windows (eg. DOS-based) program, I will probably have to print the pages on a laser printer and feed the result manually through the fax machine. Incoming faxes are printed on our plain paper fax machine and distributed by hand. The only messaging medium which we use which is close to being fully automated and paperless is e-mail: and even e-mail requires considerable manual labor when we have to communicate with someone outside our company who uses an e-mail system for which we don't have a pre-programmed "gateway".

We can easily imagine an office in the not too distant future where different messaging media: voice, text, graphics images and video are integrated and controlled from the desktop PC. To make a phone call, you would pull up a phone book on the screen, highlight the person or company you want to call, and click on the "Call" button with your mouse. To make a conference call with two of your colleagues, you would pull up the company directory and drag the parties' names into the "Conference" box. To check for messages, you would open your on-screen "in-box" to see a list of new and saved voice, text, image and video messages. To reply, you would compose a document in any application, and drag the document to the "Reply" button. The PC might determine the medium and means of communication required to deliver the reply.

Devices such as PBXs, fax machines, desktop telephones and printers have traditionally been built to accept and/or transmit information on

TAPI

their specialized medium, i.e. a telephone line or a sheet of paper, but not to provide intelligent, two-way communication with a controlling desktop computer or network. A printer, for example, can only transmit a few rudimentary messages back to a computer: "out-of-paper", or "on/off-line". This lack of communication is despite the presence in most of these devices of sophisticated microprocessors comparable in power at least with early PCs.

In the future, printers will have greatly enhanced abilities to communicate: from your PC, you will be able to see the number of sheets of paper in the feeder, check the level of toner in each color cartridge, send a command to switch to stationery instead of blank paper, and so on. The desktop telephone will have an RS-232 serial port or other connection to a computer which will enable two-way communication: the computer will be able to ask the telephone to dial a number, for example, and the telephone will send status indications back to the PC similar to those displayed on the phone through colored lights or an alphanumeric LED panel. Ultimately, the telephone itself may become an expansion card with a socket for a handset or headset with microphone and loudspeaker.

The company phone system will be controlled, probably via a serial link, by a PC attached to the corporate LAN instead of relying entirely on its internal CPU. When an incoming call is signaled, the PBX will notify the computer, which will reply with a command telling the PBX where to route the call. (Such PBXs, so-called *dumb switches*, are already available from companies such as Summa Four). To make a conference call, the controlling computer will send a command to the PBX. The voice mail PC will be connected to the PBX as a series of extensions and also to the controlling PC via a LAN.

Desktop Applications

Microsoft has a clear vision of how this technology should be tied together: via graphical applications running in a Windows environment. As a first step, Microsoft, in conjunction with Intel, has drafted version 1.0 of the Windows Telephony Applications Programming Interface, or *TAPI*, which was published in May 1993. This is an attempt to define a set of function calls (the API) which will be used by Windows programmers to interact with telephony devices such as telephones, telephone lines and telephone switches. TAPI is sometimes referred to simply as *Windows Telephony*.

Some important categories of application which will incorporate TAPI include:

Integrated Messaging.
A single system for receiving, archiving and responding to voice, text, graphics and video messages.

Personal Information Managers.
These will include facilities for automated dialing and collaborative computing over telephone lines.

Advanced Call Managers.
Do you really know how to use the conferencing feature on your company telephone system? How to ask for "camp on", where a call will be put through to your colleague who's currently on the phone as soon as he hangs up? How to put that "At lunch" message on the phone display panel to notify callers to your extension that you're away from your desk? I don't know how to use half the features on our digital phone system, and I know I'm not alone. I'm looking forward to an application that gives me access to the features of our phone system through graphical menus on my PC. Through *advanced call managers,* Microsoft is envisaging control of your phone system through Windows.

Icon-Driven Data Transmission.
By "dragging and dropping" icons representing documents to a disk drive or other target representing a remote computer, an application will be able to determine whether the document should be copied directly over a LAN or WAN, transmitted over an ISDN digital link and choose an appropriate medium such as e-mail (for text) or fax (graphics). With widespread use of ISDN, data transfer over the same link as a voice call will make whole new categories of applications possible which we can only dimly imagine today.

Remote Control.
Software to control a PC over a phone line is well-known today. Creating such software requires detailed understanding of some DOS or Windows

internals and lots of fancy programming which requires detailed code to control modems and other devices. Using TAPI, both the host and caller modules can be written in a device-independent manner.

Information Services.
TAPI will allow access to on-line services such as news, database retrieval, financial information, weather, sports etc. Developers will be able to write new interfaces for specific environments.

TAPI Structure

Windows Telephony is a particular instance of Microsoft's *WOSA* (*Windows Open Services Architecture*). The key idea in WOSA is to provide a device-independent API for application developers, and an application-independent driver specification for peripheral developers. Since this sounds rather complicated, an illustration of this philosophy might help.

The well-known WordPerfect word processing package was a breakthrough at the time it was introduced for two main reasons: it offered very extensive formatting capabilities well beyond most of the packages available at that time, and because it supported almost all the capabilities of hundreds of different printers. The support for such a large range of printers was achieved through a Herculean effort: by a team of programmers painstakingly describing the exact capabilities of each printer (range of fonts, a bit-mapped representation of each character set for on-screen print previews, paper sizes available, capabilities such as bold, italic, overstrike..., available methods for printing graphics and for rendering characters in graphics mode not available in built-in character sets, downloadable soft fonts, font cartridges...). This detailed description of each printer was called a *printer driver*. (Other applications used a similar scheme, but with far fewer capabilities). Not only did the programmers at WordPerfect have to describe each printer in detail, it was also necessary to create code for each possible PC display type. As every PC programmer knows, there are several monitor types which are "standard" in the PC world: monochrome, CGA, Hercules, VGA and XGA. Each type accepts different commands and has distinctively different capabilities such as the number of pixels (color dots) displayed on the screen and the number of different colors which can be used at one time. When using

MS-DOS, each application developer is faced with the same issues: printing to a vast range of different printer types and displaying on a range of different monitor types.

Windows offers application developers (some) relief from these burdens. Printer and display drivers are included among the services provided by Windows to application programs. To draw a line, for example, from pixel coordinates x_1,y_1 to x_2,y_2, a Windows program could call the C functions *MoveTo(hDC, X1, Y1)*, *LineTo(hDC, X2, Y2)*. This request would be passed by Windows to the display driver for the installed display. Each vendor of a display device will provide a Windows driver written and optimized for that particular display: applications programs will be able to run on this display, and even take advantage of new capabilities such as a larger number of displayed pixels or a larger color palette. The application program will call a Windows function such as *GetDeviceCaps(hDC, nIndex)* ("Get Device Capabilities") to determine the exact capabilities of the current display device, such as HORZRES (number of pixels horizontally), VERTRES (number of pixels vertically), BITSPIXEL (number of bits per pixel for representing color, 1 would indicate monochrome), and so on. Printers are handled in a similar manner.

WOSA carries this philosophy to new types of peripherals. A standard programming interface, such as TAPI for telephony devices, is defined. This programming interface will include:

1. Functions called by application programs to query the capabilities of, and interact with, peripherals (this is known as the *Applications Programming Interface*, or *API*), and

2. Functions called by Windows to the peripheral drivers in order to implement the request (this is known as the *Services Programming Interface*, or *SPI*).

This architecture might be envisaged as follows:

TAPI

Windows WOSA architecture.
By providing a well-defined interface between Windows and the hardware (an SPI, or device driver), applications programs can be written in a hardware-independent way.

A peripherals vendor would write a device driver to accept the commands from the SPI and supply this software with the peripheral. In this way, applications programs can be written without having to take into account all the details of all peripherals to be supported, and the peripherals vendor can write one device driver conforming to a single SPI rather than separate drivers for each application program, as would be the case with an operating system such as MS-DOS.

As the TAPI programmer will quickly learn, the capabilities of telecommunications devices are many and varied. The *lineGetDevCaps* and *lineGetDevCaps* functions, which return the capabilities of "line" and "phone" devices respectively (these have particular meanings in TAPI) return extensive and diverse information to the application. Creating truly device-independent applications is more of a challenge under TAPI than when drawing graphics and text on the screen using the device-independent Graphics Device Interface (GDI) functions in Windows.

TAPI Functionality

Perhaps the most important thing to realize about TAPI is that it deals with the medium, not the message. TAPI is for controlling the peripheral devices which have phone lines attached, not for controlling the data which is transmitted or received on those lines. (There are minor exceptions: hang-up detection and digit capture may require analysis of the data stream for in-band signaling, these functions should be handled through TAPI).

Thus, TAPI functions might be used to establish a modem call, and later detect a hang-up by the called modem, but not to transfer a file using Zmodem protocol. In Windows jargon, the "message" is called the *media stream*. For example, in a fax phone call, the media stream would be the graphics image, in a data call, the media stream might be a transferred file, in a voice call, the media stream is the audio waveform describing the sound, which might perhaps be accessed as Wave data. Other Windows programming interfaces may be used to access the media stream, such as the Comm API or Media Control Interface (MCI). These other APIs are beyond the scope of this chapter, but are described by Windows programming books. Of particular relevance will be the MAPI (Messaging API) which will be included in Chicago, Microsoft's next Windows version, which is just entering beta testing at the time of writing.

The *media mode* is the TAPI term to describe the type of data being communicated in a call. While control of the data stream is not possible through TAPI, the TAPI application may be notified of the media mode due, for example, to detection of tones or voice-like audio detected on the line. TAPI defines the following media modes:

Unknown.
Self-explanatory: the mode of the call has not been classified.

Interactive voice.
A live person is at the application's end of the call.

Automated voice.
Similar to interactive voice, except that the voice is assumed to be from a mechanical source such as a voice mail system or answering machine.

Data modem.
Again self-explanatory. Modems present a special problem for automated mode detection since current protocols require the called party to initiate the handshake. A heuristic method (i.e. a guess) might be to assume a period of silence after answering the call indicates a modem at the other end.

G3 fax.
A Group 3 fax call.

TAPI

G4 fax.
A Group 4 fax call.

TDD.
A call using the TDD (Telephony Devices for the Deaf) protocol.

Digital data.
A digital data stream of unspecified format.

ADSI.
An *Analog Display Services Interface* session. ADSI telephones have a small alphanumeric display which can be updated over a regular voice telephone line and "soft buttons" analogous to function keys whose meaning can be defined by the application.

Teletex, Videotex, Telex or Mixed.
A session using an over-the-phone service of the given type.

TAPI is an ambitious and forward-looking standard: many of the applications envisioned in the design of TAPI cannot be implemented with current devices. The next generation of desktop telephony peripherals will be equipped with interfaces allowing more direct interaction with a PC.

The typical functions which can be performed through TAPI are:

Query the capabilities of a line or phone device.

Dial an out-bound call.

Receive notification of an incoming call.

Establish and control conference calls.

Detect changes in call status (such as a caller hang-up).

Collect digits from a caller.

Control a phone device (set volume, control status lights, for example).

Determine and utilize vendor-specific capabilities.

Coordinate call processing with other executing Windows applications.

Getting The TAPI SDK

The TAPI Software Development Kit (SDK) is available from Microsoft Corp. of Redmond, WA (800) 426-9400. Among other sources, it is available on the Microsoft Developer Network (MSDN) CD-ROM. The MSDN CD-ROMs are an invaluable resource for obtaining Microsoft's SDKs together with technical information, articles and the invaluable Knowledge Base which will help you implement your application.

Chapter 33

TAPI Programming

Introduction

The chapter delves further into the details of using Windows Telephony (TAPI) to program telecommunications applications. This is intended as a brief introduction only, for more details, the reader is referred to the Microsoft Windows Telephony Programmer's Guide, available from Microsoft. The reader of this chapter is assumed to be a

C programmer with at least some familiarity with Windows programming.

The main abstractions used in TAPI are:

Line devices.
An abstraction corresponding to a device which attaches a phone line (modem, fax card, telephone).

Phone devices.
An abstraction corresponding to a telephone-like device with a microphone and loudspeaker. Overlaps the definition of a line device.

Calls.
A call, as you might expect, is an active connection between two or more parties.

Line Devices

In TAPI, there are two kinds of peripherals: *line devices* and *phone devices*. There is some overlap between these two classes of device: some effects may be achieved through function calls for both classes.

A line device is a fax board, modem, ISDN card or voice processing card which is physically connected to one or more phone lines: it need not be installed in or directly connected to the computer running TAPI. A single device has one or more channels with identical capabilities. A telephone may be represented as a line device used for voice calls: it need not be opened and used as a phone device unless further capabilities are to be exploited.

TAPI allows the *service provider*, i.e. the author of the SPI device driver and probably also the vendor of the peripherals, to define several different models for abstracting particular ways that telephony devices may interact with the computer. For example, a line device may be a fax card installed in the PC, or it may be a PBX which interacts with the PC through an RS-232 serial connection.

TAPI Programming

The main requirement of TAPI with respect to a line device is that it support all of the capabilities of Basic Telephony (to be described).

Before an application can use a line device, it must first open the line (analogous to opening a file). The TAPI *lineOpen* function can be used to open a specific line (eg., get me line 4) or request a line from a particular group (eg., get me a free outside line). The function prototype is as follows:

```
LONG WINAPI lineOpen(
        HLINEAPP hLineApp,
        DWORD dwDeviceID,
        LPHLINE lphLine,
        DWORD dwAPIVersion,
        DWORD dwExtVersion,
        DWORD dwCallbackInstance,
        DWORD dwPrivileges,
        DWORD dwMediaModes,
        LPLINECALLPARAMS const lpCallParams);
```

Phone Devices

A phone device is a peripheral which includes some or all of the following elements:

Hook-Switch/Transducer.

By this is meant a means for audio input and output. TAPI distinguishes three types:

Handset.
The traditional telephone.

Speaker-phone.
A hands-free variant of the telephone with loudspeaker. The important characteristic is that it permits multiple listeners compared with the handset.

Headset.
Hands-free variant without loudspeaker.

Volume Control.

TAPI provides functions for controlling the received and transmitted volume: for example, for changing the volume or muting a speaker-phone.

Display.

An area on the phone device for showing messages to the user. The main characteristic of the display is the number of rows and columns.

Buttons.

The user of the phone device may have a number of buttons available, such as Conf, Redial and so on. These buttons will generate messages through the API just as pressing a key on the PC keyboard or pressing a mouse button.

Lamps.

There may be one or more lights on the phone device itself, such as message waiting indication, line status etc. These lights may be turned on or off, or "blinked" and "flashed" in various ways.

Data Area.

The phone device may be programmable. TAPI provides a means for downloading data (i.e., programming) the device.

Analogous to *lineOpen* is *phoneOpen*:

```
LONG WINAPI phoneOpen(
        HPHONEAPP hPhoneApp,
        DWORD dwDeviceID,
        LPHPHONE lphPhone,
        DWORD dwAPIVersion,
        DWORD dwExtVersion,
        DWORD dwCallbackInstance,
        DWORD dwPrivilege);
```

Calls

Line devices and phone devices are static: they are present or absent for the entire life of an application execution. Calls, on the other hand, are dynamic: they are created and destroyed as the application progresses.

TAPI Programming

Calls are identified by *call handles*, which are small integer numbers uniquely identifying each call in progress. These are analogous to file handles used to identify open files. Some TAPI functions create calls and therefore return call handles, other call handles may be communicated to an application through an unsolicited message (as when an incoming call is answered), or when a call is "handed off" by another application.

A call has several attributes as far as TAPI is concerned, some of the most important include the following.

Bearer Mode.
The *bearer mode* is the type of medium used: for example, voice or fax.

Media Rate.
The *rate* is the data speed: for example, 9600 or 2400 baud in the case of a fax call.

Call Origin.
Indicates whether the call originated from an internal caller, outside line or unknown source.

Caller ID.
The originating party. Might be the extension, Caller ID or ANI of an inbound call, or "unknown" if this information is unavailable.

Called ID.
The dialed telephone number.

Connected ID.
Identifies the address (essentially, the telephone number) of the party to which the call was completed. Not necessarily the same as the dialed number if the call was diverted.

The generic function used to establish an outgoing call in TAPI is *lineMakeCall*, which seizes a line, waits for dial tone and dials the number. The function prototype is as follows:

```
LONG WINAPI lineMakeCall(
        HLINE hLine,
        LPHCALL lphCall,
        LPCSTR lpszDestAddress,
        DWORD dwCountryCode,
        LPLINECALLPARAMS const lpCallParams);
```

The important parameters are:

hLine	Line handle (from *lineOpen*)
plhCall	Call handle (returned)
lpszDestAddress	Dial string
dwCountryCode	Country code (as integer)
lpCallParams	Pointer to call parameter structure filled in by the application.

Once the call is established, status messages with notifications such as caller hang-up may be relayed to the application. Like the Dialogic programming interface, TAPI is designed to operate on a state machine model. The function *lineMakeCall* returns immediately, a later message will indicate whether the call was successfully complete. A TAPI application, especially if it deals with more than one phone line, will therefore be constructed according to the state machine model.

Typical Call

A typical call made using TAPI will go through the following steps:

1. Determine the media mode.

The application, perhaps through a dialogue with the user, will determine the media mode of the call. The media mode might be voice, or data, for example.

2. Initialize TAPI.

The application establishes a means of communication with TAPI. This is done using the *lineInitialize* function.

TAPI Programming

3. Obtain a line.

This is done using the *lineOpen* function.

4. Place the call.

This is done using the *lineMakeCall* function.

5. Send data.

TAPI itself will not be involved in this step. In the case of a simple voice call, the user will be performing this function.

6. End call.

The call may be ended either because a message was received indicating that the caller hung up, or because the application requested that the call be terminated. TAPI offers several ways of terminating a call for different purposes.

TAPI Structure

Unfortunately for the applications developer, TAPI is an unusually complex API even by Windows' notorious standards. There is, however, some good news: there is a simple set of just four function calls knows as *Assisted Telephony* which is provided for applications for which telephony is a secondary feature. An example of an application using Assisted Telephony might be a database manager which allows the user to highlight a telephone number field and select a "Dial" option to initiate an outbound voice call. If Assisted Telephony (described in more detail in the next section) is all that you need, you're in luck, and you don't have to learn much more about TAPI.

The programming model is fairly typical for Windows in some respects: in the naming of devices and in function calling conventions. However, an important deviation is what Microsoft terms the synchronous/asynchronous operational model which notifies the application of the result of function calls and of unsolicited events.

The mechanism used by TAPI is similar to the asynchronous call-back scheme proposed for the SCSA API (see the SCSA chapter). When an event occurs which must be passed to the application, TAPI invokes a "call-back" function. The address of the call-back function is passed to

TAPI when the initialization function *startTAPI* is called. The call-back is only made when the application calls the usual Windows *GetMessage* function. Since *GetMessage* suspends execution of the application until a message or TAPI call-back is pending, this method is more efficient than the polling mechanism used by the Dialogic MS-DOS driver. However, the programming techniques required are very similar. (Programmers should not interpret this remark as implying that Windows 3.1 might provide a good platform for developing a voice processing server. On the contrary, the cooperative multi-tasking scheme of Windows 3.1 renders it unsuitable for building voice processing servers, in this author's opinion. Future versions of Windows with pre-emptive multi-tasking may provide a more suitable operating system).

When the call-back function is invoked, it is passed a *Request ID* and an error/success code. When a TAPI function is invoked, it may complete immediately and return zero (like a Dialogic "non-multi-tasking function") or return a Request ID indicating to the application that a later call-back will be made with the result of the call (like a Dialogic "multi-tasking" function which will trigger a later terminating event). A given function will always operate in one of these two modes: synchronous (immediate return) or asynchronous (later call-back).

Assisted Telephony

Microsoft has provided four simple TAPI functions for the telecom-impaired programmer:

tapiRequestMakeCall	Requests a voice call between the user and another party.
tapiRequestMediaCall	Requests a call with a specific media mode.
tapiRequestDrop	Disconnects a media mode call.
tapiGetLocationInfo	Returns the country and city code of the user, as set in the Telephony Control Panel.

Status messages about the call (generated, for example, when the called party hangs up) are sent to the application as a standard Windows

TAPI Programming

message with type *TAPI_REPLY* rather than through the TAPI callback mechanism.

To make a voice call:

```
LONG WINAPI tapiRequestMakeCall(
      LPCSTR lpszDestAddress,
      LPCSTR lpszAppName,
      LPCSTR lpszCalledParty,
      LPCSTR lpszComment);
```

All the arguments are character strings, and all are optional (may be specified as NULL) except the destination address (dial string). No messages are passed back to the application for notification of the call status. The request is passed to a call-control application such as the Dialer applet provided in the TAPI SDK.

To make a media mode call:

```
LONG WINAPI tapiRequestMediaCall(
      HWND hWnd,
      WPARAM wRequestID,
      LPCSTR lpszDeviceClass,
      LPCSTR lpDeviceID,
      DWORD dwSize,
      DWORD dwSecure,
      LPCSTR lpszDestAddress,
      LPCSTR lpszAppName,
      LPCSTR lpszComment);
```

A handle to a Window and Request ID is needed so that *TAPI_REPLY* messages regarding the state of the call can be sent to a private window defined by the application. The destination address is again the dial string. A "secure" call is one which suppresses features like call waiting, which can disrupt a modem or fax call.

The DeviceClass string identifies the API which will process the media type for the call, and is really an API name. Names are not case sensitive. Some standard names are:

"comm"	Generic serial-device API for a COM port.
"comm/datamodem"	Reserved for use in future Windows versions.
"wave"	The Wave audio API. Treats audio as a single wave-form.
"mci/midi"	MIDI sequencer. MIDI is a higher-level description of audio than Wave.
"mci/wave"	High-level Wave device control.
"tapi/line"	TAPI line device.
"tapi/phone"	TAPI phone device.
"ndis"	Network driver interface services.

A service provider (vendor of a TAPI SPI and probably of the peripherals which are supported by that driver) may define new media modes, and will in that case need to define a new API to manage devices supporting that medium. To avoid collisions between new APIs, Microsoft recommends a standard naming using "vendor/medium", such as "dialogic/video".

To disconnect (hang-up) a media mode call:

```
LONG WINAPI tapiRequestDrop(
      HWND hWnd,
      WPARAM wRequestID);
```

This will generate a *TAPI_REPLY* message. The Request ID is that which was returned by the *TAPI_REPLY* message following the invocation of *tapiRequestMediaCall* which established the call.

To get the user's country and city code:

```
LONG WINAPI tapiGetLocationInfo(
      LPCSTR lpszCountryCode,
      LPCSTR lpszCityCode);
```

The country code and city codes may be useful to the application in forming dial strings. These country and city are retrieved from the TELEPHON.INI text file. The codes are generally set by the user through the Telephony Control Panel.

TAPI Programming

Assisted Telephony is implemented by passing requests through to a call manager application, which may be based on the Dialer applet in the TAPI SDK. The architecture might be visualized as follows:

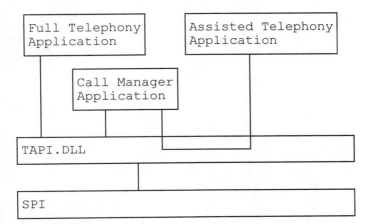

Assisted Telephony architecture. Requests (generated by Assisted Telephony function calls) are passed through to the call manager application.

The call manager registers with TAPI that it is ready to receive Assisted Telephony requests.

Chapter 34

TAPI, Wave And The Dialogic API

Introduction

Developers are faced with a bewildering choice of different platforms and APIs to choose from. In this chapter, we will compare some important areas of the Dialogic proprietary API for Windows with Microsoft's TAPI API. By the "Dialogic API", we mean the Dialogic System Release 4.x driver series for Windows.

TAPI drivers are technically known as TAPI Service Providers, abbreviated TSPs. When we refer to Dialogic's TAPI driver, we will explicitly say "Dialogic TSP". "Dialogic API" will always refer to their proprietary drivers in the Dialogic System Release 4.x series.

TAPI addresses many different types of hardware configuration. Currently available TAPI devices fall into two main categories: voice cards like Dialogic, and C-T links to PBXs and ACDs. Voice cards, of course, have a long history. As the dominant vendor of voice cards, Dialogic has long set the de-facto standard for PC call processing APIs. Computer-Telephony Integration (C-T or CTI) links are a newer phenomenon. The basic idea is that a PC (the C-T server, which might or might not be the same PC as the disk server) is hooked up to a switch via a serial line, Ethernet or other type of connection called a C-T link. The link allows the PC to receive messages from the switch (incoming call, caller hung up...) and send commands to the switch (answer call, put on hold, transfer to a given extension...). The typical use for C-T APIs is in call centers. A client application on an agent's desk could use TAPI to implement screen pops and manage ACD queues. The same call center might include a voice processing PC which is connected to the same switch as a group of analog extensions. This PC might contain Dialogic cards used to provide auto-attendant and voice mail features, and could be programmed using TAPI or the native Dialogic API. The voice processing PC can be programmed quite independently of the C-T link to the switch, which might be programmed using TAPI or an alternative C-T API such as Novell's TSAPI. So, a call center might use TAPI both to control an ACD and also to control a voice response unit, but there would not necessarily be any connection or coordination between these two different uses of TAPI.

Computer telephony developers are looking towards what I call "HAPI Days", the advent of Hardware-independent APIs. TAPI provides some promise, other initiatives such as the ECTF's S.100 specification are also under way. In the not too distant future, it may be possible to write applications which will run unmodified on many different vendors' products, helping perhaps to fuel an even greater growth of the computer telephony market. It is still too early to say if TAPI will be the HAPI of choice, if for no other reason than there are still only a few voice boards and C-T servers which support TAPI. Dialogic, for example, supports only a few entry-level boards with its TSP, high-end boards for T-1, E-1 and ISDN are not supported.

Writing A TAPI Application

Windows programmers will find the basic concepts of TAPI quite easy to understand since the programming model is similar to other Windows APIs. For example, to create a window, a C or C++ program calls *RegisterWindowClass* and *CreateWindow*, which (in addition to a lot of other information) gives Windows the address of a "call-back" function known as the window procedure. Windows calls the call-back function whenever there is a message for the window. A typical message is *WM_PAINT*, which asks the window procedure to draw the text and graphics to be displayed in the client area of the window. TAPI operates in a similar fashion. The *lineInitialize* function gives TAPI the address of a call-back function which is called by TAPI whenever there is a telephony-related message. Just as a window handle in a message identifies the particular window which is to receive the message, TAPI has device handles, which identify telephony hardware devices, and call handles, which identify telephone calls in progress. Device handles are obtained by calls to the TAPI functions *lineOpen* and *phoneOpen*, call handles are generated internally by TAPI and sent to the application via messages. Typical TAPI messages are *LINE_DEVSTATE*, which report events from a device such as when a line is ringing, and *LINE_CALLSTATE*, which informs the application when the state of a call changes, for example when a caller hang-up is detected.

Functions which take a relatively long time to complete, such as *lineGenerateDigits* which dials digits on a line, return an indication of whether the operation was successfully started, and later generate a *LINE_REPLY* message to report that the operation completed. TAPI functions never "block". In other words, there is no equivalent in TAPI of the Dialogic "synchronous" mode where time-consuming functions such as playing, recording, dialing etc. are made with a single function call which puts the application thread "to sleep" and only returns when the operation is done. Synchronous code is often be easier to write but requires separate threads or separate EXEs for each line, introducing significant overhead for some types of application. It is a limitation of most TSPs we tested that two functions which result in *LINE_REPLY* messages cannot be in progress at the same time. (The application should be able to use the *lineGetCallStatus* function to check the *LINECALLSTATUS.dwCallFeatures* bit-map to check if a second function call may be made while a function is already in progress, providing that the TSP vendor has correctly implemented this feature). We found in Hawaii that some TSPs would simply if a second function was invoked, this is a bug: a TSP should fail gracefully with an error indicating that the device is busy.

TAPI devices are divided into two types: "line" devices and "phone" devices. Line devices are channels on modems or voice cards, or PBX/ACD stations. Phone devices are intended for telephone- or headset-type hardware: they have microphones and speakers of some kind, and they may have feature lights and buttons and LED. Here, we will concentrate on line devices, for two reasons. There are few or no TAPI phone devices available (we saw none at the last TAPI Bakeoff in Hawaii, in fact we have never seen such a device: vendors please send me your literature!). Also, line devices are probably of most interest to readers since TAPI drivers for voice cards such as Dialogic are implemented as this type.

Line devices are given device identifiers 0, 1 ... up to the number of installed TAPI devices, which is returned by *lineInitialize*. A device may be opened in "monitor" or "owner" mode, which affects what the application is allowed to do with the call. A call may have several owners. A line device must be opened in order for the application to receive messages relating to the device, or to invoke API functions for the device, such as *lineMakeCall* to dial an out-bound call.

Answering In-Bound Calls

When an in-bound call is detected by a TAPI line device, a *LINE_CALLSTATE* message is sent to all applications which opened that device. As with most TAPI messages, there is a message sub-type, a numerical value in one of the parameters of the message, which in this case will be *LINE_CALLSTATEOFFERING*. A new call handle will be created by TAPI and this handle is also passed to the application as a message parameter. The application can choose to answer the call with *lineAnswer*, or reject the call with *lineDrop*.

Here TAPI gives the developer a significant advantage over the Dialogic API. Using TAPI, the programmer doesn't have to know anything about the type of trunk and protocol required to answer the call. Using Dialogic, the API calls needed are quite different depending to the type of board and trunk used. For example, on a classic D/41D board, the application waits for a *DE_RINGS* event and answers the call with *dx_sethook* to take the channel off-hook. On a D/240SC-T1 board, the application must wait for a *DTG_SIGEVT* event and answer the call using *dt_settssig*. And so on.

On the other hand, TAPI can sometimes fail to meet the needs of an application for the very same reason that it makes life easier. Sometimes voice boards answer incoming calls in unconventional ways. For example, an interactive information system which provides a wall phone at a bank or museum may be connected directly to a computer via a small box which provides loop currently only. When the user picks up the phone, loop current starts to flow and a Dialogic board can respond to this event by starting the application and playing a message. Using TAPI, there is no way to do this unless the TSP provides a setup option to start an incoming call on loop current (no TSP that we have seen does this, including Dialogic's).

Using TAPI, getting ANI/Caller ID and DNIS information is as simple as calling *lineGetCallInfo*. Again, with a low-level API like Dialogic's, the protocol for retrieving this information is left up to the application and varies according to the board type and trunk type used.

Monitoring For Hang-Up

The situation with a caller hang-up is similar to that for in-bound calls: TAPI usually sends a *LINE_CALLSTATE* message with sub-type *LINECALLSTATE_DISCONNECTED* to indicate that the caller hung up. To disconnect the call, the application uses *lineDrop*. Using the Dialogic API, the application must know the type of board and trunk and must make quite different API calls based on how a hang-up is signaled (by a drop in loop current, by a change in signaling bits on T-1, by a D-channel message on ISDN, by a PBX hang-up tone, etc.). Unfortunately, we were unable to determine a method of fully releasing a call which would work reliably on all TSPs. The problem was that different TSPs would send different messages and require different API calls in different circumstances. The TAPI specification is vague when it comes to states and calls which must be supported for situations such as a call tear-down. The only solution we could find was to provide specific patches based on the behavior of specific TSPs, not the ideal situation for an API which is designed to be vendor-independent.

Detecting a hang-up when a voice card is hooked up to a PBX or ACD is a also practical problem with current implementations of TAPI. A common configuration is to have a voice board, such as a Dialogic D/41D, hooked up to a PBX extension in order to provide auto-attendant or voice mail features. Most PBXs signal a hang-up via a proprietary tone, which varies from vendor to vendor and even from switch model to switch model. TAPI does not define any "hooks" for the application or TSP to customize the PBX hang-up tone specification. According to the TAPI specification, TSP vendors can (should?) provide a setup dialog where the user can enter information about hang-up tones. In practice, none of the TSPs we have seen at Parity Software offer this feature. This means, for example, that there is no way for an application using TAPI and Dialogic's TSP (as shipping at the time of writing in June 96) to detect the hang-up tone of a PBX, rendering this combination effectively useless for most systems which require PBX integration. (Strangely, the Dialogic TSP Advanced Setup dialog offers several options for PBX integration, but nothing to specify a hang-up tone).

Making Out-Bound Calls And Transfers

The *lineMakeCall* function is used to make a call. (Programmers note: the documentation does not make this clear, but TAPI saves the address you pass for the call handle and writes the value of the handle to this address later, when the LINE_REPLY message indicating that the dial has been completed is sent; this means that you can't pass the address of a local variable, and the value of the handle is not set at the point when *lineMakeCall* returns. Don't make the mistake we did when we started out with TAPI and pass the address of a automatic variable to *lineMakeCall* and then copy this result into a global variable: TAPI will later overwrite your stack, with unpredictable and probably disastrous results). The arguments to *lineMakeCall* are the phone number to dial, the handle to a line device and other parameters.

TAPI, Wave And The Dialogic API 529

LINE_CALLSTATE messages are sent to indicate the progress of the call, typical messages would indicate states of dialing, proceeding, ringback, and connected. The number and types of the messages will vary according to the TSP and type of call being made, applications should not be designed depend on a particular sequence. If the call is successfully completed, a *LINE_CALLSTATE* message of sub-type *LINECALLSTATE_CONNECTED* will be generated. The TAPI specification is not very clear about calls which are not completed. Transferring is accomplished by *lineBlindTransfer*, which does not attempt call progress, or by *lineSetupTransfer* and *lineMakeCall*, which does attempt call progress.

As with detecting and answering an in-bound call, TAPI is nicely designed to hide the lower-level details of the protocol required to dial and monitor call progress. However, as with hang-up detection, there is a practical problem: again, TAPI provides no "hooks" for the application or TSP to specify or customize the environment. For example, a voice board hooked up to a PBX may need to do internal transfers to another PBX extension, and to dial local, long-distance and international calls. Each of these types of call require different call progress parameters (what types of tones to monitor for busy or ringback, for example), and in some cases, such as international calls, may require many different sets of parameters depending on the number dialed (call progress tones from England are quite different from Italy's). Each type of PBX has its own set of proprietary call progress tones, and these are rarely documented by the PBX vendor. In theory, and according to the spirit of the TAPI specification, the TSP could analyze the digits of the phone number to dial to see what type of call is being made. For example, a 7-digit number might be a local call, a 10-digit number a long-distance call, a number starting with 011 would be international, a table lookup would then be done for the country code. The TSP would then adapt its internal call progress tables according to the target environment being dialed. This would require the TSP vendor to provide a sophisticated database of PBX types and of public phone network call progress signals for all countries in the world, and a setup dialog allowing the user to define the details of the current installation (dialing prefixes used to reach different long-distance carriers, for example). In practice, nobody in the computer telephony business has a complete database (to the best of my knowledge), and it would be a major investment to develop and maintain.

Dialogic's TSP supports only a limited amount of customization for call progress tones, which can be set in a configuration dialog box which is accessed through Control Panel. There is no ability to change tone definitions for call progress to adapt to different environments while the application is running (e.g., for dialing both within the US and to England).

This is where the hardware-independent design of TAPI breaks down in the face of computer telephony reality. Lower-level APIs like Dialogic's "solve" the problem by leaving the details up to the applications programmer: there are well over 100 parameters describing call progress details which the Dialogic programmer can set and change as the application runs. The big problem with this approach is that the investment in determining good values for the parameters is left up to the developer, and it is rarely easy to find or measure all of the tones and signals involved.

Microsoft's position on this issue is to point to TAPI's WOSA (Windows Open Services Architecture) design philosophy, which is intended to hide hardware implementation details from the applications programmer, making it the responsibility of the TSP vendor to provide for call progress and PBX disconnect customization through setup dialogs or intelligent adaptation to the telephony trunk environment. In the opinion of this author, Microsoft will need to provide more leadership in order for good solutions to these problems to evolve under TAPI because TSP vendors are unlikely to have the expertise or motivation to provide comprehensive, hardware-independent solutions to these issues without Microsoft's help.

If TAPI is to be a credible alternative to the low-level voice board APIs, the call progress parameter issue must be solved, either by improvements to the TAPI specification or by improvements in TSP implementation.

Getting Digits

TAPI offers two methods for getting digits from the caller for menus and data-entry. The *lineMonitorDigits* function asks TAPI to send a message to the application each time a digit is detected. The application can ask for DTMF or pulse detection, although most TSPs will not support pulse recognition over analog lines (this is a difficult recognition problem somewhat like voice recognition). Alternatively, *lineGatherDigits* requests the TSP to wait for one or more digits from the caller. This does not add much that could not be accomplished by processing messages enabled by *lineMonitorDigits*, except for finer control of time-out and the convenience of allowing TAPI to worry about some of the logic (for example, terminating a 16-digit credit card entry early if the # key is dialed. We have found that several vendors have so far chosen to implement *lineMonitorDigits* or *lineGatherDigits*, but not both, in their TSPs. An application should be designed to test a TSP extensively to determine which functions are and are not supported, it cannot rely on the complete TAPI function set being available.

Playing And Recording Messages

TAPI itself includes no "media transport" functions. It is left up to other Windows APIs to deal with voice or other data transmitted in the call.

Neither the older 16-bit Windows 3.x API nor the new 32-bit APIs for Windows NT and Windows 95 provide a standard set of functions for fax transmission.

Playing and recording sound is done via the Wave API. The TAPI *lineGetID* function finds the Wave devices associated with a given line device, the device class parameter *wave/in* requests the Wave device id of an input device (recorder), *wave/out* for an output device (player).

Many TAPI TSPs, especially those for CTI integration with PBXs and ACDs, and also those for basic data modems, will not support playing and recording, in other words there will be no Wave driver provided to work with their TSPs.

Unfortunately, the Wave API was designed with sound cards in mind, not telephony devices, and no new extensions have been added to help with TAPI/Wave applications. Even basic operations such as playing and recording Wave files can demand a considerable programming effort. You might expect to get a few simple functions, say *wavePlayFile* and *waveRecordFile*, but these simply don't exist. A near miss is *sndPlaySound*, which can play a complete Wave file, but this function doesn't accept a Wave device as a parameter: Windows chooses the first device it can find which is capable of playing a file of that type, clearly unacceptable for a multiple-line telephony system. To play a Wave file, an application must set up a multiple buffering scheme and ensure that buffers are kept full by doing "read-ahead" in the Wave file, reading one buffer at a time. A window will be needed to receive the multi-media messages required by the play process (a call-back function may also be used, but there are many restrictions on API calls which may be made when a multi-media call-back is invoked, so call-backs are not usually convenient). The Wave device is opened by *waveOutOpen*, a Wave file is opened by *mmioOpen*. The format information is found by *mmioDescend* looking for a "fmt" RIFF chunk, and read by *mmioRead*. The Wave "data" chunk in the file is located by *mmioDescend*. Buffers are read from the file using *mmioRead*, processed for mysterious reasons by *waveOutPrepareHeader* and finally sent to the driver for playing by *waveOutWrite*. The application must react to MM_ (multi-media) messages by freeing used buffers and feeding new buffers to the driver until the play is completed. If this sounds like a lot of work just to play a file, it is. Using the Dialogic API, a single function *dx_play* takes care of all of the details. For once, the roles are reversed: the Dialogic API looks higher-level and more convenient. Recording a Wave file follows a similar sequence of steps to playing.

Not only is the Wave API difficult to use, it is hard to find well-written code illustrating how to implement a multiple buffering scheme. At Parity Software, it took us several days of painful trial and error to figure it all out. We couldn't find any samples showing how to do this with the Microsoft Windows SDK or Microsoft Developer Network CDs. We did find several unhelpful examples. A typical example of poorly implemented Wave code is Dialogic's TAPI/Wave sample program called TALKER32 which ships with their TSP. (Not to knock Dialogic in particular, this just happens to be code that many computer telephony developers are likely to see). TALKER32 plays a Wave file by finding the size of the Wave data, allocating a single memory buffer

TAPI, Wave And The Dialogic API 533

big enough to hold the entire file, and then writing this one buffer to the driver, thereby avoiding all of the code needed to implement a multiple-buffer scheme. This is clearly not a suitable approach for multi-line systems, especially if large files may be played: quite apart from the memory overhead required, reading the complete file into memory before play starts is likely to introduce an unacceptable delay before the caller hears the beginning of the message.

There are many different ways to digitize and store audio data, the proliferation of sound file types can present new challenges to the TAPI developer who needs to include interactive voice menus to the caller. The most important audio data types are:

- 8-bit, linear PCM. The data stored in each 8-bit sample represents directly the amplitude (loudness, volume) of the sound at the time the sample was measured. Standard Wave files use this type of encoding.

- 8-bit companded PCM. The telephone network uses companded PCM to transmit data internally between switches. Companded PCM uses a non-linear scale to convert the sample value to amplitude, allowing more precise rendition of sound at lower volumes where the human ear is more sensitive, sacrificing accuracy at higher volumes. Two scales are in common use: the mu-law scale in North America, and the A-law scale in much of the rest of the world. Most Dialogic boards support either mu-law or A-law PCM files sampled at 6 kHz or 8 kHz.

- ADPCM. This is a compressed form of PCM which comes in many flavors. ADPCM files generally compress the data by storing fewer bits per sample, often 4 instead of 8 bits. One vendor's ADPCM is generally incompatible with another's. Dialogic's default sound file format is a type of ADPCM originated by Oki which has 4-bit samples at a rate of 6 kHz.

- VOX. This is a generic name for the 6 sound file formats traditionally supported by Dialogic. There are two 4-bit ADPCM types, recorded at 6 and 8 kHz, and four 8-bit companded PCM types: 6 and 8 kHz, mu-law and A-law PCM.

Standard Wave files are recorded at 11, 22 or 44 kHz with 8- or 16-bit linear PCM data samples with one (mono) or two (stereo) channels. This makes a total of 12 standard Wave file formats. The most high compressed (and therefore lowest quality) is 11 kHz, 8-bit mono, with a data rate of 88 kbps (thousand bits per second). By comparison, Dialogic VOX files are recorded at 6 kHz or 8 kHz with 4- or 8-bit samples. The least compressed, highest quality Dialogic format is therefore 8 kHz, 8-bit VOX, with a data rate of 64 kbps. The best quality/highest data rate VOX format therefore has a lower data rate than the lowest quality/slowest data rate standard Wave format. The rate of sound data is very important for multi-line telephony applications. If there are many lines in a single PC, the limiting factor determining the maximum number of lines is often the throughput required from the hard disk when playing or recording on many lines simultaneously. The Dialogic default format, 6 kHz 4-bit ADPCM, with a rate of only 24 kbps, gives sound quality adequate for many applications and can support approximately a factor of about $88/24 = 3.7$ times more lines than a device using 11 kHz, 8-bit Wave files.

Dialogic has made a long-overdue improvement to the VOX file format (which has no header information to specify the type of data encoding used) and added a Wave header so that all VOX formats are now supported as files which follow the Wave file layout specification. However, this does *not* mean that most sound cards and Wave utilities can play, record or edit Dialogic Wave files since the encoded audio data inside the file is still in VOX format. Products such as VOX Studio (available from Parity Software) can be used to convert to or from standard Wave file types. Real-time conversion is also possible (Parity's VoiceBocx ActiveX is able to play VOX files in real time on Wave boards), but the quality of the result can be quite variable, depending on the Wave hardware since the software conversion is not able to tweak the output as much as an off-line tool. The Dialogic TSP does support 11 kHz 8-bit standard Wave, 6 kHz and 8 kHz, 8-bit versions of standard Wave, and all four VOX file types with Wave headers. Many, but not all, Wave-compatible sound cards are able to play and record the 6 kHz and 8 kHz, 8-bit linear PCM Wave. Note that the 6 kHz and 8 kHz, 8-bit mu-law and A-law PCM Wave files also supported by Dialogic, using VOX format data, will not be supported by standard desktop Wave cards. This may all be rather confusing, the Voice File Support table summarizes the details.

Voice File Support

File Type	Dialogic API Support[1]	Dialogic TSP Support	Desktop Sound Card Support	Data Rate
4-bit, 6 kHz ADPCM VOX	Yes	No	No	24 kbps
4-bit, 8 kHz ADPCM VOX	Yes	No	No	32 kbps
8-bit, 6 kHz mu-law PCM VOX	Yes	No	No	48 kbps
8-bit, 6 kHz A-law PCM VOX	Yes	No	No	48 kbps
8-bit, 8 kHz mu-law PCM VOX	Yes	No	No	64 kbps
8-bit, 8 kHz A-law PCM VOX	Yes	No	No	64 kbps
4-bit, 6 kHz Dialogic ADPCM Wave	Yes	Yes	No	24 kbps
4-bit, 8 kHz Dialogic ADPCM Wave	Yes	Yes	No	32 kbps
8-bit, 6 kHz mu-law PCM Wave	Yes	Yes	No	48 kbps
8-bit, 6 kHz A-law PCM Wave	Yes	No	No	48 kbps
8-bit, 8 kHz mu-law PCM Wave	Yes	Yes	No	64 kbps
8-bit, 8 kHz A-law PCM Wave	Yes	No	No	64 kbps

File Type	Dialogic API Support[1]	Dialogic TSP Support	Desktop Sound Card Support	Data Rate
8-bit, 6 kHz linear PCM Wave	Yes	Yes	Some, not all	48 kbps
8-bit, 8 kHz linear PCM Wave	Yes	Yes	Some, not all	64 kbps
8-bit, 11 kHz linear PCM Wave	Yes	Yes	Yes	88 kbps
8-bit, 22 kHz linear PCM Wave	No	No	Yes	176 kbps
8-bit, 44 kHz linear PCM Wave	No	No	Yes	352 kbps

[1] *Not all formats marked "Yes" supported on all boards.*

So when you hear "Dialogic supports Wave", remember that things are not as simple as they might sound. Out of the 11 Wave formats in the table, there is only one (11 kHz linear) that is supported by the Dialogic API, the Dialogic TSP and all Windows sound cards.

Applications for TAPI/Wave have a special problem when it comes to this proliferation of sound file types. The Wave API does not allow an application to ask for a list of encoding formats which a device supports. The *waveInGetDevCaps* and *waveOutGetDevCaps* functions return only a bitmap which specifies which of the 12 standard Wave formats are supported by the device. It is possible to ask if a given format is supported, by using *waveIn/OutOpen* with the *Wave_FORMAT_QUERY* parameter, but this means that the application must know about proprietary formats, such as the Dialogic Wave files with VOX encoding, in advance, which makes writing hardware-independent TAPI/Wave applications that much harder.

Speaking Phrases

Phrases involving variables, such as "your balance is one thousand and five dollars and six cents" are constructed by playing pre-recorded pieces in the appropriate order. This example phrase might be spoken by playing the following segments: "your balance is", "one thousand", "and", "five", "dollars" "and", "six", "cents". A typical system might have one or two hundred pre-recorded segments stored in sound files ready to play this kind of phrase. An application using Wave might choose to store each segment in a separate Wave file, but this would have several disadvantages. There would be significant overhead involved in continually opening, seeking and closing this large number of files. This solution is also more difficult to administer for the application designer and developer since the number of files to be installed and distributed becomes much greater.

The Dialogic API allows an application to build a phrase by providing an array of file handles and file positions (which might or might not all be in the same file), a single function call to *dx_play* can play the whole phrase in a seamless fashion. The Dialogic driver takes care of the required buffering and read-ahead to accomplish this. So-called "Indexed Prompt Files" or "VBASE40" files for Dialogic data formats store many segments inside a single file for optimal generation of phrases. Neither Microsoft nor Dialogic appears to have recognized the need to standardize a new Wave-derived file format to define a Windows standard for this type of file. We believe that a new Windows multi-media file format should be defined for this purpose so that Wave data, a title ("Main Menu"), the full script for each segment, and other information (language, dialect, male/female voice, name of person recording...) can be stored in a standardized fashion. This means that applications from different vendors and third-party utilities such as VOX Studio and Voice Information Systems' VFEdit would all support the same file format, with benefits to developers and end-users alike. Current TAPI/Wave tools (including Parity's) use a crude derivative of the VBASE40 format, which is in our opinion inadequate as a long-term standard.

Speed And Volume Control

TAPI does provide one special hook to the Wave API, the *lineSetMediaControl* function. The application specifies a list of digits which, if detected on the line, are to control the speed and volume of the sound data being played (if supported by the Wave driver). The following features may be assigned to digits: rewind, fast-forward, speed-up, slow-down, volume-up, volume-down, reset speed, reset volume, pause and resume. These could be done by trapping digit messages and using Wave API functions such as *waveOutSetPlaybackRate*, but having the TSP implement this directly allows better response times, there might be unacceptable latency in a busy system if the application were responsible for implementing these features. The Dialogic API provides for similar features. If supported by the TSP, the *lineSetMediaControl* function provides a (not very obvious) means to implement a common feature: touch-tone interruption and type-ahead through menus.

Generating And Detecting Custom Tones

To generate a tone, TAPI offers *lineGenerateTone*, which can accept a request to generate a tone with any number of frequency components, each with their own cadence pattern (length of on and off periods), and volume. Most TSPs are likely to support only tones with one or two frequencies. Here, TAPI goes in some respects one better than the Dialogic API, which only supports single- and dual frequency tones, and does not support cadence patterns when generating tones. This can result in variable silence periods (making the tone sound irregular) because the application has to "sleep" for a timed period, which is hard to do accurately under Windows. Neither TAPI nor the Dialogic API allows for generation of tones with a cadence pattern more complex than a single on and off period each of fixed length. For example, the ring-back tone in the United Kingdom is a "double-ring" (ring, short pause, ring, long pause), this must be simulated in application code, again with irregular-sounding results. A work-around is sometimes to record a sound file with the required tone, but this requires a Wave driver and also introduces more overhead because of the increased disk i/o.

The TAPI function *lineMonitorTones* requests that a TSP should send a message each time a custom tones is detected, in a similar way to *lineMonitorDigits*. Tones to be detected can have up to three frequencies and a minimum duration (the duration is a feature mainly to prevent false-positive detection). TAPI, unlike the Dialogic API, does not provide for a cadence pattern. Sometimes tones are distinguished only by cadence (for example, some types of busy and fast-busy), so this is a limitation of TAPI which needs to be corrected. Neither the Dialogic API nor TAPI allows the application to specify the "twist" (allowable ratio in volume of the two frequencies), which can be helpful in avoiding false-positive detection of dual tones ("talk-off" and "play-off").

Configuration Information

TAPI provides extensive configuration information through the *lineGetDevCaps* function, which provides a wealth of detail about the capabilities of a device. Providing complete and accurate configuration information has not been a strong point of the Dialogic API. As mentioned earlier, the Wave API, which will be as important as TAPI for many telephony applications, does not provide full information on the types of Wave data which a telephony card supports.

C-T and PBX Control

TAPI provides a number of functions which are targeted at C-T applications where a TAPI client needs to interact with a PBX or ACD across a network. These functions are typically used for call center systems. These functions allow calls to be transferred, put on hold, parked and so on. Microsoft has put some effort into improving these functions to compete with Novell's TSAPI, which is mainly targeted at call center networked solutions. These functions, *lineHold, linePark* and so on are not supported by voice board TSPs such as Dialogic's.

API Design

Microsoft, as an operating system vendor, clearly understands many of the issues facing application developers who want to write hardware-independent, forwards- and backwards-compatible applications. Version information can be queried, the application can negotiate with TAPI and the TSP to agree on a level of TAPI support and any vendor-specific extensions which may be known to the application. All data structures which are input or output by TAPI have size fields which allow new structure members to be added to future TAPI specifications without breaking existing applications. The higher level of abstraction provided by TAPI is often convenient for the application developer, but sometimes presents serious problems, as with PBX hang-up detection and call progress parameterization. The Wave API is unnecessarily complex for simple chores like playing and recording voice files, where the API should provide higher-level functions to take care of the details.

Conclusions: TAPI/Wave Pros And Cons

Pros:
- Powerful standards-maker, Microsoft, promoting API.
- Applications can relatively easily be designed to run on hardware from multiple vendors.
- High level of abstraction hides details of trunk types and protocols.
- Solid function set for phone-line devices and C-T links, single API covers several classes of computer-telephony configurations.

Cons:
- As yet, not many voice cards with TAPI drivers available.
- Only available for Windows: DOS, UNIX and OS/2 not supported.
- Call progress and hang-up tone customization issues not solved in practice.
- Specifications not fully established for how to implement certain operations, such as releasing a call, independent of the TSP vendor.
- High level of abstraction sometimes makes it impossible to implement specialized applications requiring device-level control.
- Facilities for sharing resources such as line interface devices, players and recorders between applications are limited compared with more advanced APIs such as ECTF S.100.
- Only programming model is asynchronous, no "blocking" model.
- Wave API makes simple functions (play or record file) difficult to implement.
- Wave API and file format lacks features for hardware-independent support of highly compressed voice files and for efficient and flexible building of phrases.
- Fax, voice recognition and other collateral APIs not yet available for telephony boards.

Chapter 35

SWV: A Proposed Sound File Format

Improving On Wave

As discussed in the last chapter, the Wave API and sound file format have several limitations.

In this chapter, we will propose enhancements to the Wave file format designed to address the requirements of telephony applications. We expect that this format could have uses in other areas also. If you have comments or suggestions for this proposal, please e-mail me c/o tech@paritysw.com. This specification is offered for use by any third party royalty-free. (Note however that this chapter, like the rest of this book, is protected by copyright law and may not be duplicated in any way without express written permission).

Our proposed format is based on the Microsoft Multimedia Standards Update. This document is provided as part of the Windows SDK. For example, on the MSDN CDs, this document may be found under Product Documentation | SDKs | Multimedia Standards Update. We prefer to base a file format on known standards rather than introducing new proprietary formats such as the "VBASE40" indexed prompt file.

The main objectives of the new format are as follows:

- Extend the Windows Wave sound file format to allow files with two or more Wave data chunks, called *segments*. The main uses envisaged for this format are:

    1. to prepare phrases by concatenating segments, as for example "your balance is" "twenty" "three" "dollars" "and" "seven" "cents". Applications are anticipated in computer telephony (TAPI/Wave) and possibly for some types of simple desktop speech synthesis, and

    2. to contain many short prompts as needed by an automated system: the main greeting, menu prompts, error messages, and so on; thereby limiting the number of files which are needed in a system.

- Utilize a Windows multimedia (RIFF) file format as close to Wave as possible.

- Accommodate proprietary audio encoding formats, especially from telephony card vendors.

SWV: A Proposed Sound File Format 543

- Avoid any mandatory data in the file which is not present in a Wave file. (It is then guaranteed that a utility can create a valid SWV file from a set of Wave files with the same Wave format).

- Provide backwards-compatibility with Wave so that existing Wave utilities, file viewers, etc. can at least play the first Wave segment in the file, display the audio data format, copyright information, etc. It is expected that SWV software will accept a standard Wave file as an SWV file with one segment.

- Allow an application to retrieve an index to all segments efficiently. For example, traversing all chunks in a RIFF file to build a table would not be acceptable.

- Standardize optional information in the file which will be of use to application development tools and other software. Allow for future extensions to this set of information. Among the information most likely to be useful:

Information	Per File or Per Segment	Illustrative Example
File title	File	Information System Prompts
Copyright	File	Copyright 1996 Parity Software Development Corporation
Creation date	File	1996-07-03
Language	File	English
Dialect	File	American
Speaker	File	Jane Doe
Gender	File	Female
Code page	File	Code page to use for ASCII text encoding.
Segment title	Segment	Main menu
Segment script	Segment	For this choice press 1, for another choice press 2.

Segments

The basic element of sound to be stored in the file is termed a *segment*. A segment contains audio data which can be played back in one uninterrupted sequence stored as a contiguous array of bytes.

Some audio data compression schemes result in samples of varying bit length within a segment and where the state of the decompression algorithm can be highly dependent on preceding data. It can therefore be computationally difficult or impractical to concatenate all segments in such a way that they can be played as a single play-back. This eliminates the possibility of a standard which uses a single Wave data chunk and an index into this data, the result would be that existing Wave software would attempt to play the data incorrectly. Segments must therefore be stored as separate data chunks, at least in some types of segmented file. We choose for simplicity to make all SWV files segmented.

The proposed standard imposes the following restrictions:

- All segments are encoded using the same format.

- There is no provision to specify more than one language, dialect or speaker. It is anticipated that there will be little need for concatenating segments with different languages, dialects or speakers.

File Extension

Since an objective is to allow backwards compatibility with existing Wave software, the recommended file extension should be .wav or .WAV (software should accept either). If the application developer wishes to reduce the emphasis on Wave compatibility, we recommend the alternative file extension of .swv or .SWV. A conforming file may have any file name permitted by the host operating system, the extensions here are suggested, not mandatory.

SWV: A Proposed Sound File Format

File Format

The Segmented Wave (SWV) file format is a superset of the Wave file format. Any Wave file is also an SWV file, the reverse is not necessarily true.

An SWV file is a RIFF file with Wave form.

One Wave fmt chunk and one Wave data chunk are mandatory.

Multiple Wave data chunks may be present.

One Wave data chunk is required for each segment which has a size greater than zero.

Segment Index Chunk

If there is more than one Wave data chunk, there must be one segment index chunk.

The segment index chunk has id 'swv'. The chunk is organized as follows:

Header
Segment Table
String Table
String Area

```
// One entry in Segment Table
typedef struct tag_SWV_SEGMENT
    {
    // String id of name
    DWORD dwStrName;
    // String id of script
    DWORD dwStrScript;
    // String id of info "name=value;name=value;.."
    DWORD dwStrInfo;
    // Offset of Wave data (0=start of file)
    DWORD dwDataOffset;
    // Bytes in Wave data (0=not present)
    DWORD dwDataSize;
    } SWV_SEGMENT;
```

```
// One entry in String Table
typedef struct tag_SWV_STRING
    {
    // Offset of start of string (0=first byte in
    // string area)
    DWORD dwOffset;
    // 0=no string, > 1 Bytes in string
    DWORD dwSize;
    } SWV_STRING;

    // SWV chunk header
typedef struct tag_SWV_HEADER
    {
    // Bytes in SWV_HEADER stucture
    DWORD dwHdrSize;
    // Bytes in SWV_SEGMENT structure
    DWORD dwSegSize;
    // Bytes in SWV_STRING structure
    DWORD dwStrSize;
    // Number of segments (>= 0)
    DWORD dwNrSegments;
    // Number of strings in string table (>= 1)
    DWORD dwNrStrings;
    // Code page or SWV_UNICODE
    DWORD dwCodePage;
    // Language id, 0=not specified
    DWORD dwLanguage;
    // Dialect id, 0=not specified
    DWORD dwDialect;
    // Gender: 0=not specified, 1=male, 2=female
    DWORD dwGender;
    // String id of speaker name
    DWORD dwStrSpeaker;
    } SWV_HEADER;
```

The *SWV_HEADER* structure is followed contiguously by the Segment Table, an array of *SWV_SEGMENT* structures with *SWV_HEADER.dwNrSegments* entries. The offset of the n'th *SWV_SEGMENT* entry from the start of the swv chunk is

```
SWV_HEADER.dwHdrSize + n*SWV_HEADER.dwSegSize
```

Contiguously following the segment array is the String Table, an array of *SWV_STRING* structures with *SWV_HEADER.dwNrStrings* entries. The offset of the n'th entry in the String Table from the start of the swv chunk is:

```
SWV_HEADER.dwHdrSize +
  SWV_HEADER.dwNrSegments*SWV_HEADER.dwSegSize +
  n*SWV_HEADER.dwStrSize
```

Following the String Table is the String Area, an array of bytes. String Table entries point into this area. The value in *SWV_STRING.dwOffset* is the offset from the start of the String Area, not the offset from the start of the chunk. The length of the string in bytes is *SWV_STRING.dwSize*; if this size is zero, this indicates that no string is stored and the *dwOffset* member should be ignored. Strings are referenced by a string id, for example *SWV_HEADER.dwStrSpeaker*. String id's are 0, 1 ... *SWV_HEADER.dwNrStrings*. The first entry in the string table must always be a string of length 0, i.e. with *dwSize = 0*. This convention means that string id 0 always refers to a not-present string and allows software to check for a not-present string without referencing the string table.

If *SWV_HEADER.dwCodePage* is set to *SWV_UNICODE*, all strings in the String Table are to be interpreted as *UNICODE*.

All structure members are stored as 32-bit integer values with Intel byte ordering (LSB is first byte).

Applications should always use the *dwHdrSize*, *dwSegSize* and *dwStrSize* members to determine the size of the structures. Future enhancements to the SWV format may mean that new structure members are added. Applications should be designed to accept files with more or fewer structure members than are in the structure definition used to compile the software.

The language and dialect id's are as defined in the Multimedia Standards Update, for example *dwLanguage=9*, *dwDialect=2* refers to UK English.

The *SWV_SEGMENT.dwStrInfo* string, if present, is a list of *attribute=value* pairs, separated by semi-colons (use \; to indicate a semi-colon in a value). Attribute names are case-sensitive. Upper-case names will be standardized, lower-case names will be application-specific. This member is provided to allow for persistent per-segment data required by particular applications which are not anticipated elsewhere by this standard. (Per-file data can be accommodated by application-specific fields in the *LIST INFO* chunk or elsewhere.) Proposed names include the following:

Attribute Name	Valid Values	Meaning
APPROVED	yes, no	Segment has been approved for use?

The *APPROVED* attribute anticipates a common evolutionary process for an SWV file. First the script will be created, but no segments will be recorded. Then segments will be recorded, and finally approved for publication.

LIST INFO Chunk

A *LIST* chunk of type *INFO* is optional but recommended. The following chunks (described in the Multi-Media SDK) are suggested for SWV files. All are stored in ZSTR format (which will not respect *SWV_HEADER.dwCodePage* since this information will be unknown to software which is not SWV-aware). This information could be folded into the swv chunk, but since the *LIST INFO* chunk id's are standardized, some software which is not SWV-aware may be able to display these fields.

Chunk ID	Description
ICOP	Copyright message.
ICRD	Creation date
ISFT	Name of software used to create file.

Sample Files

Since this is a first draft, we have not prepared examples of binary files. After collecting initial feedback and making any required modifications to the proposal, we will create some example files to assist developers create and test tools and applications using this file format. If you are developing an application based on SWV, please send us some example files to tech@paritysw.com.

Future Extension: Phrase Algorithms, RFCs

Two main uses are envisaged for SWV files:

- Consolidating voice prompts and messages into a small number of files, simplifying application distribution and maintenance.

- Creating phrases by concatenation of voice prompts.

Creating phrases is a complex process which requires significant effort on the part of the application developer.

It is possible to store the complete algorithm for a phrase within the SWV file itself, so that the file "knows how to speak itself". While creating a flexible standard would be quite a significant effort, we suggest that this would benefit all vendors (hardware, software tools, systems integrators) and might be worth considering: please send your comments.

De-coupling the phrase specification completely from the application would have a number of advantages:

- Specialized Wave and SWV editors could be extended to include phrase algorithm builders and debuggers, a process which today must implemented by each application development tool or application software in a proprietary fashion.

- The phrase quality could be improved in non-trivial ways simply by exchanging the SWV file(s) used by the system. For example, if a system speaks 7-digit phone numbers to the caller by concatenating the digits "zero" .. "nine" from a set of ten segments, this can be improved by selecting from an inflected matrix of 7 x 10 recordings where each digit is selected based on its position in the phrase, special cases like speaking 555 4000 as "triple 5 4 thousand" can be built into the file rather than the application.

- The application can be localized (translated into a different language) simply by replacing the SWV file(s). Today, due to the different ways of speaking numbers in different languages, the application software must be changed in non-trivial ways to move from English to French, say.

- Phrase files can become separate products, making much higher quality phrases available to a much wider range of applications and much lower expense. Recording studios can compete to offer better phrase files.

- Interpretation of phrase algorithms can potentially be moved from application software into firmware, becoming part of the voice board API (TAPI/Wave, ECTF S.100, etc.), increasing efficiency and simplifying application and application tool development. This might allow firmware to create phrases using more sophisticated techniques than simple concatenation of segments, this would allow firmware and voice board vendors to compete on the basis of phrase generation technology in addition to other features.

- Ideally, a high-level API for specifying a phrase should be added to the media API, such as Wave in the case of Windows. The phrase would be specified in a high-level form e.g., *YOU_HAVE(23)*, not including the algorithm to speak it. The algorithms would be determined by the enhanced Wave or other driver, perhaps on the basis of algorithms read from SWV files, or perhaps passed through to the voice board firmware. These ambitious features are more a long-term vision than a proposal at this stage, it will be sufficiently challenging merely to embed a notation for phrase algorithms into an SWV file.

Some issues which must be addressed:

- The algorithm to build a number from a string of digits may be quite complex. A fairly sophisticated language of some kind (a simple p-code?) will be needed to specify the algorithm and it must be reasonably fast to interpret the language.

- Algorithms will need to reference "sub-phrases" from the same or another SWV file. For example, an SWV file containing the phrase "You have <number> new messages" will usually need to reference a separate SWV file to play <number>.

Please send comments and suggestions on the SWV proposal to me c/o tech@paritysw.com.

Chapter 36

The User Interface

Introduction

It has become fashionable to dislike "voice mail". Newspaper columnists bemoan their recent imprisonment in "voice mail jail", comedians and comic strips poke fun. Many VRU systems are frustrating to use, for a variety of reasons.

The voice processing industry is gradually learning how to make these systems more pleasant, efficient and responsive.

This chapter explains some of the pitfalls to avoid, and gives some rules of thumb which may help make your application a success.

Menus

Among applications for personal computers, where the user interacts with a program through a screen and keyboard, one principle has become dominant: *show the options*. Whether the application is GUI-based on the Macintosh or under Windows, or a character-based DOS program, it should show the user a list of the available commands on the screen. The term *menu-driven* has been coined to describe this technique: the user can reach any command by selecting from lists of options displayed in on-screen menus. Of course, there may be "short cuts" for the experienced user, meaning control or function-key combinations which will activate the same command; these will often be shown next to the menu selection.

The same principle applies to VRU user interfaces: give the caller menus which can reach any function in the system. A typical menu is a pre-recorded voice prompt which sounds something like this:

> "To access this feature, press 1, to access another feature, press 2, to do this press 3, or to return to the main menu, press the star key."

Simple enough, you'd think, but even here there is potential trouble lurking.

Most important is to allow the caller to interrupt the menu at any time by pressing the requested digit. This saves time, since the caller doesn't have to listen to the whole menu, and is also easier to use, since the caller doesn't have to remember the choice through the rest of the explanation. To emphasize to the caller that he or she may interrupt, it may help to say "now" after each option, along these lines:

The User Interface

> "To access this feature, press 1 now, to access another feature, press 2 now, to do this press 3 now, or to return to the main menu, press star."

The drawback is that some callers may infer from this wording that they must wait for the "now" before pressing a digit. Some designers prefer an initial message that explains the type-ahead feature, along the lines of:

> "When using this service, you may make a selection at any time by dialing a digit."

In any event,

> *Allow the caller to interrupt a menu at any time by making a selection.*

Menus which say the number to press before the choice are harder to use, because the caller must remember what the number was, and then decide whether to select it. All too common are menus phrased like this:

> "Press 1 to access this feature, press 2 to access another feature, press 3 to do this, or press the star key to return to the main menu."

> *Say the digit to press after explaining the choice, not before.*

Another common problem is making the menu too long. The inexperienced caller, having heard the first selection, will often be unsure whether this is really the right one, so decides to listen to the whole menu before making a selection. If there are too many options, it is easy to forget the original choice. Therefore,

> *Don't have too many options in one menu.*

A rule of thumb favored by many experienced designers is

> *Don't have more than three options in one menu.*

It is frustrating to listen to a long list of choices before getting to the one that you want. Listening to a series of irrelevant options also adds to the average length of your call which (unless you have a pay-per-call application) is surely undesirable because, in addition to the irritation this may cause your users, it reduces the number of calls per hour which your system can process. You can minimize the number of options the average caller will hear by giving options in order of decreasing popularity: describe the most used option first, the least used option last.

Give the most popular options first.

Some pay-per-call systems exploit this trick in reverse to increase the length of the average call.

Mnemonic Menus

"Mnemonics" (don't pronounce the initial "M") is the art of aiding memory and developing mental skills through special tricks. Many software packages have a short-cut to get to the printer menu by pressing [Ctrl]+P, which is easy to remember because P stands for Printer. This is an example of a mnemonic.

There is a school of thought which believes that VRU menus can be made easier to use and remember by using the letters written on most tone pads: ABC on the [2] key to WXY on the [9] key. For example, when you have listened to a voice mail message, the menu options might be:

>   Save the message
>   Delete the message
>   Forward the message to another mailbox

It would be nice to present these using letters, using "S" for Save, "D" for Delete and "F" for Forward. Since "S" is on the [7] key, you might consider menu options like:

>   "To save, press 7"
>   "To save, press S"
>   "To save, press S, the 7 key"

The advantage is obvious: the caller can hopefully understand and remember that the Save option is [7] in future.

There are a number of problems with mnemonic menus, however. Unfortunately, the alphabet is shared among only eight keys, which causes conflicts. The natural choices "D" for Delete and "F" for forward, are both on the [3] key. Another description must be found: "Give to another user", for example, which works since "G" is on the [4] key. Less natural, less memorable.

Another problem is that the alphabet contains several letters which sound similar, particularly in a noisy environment or on a poor-quality phone connection. The letters "S" and "F" are often confused, so "To save, press S" may sound like "To save, press F". On a good day, if English is our native language, we are unlikely to misunderstand, but on a bad day, listening to a language which is not our first, we could easily make an error. This objection may be alleviated by referring to the letter codes in written instructions, such as user guides and quick-reference cards, sticking to numbers only in the spoken prompts.

Using numbers and letters together in the prompts is likely to make the menu more confusing since more information is being transmitted.

Remember also that most people find letters harder to dial than numbers. I can remember 800-LIBRARY as the number to call the publisher of this book (if you want extra copies), but I find it painfully slow searching for each letter, and I'd rather read off 800 542 7279. The letters are harder to read, since they are usually in smaller print on the phone buttons, and we use them much less often, so we're going to have to search more for a given letter than for a given number.

A further pitfall, if numbers alone are used in the menu, is that the menu is now likely to be in a "random" numerical order. If the principle of "most popular option first" is followed, the numbers are likely to come out of order. Returning to our voice mail menu, suppose we choose, in order of decreasing popularity:

Delete message   D=[3]
Save message   S=[7]
Give to another box   G=[4]

the spoken prompt might then be:

"To delete this message, press 3, to save this message, press 7, to give this message to another mailbox, press 4."

Not too bad, but each time the caller hears a number, it comes as a surprise, compared with the menu:

"To delete this message, press 1, to save this message, press 2, to forward this message to another mailbox, press 3."

and this element of surprise and lack of order makes it that much harder to remember the selection if the caller listens to the whole menu.

An important element of any user interface is consistency — the interaction should be as "obvious" as possible, intuitive and free of surprises. This suggests that *all* your menus should be mnemonic if some of them are, and this leads to yet another problem: some menus are just not naturally suited to the mnemonic approach. Do you prefer:

"Press 9 for Yes, 6 for No"

or

"Press 1 for Yes, 2 for No?"

Think carefully before using mnemonic menus.

Entering Digit Strings

Many systems require the caller to enter strings of digits: an extension number, a credit card number, a PIN code, a stock number, a date of birth. This is an error-prone process, especially since the caller, unlike when using a keyboard, gets no visual check that the correct digits have been entered. Where possible, the VRU system should validate the digits entered and give an audible message to the caller that the correct digits were accepted.

The User Interface

One simple interface convention is now sufficiently widespread that all VRU systems should follow suit: the pound key (#) erases the digit string being entered and allows the caller to start again from the first digit. This is, for example, allowed by calling-card services which expect a card number following a "bong" tone: if you press pound, you will receive a new bong and can begin your input over. Make sure your software development tools allow you to cut short expected input of a string of digits in this way without resorting to time-outs.

Allow re-entry of a digit string by pressing pound.

Where possible, design your system so that you know in advance the number of digits that the caller is going to enter. This is so that you can respond immediately when the last digit is dialed. The alternative is to *time-out*, in other words to wait until no digit has been dialed for a set period of time, and then to react. The disadvantage of using time-outs is, of course, that the system becomes less responsive because there is an inevitable pause when all the necessary digits have been dialed.

Consider a situation where the caller will dial a telephone number within the US. The number could have seven digits (local call, or possibly also long-distance in some areas), or ten digits (long-distance, with a three digit area code followed by a seven digit local number). Most areas require a long-distance number to be prefixed by a "1" (this is called "1+" dialing). This allows the phone company switch to determine in a simple way whether to expect seven or ten digits: look at the first digit dialed, if it is a "1", expect ten more digits, otherwise expect six more. (Areas which have seven-digit long-distance numbers will need an additional check on the three-digit prefix of the seven-digit number, called the *NXX* code). Contrast this with the system in my native England, where long-distance calls are prefixed by a "0". The British system has variable length area codes and variable length local numbers — sometimes even within the same area code. For example, Central London has area code 71, the local number will have seven digits, while the village of Llandeilo, Wales has area code 558 and a local number will have four digits. A VRU program which can determine the number of digits to expect in an English long-distance number as it is being dialed will need extensive look-up tables.

A common variable-length string is a credit card number. My American Express card number has 16 digits, while my Visa has only 13. The trick here is to look at the first four digits, which determines the type of the card and the total number of digits to expect.

Determine the number of digits and avoid time-outs.

If you really can't determine the number of digits before the end of the string, allow the caller to indicate the end of the string by pressing a special digit, preferably star, as a "period" at the end of the string (many systems use pound for this purpose, but this conflicts with the convention that a pound may be used to re-enter a string when making an error). Again, be sure that the software tool that you are using allows you this option.

Use star to terminate a digit string of unknown length.

It is sometimes considered helpful to re-enforce the number of digits to help the caller feel confident that the correct information is being entered. Consider using:

"Please dial the three-digit extension now"

rather than just:

"Please dial the extension now"

Another example which requires careful thought is asking the user for a date, such as a birthday or delivery date. It is not a good idea to ask for a six-digit format like month-day-year (MMDDYY), this is going to confuse a lot of people. Here is a suggested method which avoids some of the common problems.

Prompt for the month first, requesting a two-digit format:

"Please enter the month as two digits now, for example zero two for February or one one for November."

Since some European countries, including those of the United Kingdom, write dates as "DD/MM/YY", international applications may wish to reverse the order of month and day for users from those countries.

Prompt for the year, explaining that two digits only are required,

> "Please enter the year as two digits now, for example nine seven for this year."

The day can be requested as for the month:

> "Please enter the day as two digits now, for example two three for the twenty-third or zero one for the first of the month."

Notice how the caller who is unsure gets more explanation if he or she hesitates following the "now". If the caller in fact enters "6" as the first digit, of the month you might choose to accept that immediately as "June" without waiting for a second digit, otherwise you could assume that this was mis-dialed and return to the prompt; similarly if a digit greater than three is dialed for the day.

Don't even think of "spelling" the month using the first three letters, by the way — quite apart from being hard to dial, both "FEB" and "DEC" are "332".

This scheme in fact allows the experienced user to "type-ahead" a six-digit MMDDYY date.

For some applications, short-cuts like "star for today", "star for this month" or "star for this year" may help save time.

The Overall Structure

Most VRU systems answer a call with an initial *greeting*, followed by a *main menu*. While these may in some cases be combined into a single recording, there are usually two distinct functions to be performed. The initial greeting will introduce the application,

> "Hello, and thank you for calling XYZ Corporation."

A pay-per-call application may have a mandatory greeting which gives the cost of the call and gives the caller the option to hang up without being charged,

> "You have reached 1-900-PIGLETS, your complete source for piglet news and views. This call will cost you 95 cents per minute, you will not be billed if you hang up now."

The main menu, on the other hand, lists the options that may be selected:

> "For the latest piglet news, press 1 now, for general information on piglet care and maintenance, press 2 now, to leave a message for a fellow piglet-lover press 3 now, for information on using this system, press 4 now. You may return to this menu at any time by pressing the star key."

The greeting will only be heard once, but it will often be possible to return to the main menu. It will often make the caller's life much easier if there is a simple way to get back to the main menu from any place in the system, for example by dialing star. This means that if the caller becomes confused, or selects an incorrect option by mistake, there is a simple way to get back on track.

> *Make it easy to get back to the main menu from <u>anywhere</u> in your system.*

It can be surprisingly difficult to implement this feature because of the hierarchical nature of many software tools. The structure of most VRU systems can be viewed as a tree looking something like this.

The User Interface

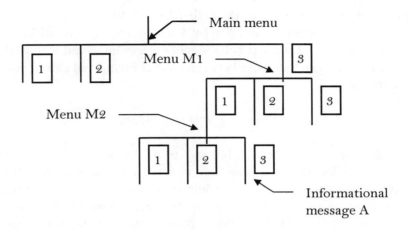

The caller reaches selection "A" by selecting "3" in the main menu to reach menu M1, then selecting "1" to reach menu M2, then selecting "3". Using many programming tools, the structure of the tree will be mirrored by the structure of the program — as a hierarchy of function calls perhaps, or as nested *switch/case* statements. These types of programs make it relatively easy to go up one level, for example from selection "A" back to menu M2. However, to jump from point "A" directly back to the main menu does not follow the tree structure, and may require some fancy programming techniques and careful advance planning.

It is usually not a good idea to explain every option and short cut in every menu. A main menu selection "for information on using our system", probably given as the last option, could be an opportunity for a new caller to listen to longer explanations at his or her leisure, or for the more experienced caller who wants to get more out of your system. This option could also allow the caller to leave a name and address to receive a printed user's guide to the application, or to leave a message with comments for the system owner or administrator.

Provide a way for the caller to get more information and to comment on your system.

If your system is an auto-attendant or other application which has options to get to a live operator, then be sure to allow the caller to reach a live operator *at any time* by dialing zero (marked "OPER" on most phones). This is a simple, friendly and often absent feature that can make all the difference between a caller loving and hating your software.

> *Allow access to a live person <u>any time</u> by pressing zero.*

Following this rule, it is all too easy to create *voice mail jail*. Voice mail jail works like this: you press zero to reach a live operator, and the system responds "please hold while I transfer you." After four rings, the receptionist fails to pick up, and the system returns you to the menu you started from. Don't let this happen to your callers; there is nothing more frustrating.

> *Avoid voice mail jail.*

There are a number of things you can do to avoid getting your callers into jail. The most obvious is to track whether there is a live person available — for example, you should allow for a "night mode" where the system knows that there is no receptionist. Also, you can make your system *context sensitive* by remembering the most recent menus and choices the caller made. This allows the software to detect the situation where the caller has got "into a loop", and provide some different options, such as:

> "We are sorry, but our receptionist is not available at this time. To return to the previous menu, press 1, to leave a message, press 2, or to return to the main menu, press star."

Damage Control

Certain options presented to the caller will probably result in irrevocable "damage" being done: for example, a voice mail message being deleted. You should design your interface so that choices of this type are distinctively different from the usual choices. For example, if most of the menu choices in your system are 1, 2 or 3, use 9 for "delete a message". This avoids catastrophic consequences if the user makes an error typing ahead through menus, or if the voice board detection algorithms fail to correctly recognize the digits entered.

Make "damage" choices different from common menu choices.

For options which result in major changes to the system, the caller should be prompted with a verification of the operation to be performed, giving an opportunity to proceed or abandon the attempt. For example, if there is a choice to delete all saved messages in a mailbox, the caller could be prompted with:

"This operation will delete all saved messages in your mailbox. To proceed with the deletion, please press the star key, to return to the last menu without deleting messages, please press 1."

As a general rule,

Verify major operations and give an opportunity to quit.

Reporting Information

Voice processing systems often convey variable information to the caller. A typical example would be a voice prompt like:

"You have six new messages."

or

"You have no messages."

In this example, the caller cannot know whether the mailbox is empty without listening to the third word of the sentence. An empty mailbox is an important special case: the experienced caller will probably want to hang up or select another option as soon as this has been reported. The wording for this type of special case should therefore be changed to something like:

"This mailbox is empty."

or some other choice so that the experienced user will be able to recognize it without listening to the whole prompt.

Change wording of reported information to allow early recognition of important special cases.

If the mailbox is empty, you don't need to waste the caller's time with a standard menu such as:

"To listen to your messages, press 1, to change mailbox options, press 2..."

This may also confuse the caller, who may interpret the first option as implying that there are in fact messages to be heard. Much better is to skip the options which do not give a useful result. Consider a solution such as starting the menu with:

"To change mailbox options, press 2..."

when the mailbox is empty, with perhaps a confirmation:

"This mailbox is empty."

to be played at the end if the caller makes no selection before the end of the menu. This will confirm the situation for the caller who perhaps is expecting to hear the first choice. For callers who type ahead, the option of pressing 1 should still be available, and should result in a message such as "This mailbox is empty."

Don't offer menu options which give no new information when selected.

The experienced caller will also be grateful for a wording which contains the variable information early in the message. For example,

"Your balance was 123 dollars at the close of business today"

might be preferred compared with:

"Your account balance at the close of business today was 123 dollars" since the first wording gives the caller the required information more quickly and gives an opportunity to move to the next transaction or hang up: improving the through-put of your application.

Give variable information as early in the message as possible.

Field Trials

The only way to evaluate the design of your user interface is through field trials. Find a group of people who have never heard your system, and record some sessions with them — make sure that your software has an option to log the menus played and responses given to a file for later analysis. Ask them for their impressions: was the system easy to use, entertaining, efficient? Make sure that your interface can accommodate both the first-time user and the experienced caller who will want short cuts and will become frustrated with having to type many touch-tones or listen to well-known menu prompts.

Be aware: a successful voice processing systems always grows. More people come to use it. More functions get added to it. Life also changes. What worked so well last month doesn't work so well this month. The most successful voice processing systems are constantly changing. Don't ever expect to design a voice processing system and walk away from it. The best operators of voice processing systems are always surveying their customers for feedback.

Chapter 37

Speaking Phrases

Introduction

Playing a complete pre-recorded message file is a simple operation. Constructing phrases from smaller pre-recorded segments of speech (a *vocabulary*) can present more of a challenge if the results are to sound good. Some examples of phrases which are constructed "on the fly" in voice processing systems are:

Mailbox Status.

For example, "You have twenty-three new messages and five saved messages in your mailbox".

Date And Time Stamps.

For example, "This message was recorded at four thirty p.m. on July second."

Time, Temperature And Weather Readings.

"The time is now eight twenty seven a.m. The temperature is sixty-eight degrees. The wind is ten miles per hour from the East, and the surf is light to moderate."

Account Balance Information.

"Your current account balance is two thousand, five hundred and twenty three dollars. Your most recent cleared check was number one hundred two in the amount of thirty seven dollars and six cents."

Telephone Numbers.

"The number of Perry's Pizza Parlor is 555-1234".

Phrase Components

Typically, the phrases which are constructed are composed of sentences with "blanks" where a few common elements are inserted. For example,

"You have ___ new messages."

is a sentence where the "blank" is filled by a whole number (zero, one, two, ...). The phrase is constructed from the partial sentence:

"You have..."

followed by the whole number, followed by:

"..new messages."

The recording "You have.." would be made with a slight upward intonation so that the listener expects more to follow, and "..new messages" would be recorded with an intonation indicating the end of the sentence.

Fitting these "sentence" pieces together can produce excellent results, but even the most careful production will have a slightly artificial "feel" to it. If there are only a limited number of possibilities, it may give a more natural-sounding result to record all possibilities as complete sentences. Suppose, for example, that your voice mail system cannot store more than 99 message. In that case, it might be worth recording 100 complete sentences:

> "You have no messages."
> "You have one new message."
> "You have two new messages."
> ...
> "You have ninety-nine new messages."

The appropriate phrase to play would then be selected based on the number of messages. Note that, as this example shows, there are in fact two or more fragments which need to be recorded for the end of the phrase: it should be "you have one message", not "you have one messages". The complete phrase "you have no messages" should probably be recorded as a special case.

Even better would be to select a different wording for the case of an empty mailbox, for example: "this mailbox is empty." This allows the experienced caller to get the information without having to listen through "you have..", he or she will recognize the start "this mail..." and have the option of immediately selecting a different menu item or hanging up.

Commonly occurring phrase components are:

Whole numbers.
One, two, three ...

Ordinal numbers.
First, second, third...

Dates.
January 1st, or perhaps January 1st 1993.

Times.
Eleven thirty five p.m.

Date/Time stamps.
January 1st at eleven thirty five p.m.

Money.
Sixteen dollars and fourteen cents.

Digit and alpha-numeric strings.
For example, account numbers, where each digit is pronounced separately (so 1234 is pronounced "one two three four" rather than "one thousand two hundred thirty four").

Phone numbers.
These are a special case of digit strings. Phone numbers often have fixed numbers of digits (say, three for a PBX extension, seven for a local number and ten for a long-distance number).

There are special considerations which may apply to each type of phrase component.

Whole Numbers

In this section we will explain in detail how computer software can generated whole numbers from a small, pre-recorded set of vocabulary files.

All whole numbers (*integers*) in English can be spoken with the following vocabulary:

Speaking Phrases

N(n)	n = 0 .. 9	Zero One Two .. Nine
N(n)	n = 10 .. 19	Ten Eleven .. Nineteen
TENS(n)	n = 2 .. 9	Twenty Thirty .. Ninety
HUNDRED		Hundred
THOUSAND		Thousand
MILLION		Million
BILLION		Billion

This is a total of only 33 vocabulary files.

An algorithm to construct the phrase can be described as follows, by building up a hierarchy of routines which can speak small numbers to those which speak bigger numbers and which call on the lower level routines. Each part of the algorithm will be given a name, which might correspond to the name of a subroutine in a program.

The hierarchy of routines is as follows:

Spk9(x)	Speaks any number x = 0 .. 9.
Spk99(x)	Speaks any number x = 0 .. 99.
Spk999(x)	Speaks any number x = 0 .. 999.
SpkNum(x)	Speaks any number x less than one billion.
SpkBig(x)	Speaks any number x.

We will use the symbol "%" to represent the modulo operation, as it does in the C programming language. "Modulo" means "the remainder after dividing by". For example, 5 Modulo 2 is 1, and 18 Modulo 4 is 2. We write this as 5 % 2 = 1, 18 % 4 = 2. The modulo operation is useful for getting decimal digits out of a variable. For example, 14 % 10 is 4. We will represent multiplication by "*". Division is represented by "/", meaning whole number (integer) division where the fractional part is thrown away. For example, 4 / 2 = 2, and 5 / 2 = 2 also.

Spk9(x)
    speak N(x).

Spk99(x)
   If x < 20 then speak N(x) and finish.
   Let t = x / 10 (now t is the "tens" digit).
   If t > 0 then speak TENS(t).
   Let u = x % 10 (now u is the "units" digit).
   If u > 0 then speak N(u).

Spk999(x)
   If y < 20 then speak N(x) and finish.
   Let h = x / 100 (now h is the "hundreds" digit).
   If h > 0, Spk9(h), speak HUNDRED.
   Let y = x % 100 (y is now the "tens and units" part).
   If y > 0 then Spk99(y).

SpkNum(x)
   If y < 20 then speak N(x) and finish.
   Let m = x / 1000000 (now m is the number of millions).
   If m > 0 then Spk999(m), speak MILLION.
   Let t = x / 1000 (now t is the number of thousands).
   If t > 0 then Spk999(t), speak THOUSAND.
   Let y = x % 1000 (now y is the "hundreds, tens and units" part).
   If y > 0 then Spk999(y).

SpkBig(x)
   Let y = x.
   Repeat: If y < 100000000 then SpkNum(y) and finish.
   Let z = y / 100000000 (now y is the number of billions).
   Let w = the first three digits of z.
   If w > 0 then Spk999(w), speak BILLION.
   Remove the first three digits of y.
   Go to "Repeat".

As you will see, SpkBig doesn't know about terms like "trillion" or bigger, but these issues are unlikely to be important. The great majority of voice processing applications will be using something like SpkNum or simpler. Note that British English traditionally uses "billion" for a million million in contrast to US English, where a billion is a thousand million.

This method may easily be translated into code in a programming language. A useful test of the algorithm for debugging is an option which produces text output rather than spoken output. Each vocabulary file will have a corresponding text string. For example, calling:

    SpkNum(12043);

in your programming language might produce the test output:

    Twelve Thousand Forty Three.

It is quicker to produce and check output in this format, and has the advantage that a voice processing card, telephone, cables and software to take the card off-hook etc. is not required.

By adding more files, the quality of the generated numbers may be improved. For example, each whole number from zero to 99 might be recorded as a separate file: a single recording of "twenty four" will sound better than the concatenated pair "twenty" "four". The numbers 1 .. 99 represent probably the most commonly used numbers. If not all 99, perhaps the first thirty or forty numbers.

Non-English languages can be approached in a similar way—but the details of the algorithm are likely to be different. For example, French speaks 21 as "vingt-et-un" (literally "twenty and one"), Danish speaks 21 as "to-og-tyve" (literally "two and twenty"). To accommodate these languages, you might think of adding "et" or "og" as a separate vocabulary file, but then three files would have to be concatenated to speak the single number 21. This is unlikely to give good results. A better solution is either to record all numbers under 100 in separate files, or to record "vingt-et.." or "..og-tyve" as separate files, or to record "..et un", "..et deux" or "en-og..", "to-og.." separately. Experimentation in each language is advisable to determine the best solution.

Another type of difference from English is the anomalous way that French speaks the numbers 90 - 99, as "quatre-vingt dix" .. "quatre-vingt dix-neuf" (literally "eighty ten" to "eighty nineteen"), which again requires a special modification of the algorithm.

Special cases which often need to be considered are zero and one. In some situations, it might be better to say "no" rather than "zero": for example, "you have no messages" rather than "you have zero messages". As mentioned earlier, the zero case often deserves its own recording, such as "this mailbox is empty". The case of one also changes the following noun from singular to plural: "one message" but "two messages". In many non-English languages, there are two or more different varieties of "one" which depend on the noun. For example, Danish has both "et" and "en" (neuter and gendered), thus "et hus" (one house) uses a different variety than "en hund" (one dog). This can impact not only the algorithms needed to speak numbers but also database information: a database field may be needed to indicate whether a thing is masculine or feminine, neuter or gendered in the local language. Systems which cater to multi-lingual users must take several such complications into account. For example, you may be serving Swiss subscribers, where both French and German have such complications (but different in each language, of course).

Ordinal Numbers

The *ordinals* are the whole numbers as used to indicate an ordering or ranking: "first", "second", "third" ...

Adding 32 new vocabulary files is all that is required to speak all ordinals, as you might expect:

NTH(n)	n = 1 .. 9	First Second .. Ninth
NTH(n)	n = 10 .. 19	Tenth Eleventh .. Nineteenth
TENTHS(n)	n = 2 .. 9	Twentieth Thirtieth .. Ninetieth
HUNDREDTH		Hundredth
THOUSANDTH		Thousandth
MILLIONTH		Millionth
BILLIONTH		Billionth

Many applications will use ordinals only for dates, in which case it makes sense to record each from "first" to "thirty-first" as a separate file. For other applications, the algorithm described in the last section can easily be modified to speak ordinals.

The routine to speak an ordinal number x will be denoted SpkOrd(x).

Dates

Speaking dates can quickly be accomplished by adding twelve vocabulary files, one for each month, and building on the functions already created for speaking whole numbers and ordinals:

   MONTH(n)   n = 1 .. 12   January .. December

A date expressed in the form YYMMDD (year, month, day) can be spoken in the form

   January first nineteen ninety-three

by using:

   Speak MONTH(MM), SpkOrd(DD), Spk99(19), Spk99(YY).

Don't forget that the year 2000 is coming up. If you represent years as two- rather than four-digit numbers in your software, I'd recommend that you treat all years before (say) 80 as 20xx and all years ≥80 as 19xx. If you don't have some convention like that, your software will stop working correctly on Dec 31st 1999. Special cases are probably needed when speaking the years from 2000, 2001 and so on. The date 1/1/2000 will be probably be spoken as "January 1st two thousand", and 1/1/2001 will be "January 1st two thousand one". Your routine to speak a year will need to go from "nineteen ninety-nine" to "two thousand", which represent two different ways of interpreting the four digits in a year.

The issue of speaking years is a case where English, which is structurally simpler than most languages for numbers, dates etc., is more difficult than it might be. The French say (literally translated) "One thousand nine hundred..." for the year 19xx, so presumably they will keep the same scheme and use "two thousand" for the year 2000.

You might offer an option to drop the year if it is "this year", or perhaps even if the month in question is November or December last year when we are now in January. For example, if today is 8/8/93, then the date

6/6/93 would be spoken as "June sixth", but "5/5/92" would be spoken as "March fifth nineteen ninety two." If dates are going to be spoken regularly, it might also be worth dropping the "nineteen" and going with the more informal style "May fourth ninety one". Compared with information which can be instantaneously displayed and appraised on a screen, spoken information is conveyed very slowly, so cutting out unnecessary words can make your application more effective.

Times

Speaking times can again build on the components we have already discussed. There are perhaps only four new vocabulary files we need:

AM	a.m.
PM	p.m.
MIDNIGHT	midnight, or twelve midnight
NOON	noon, or twelve noon

As the above list suggests, the times 12:00am and 12:00pm might be considered special cases and spoken as "twelve midnight" and "twelve noon". Some non-English languages, eg. Danish, use a 24-hour clock in colloquial (every-day) speech; for these languages no special cases might be needed at all.

Date And Time Stamps

A date and time stamp is the moment when an event occurred, such as the recording of a message, as stored on the computer. The simplest approach is just to speak the date followed by the time (or vice versa). However, it might be annoying to get such a long phrase. For example, if I am listening to my voice mail messages, chances are that most of them were recorded today. It will significantly add to the time it takes me to listen to all my messages if I must sit through "twelve thirty five p.m. August eighteenth nineteen ninety three", or similar, before each message. In this type of situation, it might be best to check for some special cases. For example, when playing the date and time for a voice mail message, the following approach might be taken:

Today.
Play the time only. Perhaps say "today" as a confirmation.

Yesterday or previous business day.
Play the time followed by "yesterday" (much quicker than a date), or "Friday" etc., especially if a business holiday has intervened.

Within the last month.
Play the time, day of month and month.

Older than the last month.
Play the time, day of month, month and year.

If date/time stamps older than a month are going to be common, it is worth considering whether the time of day is important or whether it could be omitted. In a voice mail system, this is likely to be a rare exception, so a sensible default is to provide all the information.

Money

Money amounts can again easily be spoken by building on the SpkNum routine already developed. We just need to add:

DOLLAR	dollar
DOLLARS	dollars
CENT	cent
CENTS	cents,

and perhaps

NOCENTS	and no cents
EXACTLY	exactly.

To speak the amount $DDD.CC, we would use:

SpkNum(DDD), speak DOLLAR(S), SpkNum(CC), speak CENT(S).

Some might prefer to add the word "and" between the dollar and cent amounts.

Special cases to consider are one of each (use Dollar instead of Dollars, Cent instead of Cents), and:

$0.00
This might be spoken as Zero Dollars Zero Cents or as Zero Dollars No Cents.

$0.CC
This might be spoken as CC Cent(s), or as Zero Dollars CC Cents.

$DDD.00
This might be spoken as DDD Dollar(s) exactly, or DDD Dollar(s) and no cents.

The best solution is a matter of taste and the specific application.

Digit Strings

Digit strings are perhaps the simplest structure: the application needs only to play each digit in sequence.

To speak a long string such as a credit card number, consider inserting short pauses to group the digits. Groups of four digits might be a good choice for Visa and Mastercard numbers, for example.

English speakers use both "zero" and "oh" for the digit 0 in a string. A more formal system, such as banking by phone, might choose to speak "zero" in an account number since it sounds more official. An entertainment program might choose the more friendly-sounding "oh": "your lucky number is two oh one!".

Where critical information must be provided, such as in an emergency notification system or applications to be used in noisy environments, unambiguous pronunciation techniques might be considered. An example is the military "niner" to distinguish "nine" from the similar sounding "five". With alphanumeric strings, the internationally standardized phonetic alphabet (Alpha, Beta, Charlie, Echo...Zulu) might be used for the letters, so that "P19" could be spoken "Papa One Niner". Special cases which sound similar, such as "thirteen" and

Speaking Phrases

"thirty", might also be a consideration when accuracy is important in a less than ideal listening environment.

Phone Numbers

Phone numbers are special cases of digit strings. Again, it might be worth grouping the numbers according to the usual convention. A local number (NXX-YYYY) in the US might have a pause after the first three digits, a long-distance number (AAA-NXX-YYYY) might have pauses after the third and sixth digits.

Again, the choice must be made between "zero" or "oh" for speaking "0". In the US, "oh" is the usual choice, "601 1234" would be "six oh one...".

Phone numbers also provide opportunities for checking special cases as a person would. For example, the phone number "601 2233" might be spoken as "six oh one double two double three". This technique would not generally be applied to other situations, such as account numbers, which also speak digit strings.

Other languages have their own conventions for speaking telephone numbers. The English name each digit, but the Danes look at digit pairs so that the number "01 23 43.." would be spoken (literally translated) as "zero one, twenty three, forty three ..".

Inflection

If only one recording (vocabulary file) is used for each digit, month etc., then the resulting numbers and dates produced will have a monotone sound since the tone of voice and pitch will not vary as the phrase is spoken. Natural speech varies tone and pitch, giving the listener cues for positions in the phrase and the information being conveyed. For example, an English speaker will often raise the pitch of his or her voice at the end of a question, or lower the pitch of the voice at the end of a normal sentence. In contrast, a speaker in many Indian languages raises his or her voice at the end of a normal sentence. When speaking a phone number, an English speaker will speak the digits in groups with a slight upward intonation at the end of each group except the last, where the tone of voice will drop. These variations in the voice are called *inflection*.

Phrases generated by voice response systems can add inflection as an ingredient to make the resulting speech sound more natural.

An inflected computer voice probably familiar to most readers is used by the directory information services (411 or 555-1212) in the US to speak the number back to the caller. Both the greeting ("Hello, my name is Fred, what city please?") and the telephone number found are machine-generated, minimizing the time spent by each operator on a call. The technique used by such a system is to record each digit, month etc. with several different inflections; the particular recording chosen for each digit depends on the digit's position within the phone number.

At Parity Software, we have experimented with several different inflection methods in creating the automatic phrase generator module of our VOS development environment. We identified three different types of inflection: middle/end inflection, beginning/middle/end inflection and complete inflection.

We will introduce the term *element* to mean a single spoken unit of a constructed phrase. A typical element might be a digit, month, or word such as "midnight".

Middle/End Inflection.
This is the simplest possible class of inflection, which has just two alternative vocabulary files for each element. Consider a phrase such as "Your number is twenty three" built of the elements "Your number is" "twenty" "three". In this case, the "three" comes at the end of the sentence, and should be spoken with a slight downward ("ending") inflection. "Twenty" will be spoken with a neutral ("middle") inflection because there is more to come. If the phrase were "Your number is twenty", then "twenty" would be spoken with an ending inflection.

Beginning/Middle/End Inflection.
This extends the middle/end technique to add a third possibility: the beginning or starting inflection which an element receives at the start of a phrase or group. Consider a phrase such as "thirteen dollars twenty nine cents", composed of the elements:

"Thirteen" "dollars" "twenty" "nine" "cents".

Here, "Thirteen" would be given beginning inflection, "cents" would be given ending inflection, and the remaining elements would receive middle inflection.

Creative use of the three inflection types can give different effects. For example, a seven-digit phone number might be spoken in any of the following ways (B=beginning, M=middle and E=ending inflection):

BME-BMME
BMM-MMME
BMB-MMME

The last may seem like a violation of the rules (a beginning inflection in the middle of the phrase), however the rising effect from the beginning inflection may produce a similar effect to the rising effect spoken by people in groups which are not the last group to produce an expectation of more digits to come.

Complete Inflection.
Take the example of a seven digit phone number. Each digit zero through nine could be recorded seven different times for the first, second ... seventh position in the number. The complete phone number would be constructed by choosing digits from a 7 by 10 matrix of digits (7 by 11 if "zero" and "oh" are both offered as options). This matrix can be constructed by using a professional voice talent (radio announcer or other trained voice) to speak ten different phone numbers of repeating digits:

```
000-0000  (oh)
000-0000  (zero)
111-1111
...
999-9999
```

Each phone number can then be split up into the seven component digits by using a voice editor. Each of the 7 x 11 = 77 different voice files must be carefully matched in terms of volume and space trimming at the beginning and end. To generate a number, digits are selected from the appropriate positions. For example, to play 989-0330, the digits would be selected as follows:

```
000-000̲0̲
111-1̲111̲
222-2222
333-3̲3̲3̲
444-4̲444
555-5555
666-6666
777-7777
8̲8̲8-8888
9̲9̲9-9999
```

People pronounce words differently depending on the context. For a fairly extreme example, "seven nine" will usually be pronounced with only one "n" sound, but other combinations will produce variations also. It might be interesting to experiment in producing this same matrix from other spoken samples sets, such as:

```
123-4567
012-3456
901-2345
...
456-7890
```

This set also has each digit in all of the seven possible positions, but now spoken in a different context. When chopped up into individual digits, this will result in subtly different recordings for each position in the matrix.

This type of approach can give superb results, but requires a considerable investment of effort.

Conclusion

The preceding sections have shown that there can be a lot of work to produce the very best results from what may appear to be simple operations which speak information to the caller.

The Dialogic driver offers several methods for constructing phrases from component elements. The most obvious, simply storing each element in a separate file and making repeated *xplayf* calls to play each file, has several drawbacks. (See the chapter on the Dialogic programming interface for a discussion of *xplayf*). The Dialogic driver offers no method of "queuing" commands, so the application must wait for an indication that a file play has completed (a "terminating event") before proceeding with the next play. This will introduce a short gap between the elements, especially in a busy system. Also, there may be a large number of vocabulary files, especially when inflection techniques are used. A large number of files may produce problems in a FAT file system, large directories become increasingly slow to open files in sub-directories which have many files. The Dialogic driver offers solutions using the "indexed play" technique which address both of these issues. By using indexed play, a sequence of file fragments, each of which may be a part of a larger file, are played in sequence using a single call to *xplayf*, which guarantees that there is no gap between each component file. All elements therefore may be concatenated into one or a few vocabulary files.

Even when you've got all your software routines together, there is still the issue of recording all your voice element files. And remember that your application will evolve with time: better sign your voice talent to a long-term contract, or you'll end up with several voices in your phrases. This common malady has been described as the voice mail ransom note.

DON'T Let THIS happen TO You!

Chapter 38

ActiveX Controls

What Is A Control?

You've probably heard of controls, custom controls and ActiveX controls, but you may not know what they are even though you probably use them every day. We'll therefore start out by introducing you to the idea of a control step by step. As an example, we'll use a standard button control in Visual Basic v5.

A control is a Windows graphical user interface element which can be manipulated by the user with the mouse and/or with the keyboard. Well-known examples of controls are:

- Buttons,
- Scroll bars,
- Check boxes,
- Radio buttons,
- "Edit controls" (areas where you can type in text), and
- Toolbars.

There are many more types, but these are familiar to most users.

Here is screen shot of a running Visual Basic application which has a single button control:

To create this application, the developer started Visual Basic (VB) and created a new project. VB starts out in design mode, which allows you to "draw" controls such as buttons onto the main application window (called a *form* in VB terminology) in a similar fashion to an illustration or presentation program drawing shapes. In VB design mode, the form looks like this:

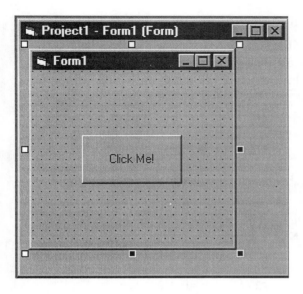

Control Properties

There are several attributes of the button that the developer can change as desired. For example, the text shown on the button face, called the Caption, the size and position of the button on the form, the font and color used to display the text, and so on. These attributes are called *properties*. In VB's design mode, the developer can change properties in the Properties Window, which looks like this:

The name of our button control is *Command1*. This is the default name which was assigned by Visual Basic. The name *Command1* is itself stored in a property called *Name* which we can change in the Properties window if we wish.

Properties may also be changed while the application is running. This is done in Basic code by using a statement like this:

   <Control name>.<Property name> = *<Value>*

As an example, suppose we wanted to change the caption from "Click Me!" to "Please Push Me Now". This can be done by the following Basic statement:

```
Command1.Caption = 'Please Push Me Now'
```

Some properties (like *Name*) can only be set at design time. Some can only be set at run time, and some can be set either at design or at run time.

Some properties can be either set by the application or read by the code in response to a user action. For example, the position of a scroll bar is determined by a property called *Value*. If the program sets *Value*, this will result in the position of the scroll bar being changed on the screen. If the user drags the scroll bar with the mouse, this will result in *Value* changing, the application can read the *Value* property to find out where the user has positioned the bar. This can be done by using *ScrollBarName.Value* as if it were a variable, for example:

```
Position = ScrollBar2.Value
```

Control Events

When the user interacts with a control, this triggers an *event*. A typical example of an event is a user clicking a button. Another example would be when a user drags a scroll bar.

ActiveX Controls

You can "attach code" to an event. When an event is *triggered* or *fired*, this code starts executing. You attach code to an event as follows.

In VB design mode, when you double-click on a control you open a code window. If we double-click on our button control named *Command1*, we get a code window like this:

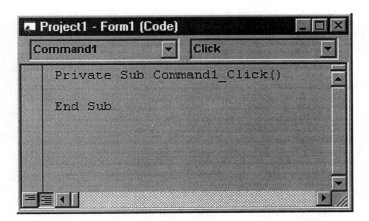

The event which is triggered by clicking on a button is named *<Control Name>_Click*. Button controls can trigger other event types, but Click is the default so this is the one which is shown when you first open the window, as the above screen shot shows. Since our button is named *Command1*, the event is named *Command1_Click*.

If the application is going to do something in response to the click, you have to write some code. This code is typed into the code window. The code that you write is put into a special type of subroutine called an *event procedure* in VB. Other development environments use different names, for example Visual C++ would call this a *message handler function*. The event procedure is the code attached to an event which is executed when the event is triggered.

In Visual Basic, the only way to get code executed is to write an event procedure and then make sure that the event is triggered. Some events are triggered automatically without user interaction. For example, each form in your application triggers a *Form_Load* event when it is first loaded, so you can enter initialization code into the *Form_Load* event procedure if you wish.

Control Methods

A *method* is a subroutine call which is executed by a control. Most common controls don't have methods – Visual Basic didn't even support methods in versions 1, 2 or 3; methods were first introduced in 32-bit controls in version 4. The button control which has provided our examples so far doesn't have any methods.

An example is the *Refresh* method which is supported by some built-in VB controls such as the FileListBox control. Invoking *Refresh* forces the control to re-read the directory and re-display its contents. This might be useful if you want to make sure that any changes to the directory are reflected in your interface, say following the deletion of some files. In Basic code, a method is invoked by a statement which is *Call <Control name>.<Method name>*, for example:

    Call FileList1.Refresh

As with subroutines you write yourself in Basic, methods may have arguments. Arguments are given in parentheses following the method name. Since method arguments are rarely used in standard Visual Basic controls, we'll leap ahead and give an example with Parity Software's VoiceBocx control. To play a Wave sound file on a telephone line, you can use the *PlayWave* method, which takes one argument, the name of the Wave file to play. In Basic code:

    Call VBocx1.PlayWave("Greeting.wav")

Invisible Controls

Most controls are user interface elements drawn on the screen. Special-purpose controls may, however, be not be shown on the screen at all when the application is run. Such controls are called, for obvious reasons, *invisible controls*. An often-used invisible control in the standard VB toolbox is the timer control. The timer control has a stopwatch icon as shown to the left. It has a property called *Interval*. If this property is assigned a value, then the timer control will trigger a *Timer1_Timer* event every *Interval* milliseconds.

Controls: A Summary

A control is defined by its appearance on the screen (if any), and its:

- Properties, which are attributes which may be readable and/or writeable at design time and/or run-time,

- Events, which are triggered by the control, usually in response to user interactions with the control but sometimes by internal actions in the control, and

- Methods, which are like subroutine calls which are executed by the control itself.

You can see the properties, events and methods of a control in the VB Object Browser or equivalent feature in other visual tools.

Custom Controls: VBXs, OCXs And ActiveX

Windows itself includes many standard controls, such as buttons, menus, scroll-bars, check boxes and radio buttons. Development environments may add further controls, such as the VB timer control, to their default toolboxes.

Visual Basic introduced an open architecture which allows third-party developers like Parity Software to add new controls to the VB toolbox. In the original, 16-bit versions of VB these were called *VBX* (*Visual Basic Extension*) controls. Other tools vendors, such as Borland with their Delphi product, reverse-engineered the VBX interface and provided support in their own tools.

For several reasons, the 16-bit VBX specification did not port well to 32-bit Windows applications for Windows 95 and NT, so Microsoft introduced a new standard based on OLE interfaces. The new type of control was a special type of OLE object called an *OLE Control* or *OCX*. Many development environments in addition to VB support OCXs, including Delphi, Visual C++, Visual FoxPro, PowerBuilder and others.

In general, a *custom control* is a control created by a third-party vendor for a visual development environment.

In 1995, Microsoft decided to introduce a new name, *ActiveX*, for many types of OLE object. This has created considerable confusion among users, especially because most of Microsoft's marketing efforts have gone towards promoting special types of ActiveX controls for Web browsers as an alternative to Java for providing downloadable, client-side browser enhancements for Web pages. By definition, an OCX control is automatically an ActiveX control, although a given OCX may not have the special OLE interfaces required to be used in a Web page, and may not have been built with the "lean-and-mean" ActiveX development kit designed to create small, easily-downloaded controls for the Web.

Chapter 39

Telephony Controls

Extending Visual Tools

"Visual" programming languages like Microsoft Visual Basic and Borland's Delphi have become enormously popular among developers of Windows applications. Telephony, however, remains a specialized niche, and none of the mainstream visual tools currently provide built-in support for Dialogic applications. Third-party vendors such as Parity

Software have therefore taken advantage of the extensibility of these tools to create ActiveX custom controls which add telephony features to these languages.

ActiveX For Telephony

Remember from the last chapter that the interface to controls is defined by:

- Properties, which are named values which can be set at design and/or run-time,

- Events, which are triggered by the control and to which you can attach programming code in your visual tool, and

- Methods, which are function calls which are processed by the control.

Telephony controls are usually invisible at run-time since they interact with a user via a telephone line, not via a screen / keyboard / mouse graphical interface.

We will show examples using VoiceBocx, Parity Software's ActiveX control for Dialogic and TAPI voice boards. VoiceBocx is a member of Parity's CallSuite family of controls which cover a wide spectrum of telephony technology. The VoiceBocx icon is a yellow-on-blue phone, as shown on the left.

Properties

The most important control property is *PhoneLine*. This is set to a value 1, 2, 3... to indicate the channel number. Other properties are used much less often. One example is *Volume*, which sets the play-back volume on boards which support variable volume. It is a property rather than a *SetVolume* method because the caller may be able to change the volume by pressing touch-tone digits, the application is then able to read the *Volume* property to get the current volume. This is a similar situation to the position of a scroll bar, which can be changed by the application code or by the user.

Telephony Controls   597

Events

The two most important events in VoiceBocx are:

- IncomingCall
- CallerHangup

As you would expect, *IncomingCall* is triggered when the control detects an incoming call on the phone line, and *CallerHangup* is triggered when the control detects a disconnect signal. Contrary to what you might expect, however, the *CallerHangup* event is not the usual method used to process a hang-up. We will return to this issue shortly.

The following screen shot shows Visual Basic with the VoiceBocx control loaded into the Object Browser. The *IncomingCall* event is highlighted, so the details of the event are shown in the lower window.

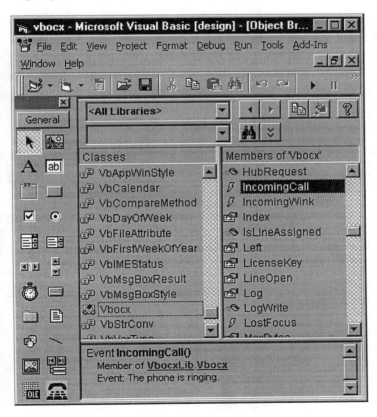

The Visual Basic object browser is a great way to get a quick overview of the programming interface of a new control. If you are using a different visual tool, you will probably find a similar feature.

Methods

Most of the work of a telephony control is done through methods. Methods are used to:

- Play messages,
- Play phrases with variable information,
- Record messages,
- Dial out-bound calls and analyze call progress,
- Get touch-tone digits from the caller,
- Play custom tones,
- Answer and disconnect calls,

and much more. For example, the *OffHook* and *OnHook* methods are used to answer and disconnect an incoming call, and the *PlayWave* method is used to play a message stored in a Wave file. The *On/OffHook* methods have no arguments, the *PlayWave* method takes a single argument which is the path name of the Wave file.

A Simple Example

We now know enough to write a complete sample program for the VoiceBocx control.

The program will:

1. Wait for an incoming call.
2. Answer the call.
3. Play a greeting.
4. Hang up the call.

To wait for an incoming call, you attach code to the *IncomingCall* event. When an incoming call is reported by the control, this code will be executed. The default name assigned to a VoiceBocx control by Visual Basic is *VBocx1*. The code will therefore look like this:

```
Private Sub Vbocx1_IncomingCall()
    Call VBocx1.OffHook
    Call VBocx1.PlayWave("Greeting.wav")
    Call VBocx1.OnHook
End Sub
```

Handling Caller Hangup

When a hang-up is detected by VoiceBocx as a loop current drop or hang-up tone detected, a *CallerHangUp* event is triggered. At first sight, you might think that this makes hang-up processing easy: when you get a *CallerHangUp* event, you hang up the call, exit the subroutine and you're done. Unfortunately, things are not that simple. The problem is that your *IncomingCall* event procedure will still be executing (though suspended), and when you exit from your CallerHangUp event procedure, the IncomingCall procedure will continue at the point where it was interrupted. This is obviously not what we would usually want: your application would continue to execute the remaining code in the *IncomingCall* procedure, even though there is no caller on the line. What we should do is hang up immediately and prepare for the next call. This means we need to find a way to stop executing the *IncomingCall* procedure as soon as a hangup is detected.

VoiceBocx solves the hang-up notification problem by generating a run-time error (called *raising* or *throwing* an *exception* in OLE terminology) when a caller hang-up is detected. If you try hanging up while a VoiceBocx application is running, you will produce a run-time error dialog box unless your code traps the error.

Processing Hangup Error Traps

To trap a caller hang-up as a VoiceBocx run-time error in Visual Basic, an *On Error Goto* label is included in the *IncomingCall* event subroutine. In other development tools such as Delphi or Visual C++, you use the exception-handling mechanism provided by that language. In Visual C++, for example, this would be done by using a *try ... catch* block. In Visual Basic, the *On Error Goto* statement looks like this:

```
' Set the error handler to catch any errors
' reported by VBocx1 or other object.
On Error GoTo ErrorTrap
```

```
    ' ...call processing code...

    ErrorTrap:
    ' If it is a hangup, exit normally
    If Vbocx1.HangupDetected Then
        Call Trace("Caller hung up.")
        GoTo Hangup
    End If

        ' Otherwise, log the error
        Call Trace("Error #" & Err.Number & ": " & _
          Err.Description)
        Beep

    GoTo Hangup
```

With this code included, you can hang up at any point in the call, and your application will respond immediately by going on-hook and exiting the event procedure, making Visual Basic ready to respond to the next call.

At the *Hangup* label, disconnect the call and perform any processing (call logging, database updates etc.) required tat the end of each call.

```
    Hangup:
    Call EndCallProcessing ' User-written subroutine
    Call VBocx1.OnHook
```

Notice that you would call *OnHook* last to avoid having an incoming call which is not answered because you are still processing inside other code such as *EndCallProcessing*.

Problems With Error Traps

The interface to custom controls was not designed with telephony needs in mind. The error trap technique itself is a kludge required to solve the issue of managing hangups. At least one other vendor offers this same trick, however further features (unique to VoiceBocx) are needed to provide fully robust applications.

Any error or exception will trigger the *Goto ErrorTrap*, so if the code at *ErrorTrap* simply hung up the phone line, there would be no notification of other errors, such as trying to play a file that does not exist. There is, however, a way to differentiate between hang-up errors

Telephony Controls 601

and "real" errors. VoiceBocx always assigns *True* to the *HangupDetected* property when a hang-up is detected. The default code tests the value of this property to see if the error that occurred was a hang-up or a "real" error.

The most important problem with error trapping is that the trap may interrupt code which is performing a sequence of related operations which must be completed in full. For example, you may have a sequence of database updates which updates a mailbox system. If the updates are not fully completed due to an *On Error Goto* trap, you may leave the databases in an inconsistent state. To solve this problem, VoiceBocx provides a *HangupDefer* property. If this property is set to *True*, VoiceBocx will remember if a hangup occurs, but will not trigger the error trap. If the application later sets *HangupDefer* to *False*, then VoiceBocx will immediately trigger the error trap if a hang-up was detected. This allows you to "protect" critical sections of code:

```
' Execute database updates without interruption:
VBocx1.HangupDefer = True
Call UpdateDatabases
' Now it's OK to jump to hang-up code.
VBocx1.HangupDefer = False
```

Another issue with error trapping is that it hides "true" run-time errors from the VoiceBocx control. For example, if you try to call the *PlayWave* method with a file that does not exist:

```
Call VBocx1.PlayWave("AWOL.wav")
```

then you would usually like to get a notification of this error. VoiceBocx provides a feature called *Error Popups* to handle this situation. If Error Popups are enabled, VoiceBocx will display an error dialog box even if error trapping is enabled in the host programming language. This allows the developer to be notified immediately of bugs in the application which might otherwise go undiscovered. Error Popups are controlled through the *ErrorPopup* property

VoiceBocx cannot generate an error trap unless a) an method is in progress, or b) your code accesses a VoiceBocx property. This is due to the architecture of OLE; it is not possible for a control to signal a container such as VB unless a control method or property is currently being accessed. If your code is performing other statements which do not reference the VoiceBocx control, you will not receive the run-time error until a later statement where VoiceBocx is accessed.

If you set the *OnHangupErr* property to False, VoiceBocx will not generate a run-time error on hang-up. The application can check the *HangupDetected* property and/or use the *CallerHangup* event to process hangups as required.

The VoiceBocx Programming Interface

In the following sections we will present tables summarizing the complete programming interface to the VoiceBocx ActiveX control. While much fuller explanations are needed for full understanding, and are of course available in the VoiceBocx documentation, this summary should help give the reader a more complete picture of the features in a comprehensive telephony control.

VoiceBocx Properties

Property Name	Usage
Action	Performs action when assigned a value.
Amp1	Amplitude (dB) of 1st freq for *PlayTone* action.
Amp2	Amplitude (dB) of 2nd freq for *PlayTone* action.
ANI	Returns ANI received at *IncomingCall* event.
AudioEventEnable	Sets *AudioOn/AudioOff* events trigger.
ConnectionLine	Specify the line to use in *Connect* and *Disconnect* actions.
DialReason	Reason for the *DialWithCPA* action *DialResult*.
DialResult	Result of the *DialWithCPA* action.
Digits	Digits retrieved by *GetDigits* action.
DNIS	Returns DNIS received at *IncomingCall* event.
ErrorNumber	Last run-time error number.
ErrorPopup	Over rides the VBOCX.INI *ErrorPopup* entry.
FileName	File used by play or record actions.

Property Name	Usage
Freq1	Frequency (Hz) of 1st freq for *PlayTone* action.
Freq2	Frequency (Hz) of 2nd freq for *PlayTone* action.
HangupDefer	Set defer of hang-up notification.
HangupDetected	True if caller hangup has been detected.
LicenseKey	Returns the current VoiceBocx *LicenseKey*.
LineOpen	Set or get the current on-hook state.
Log	Set logging to on, off or detailed.
MaxBytes	Reserved for future use.
MaxDigits	Set the maximum digits to get for *GetDigits* action.
MaxSecs	Set the maximum time for the *Record* and *GetDigits* actions.
MaxSilence	Set the maximum silence period for *Record* and *GetDigits* actions.
MessageNr	Reserved for future use.
NrDigits	Number of digits pending in digit buffer.
NrPhoneLines	Returns number of installed and licensed phone lines detected by VoiceBocx.
OffTime	Specify tone-off period for *PlayTone*.
OnHangupErr	Specify if hangup generates a run-time error.
OnTime	Specify tone-on period for *PlayTone*.
PhoneLine	Voice card channel number to use.
PhoneNumber	Number to dial for dialing actions.
Phrase	Phrase to speak for *PlayPhrase* action.
PlayCtlDigits	Control digits for *PlayFileCtl* action, *PlayVoxCtl* and *PlayWaveCtl* methods.
PlaySpeed	Set speed of audio file play in actions and methods.
PlayVolume	Set volume of audio file play in actions and methods.
PlayWait	Reserved for future use.
RecordBeep	Set beep length for record actions and methods.
Repeat	Set repeat count for *PlayTone* action.

Property Name	Usage
Reply	Reply returned from *Request* and *AsyncRequest* actions.
Request	Request used in *Request* and *AsyncRequest* actions.
RequestID	Unique ID returned from *AsyncRequest* action.
RequestType	*RequestType* used in *Request* and *AsyncRequest* actions.
Result	Reserved for future use.
RingsBeforeAnswer	Set the number of rings before an *IncomingCall* event is triggered.
SimPhoneConnected	Is true when VoiceBocx is connected to SimPhone.
SkipSecs	Skips this number of seconds on *PlayFile* action and *PlayVox* method.
StopTones	Specify the stop tones for play and record actions.
StopToneSelect	Specify the stop tone option for play and record actions.
TermTones	Specify the termination tones for the *GetDigit* action.
Type	Returns the current VoiceBocx hardware type.
Version	Returns the current VoiceBocx Version string.
VoxFileType	Specify the Vox file type for play and record actions.

VoiceBocx Methods

Method	Operation
Abort	Abort any action in progress.
AboutBox	AboutBox for VoiceBocx displays version and LicenseKey information.
AcceptCall	Pickup the current incoming call.
AddToConference(PhoneNr)	Add the current call to the conference pool.
BlindTransfer(PhoneNr)	Transfer the current call to the PhoneNumber specified without waiting for an answer.
ClearDigits	Clear the digits buffer.
ConnectCallers(ConnLine)	Connect the caller to the caller specified on ConnLine.
Dial(PhoneNumber)	Dial the specified PhoneNumber.
DialWithCPA(PhoneNumber)	Dial the specified PhoneNumber and perform 'Call Progress Analysis'.
DisconnectCallers	Disconnect the caller from the caller specified in the last ConnectCaller.
Forward(PhoneNumber)	Set/Reset the phone line to call forward.
GetCPAReason	Get the 'Call Progress Analysis' Reason from the last DialWithCPA call.
GetDigits(MaxDigits, [optional] MaxSecs, [optional] MaxSilence, [optional] TermTones)	Gather the specified number of digits.
GetNumericParameter(ParameterID)	Return the current numeric parameter value specified in ParameterID.
GetStringParameter(ParameterID)	Return the current parameter value specified in ParameterID.
Hold	Put current call on Hold.

Method	Operation
HubAsyncRequest(Request, RequestType)	Send an asynchronous query to VoiceHub.
HubRequest(Request, RequestType)	Send a query to VoiceHub.
IsLineAssigned(PhoneLine)	Return True if specified *PhoneLine* is assigned to another VoiceBocx.
MFDisable	Disable MF detection.
MFEnable	Enable MF detection.
OffHook	Set the hook state to off hook.
OnHook	Set the hook state to on hook.
Park(PhoneNumber)	Park the current call at the *PhoneNumber* specified.
Pickup(PhoneNumber)	Pickup an incoming call on the specified *PhoneNumber* or call group.
Play(FileName, FileType, [optional] StopTones)	Play a sound file.
PlayCtl(FileName, FileType)	Play a sound file with speed and volume control.
PlayDualTone(Frequency1, Amplitude1, Frequency2, Amplitude2, MaxSecs, [optional] StopTones)	Play a continuous dual tone.
PlayDualToneCadence(Frequency1, Amplitude1, Frequency2, Amplitude2, Repeat, OnTime, OffTime, [optional] StopTones)	Play a cadence dual tone.
PlayError(FileName, FileType)	Play a sound file without interruption.
PlayPhrase(Phrase, [optional] StopTones)	Play a Phrase.
PlayTone(Frequency, Amplitude, MaxSecs, [optional] StopTones)	Play a continuous monotone.
PlayToneCadence(Frequency, Amplitude, Repeat, OnTime, OffTime, [optional] StopTones)	Play a cadence monotone.
PlayVox(FileName, VoxType, [optional] StopTones)	Play a Vox file.
PlayVoxCtl(FileName, VoxType)	Play a Vox file with speed and volume control.

Telephony Controls

Method	Operation
PlayVoxError(FileName, VoxType)	Play a Vox file without interruption.
PlayVoxOnWave(FileName, VoxType)	Play a Vox file on the Wave sound board.
PlayWave(FileName, [optional] StopTones)	Play a Wave file.
PlayWaveCtl(FileName)	Play a Wave file with speed and volume control.
PlayWaveError(FileName)	Play a Wave file without interruption.
PlayWaveOnWave(FileName)	Play a Wave file on the Wave sound board.
Record(FileName, MaxSecs, MaxSilence, FileType, [optional] StopTones)	Record a sound file.
RecordVox(FileName, MaxSecs, MaxSilence, VoxType, [optional] StopTones)	Record a Vox file.
RecordVoxOnWave(FileName, MaxSecs, VoxType)	Record a Vox file on the Wave sound board.
RecordWave(FileName, MaxSecs, MaxSilence, FileType, [optional] StopTones)	Record a Wave file.
RecordWaveOnWave(FileName, MaxSecs, FileType)	Record a Wave file on the Wave sound board.
Redirect(PhoneNumber)	Redirect the current incoming call to the *PhoneNumber* specified.
RejectCall	Reject the current incoming call.
SecureCall	Secure the current call from interruption.
SetNumericParameter(ParameterID, ParameterValue)	Set the numeric parameter specified in *ParameterID* to ParameterValue.
SetStringParameter(ParameterID, ParameterValue)	Set the parameter specified in ParameterID to ParameterValue.

Method	Operation
SwapHold	Swap the on-hold call with the current call.
Transfer(PhoneNumber)	Transfer the current call to the PhoneNumber specified and wait for the party to answer.
Unhold	Make the on-hold call the current call.
Unpark(PhoneNumber)	Retrieve the call from the PhoneNumber specified.
Wink	Transmit a Wink.

VoiceBocx Events

Event	Notification
AsyncRequest(RequestID, Request, RequestType, Reply)	VoiceHub is making an *AsyncRequest.*
AudioOff	Silence is detected.
AudioOn	Sound is detected.
CallerHangUp	The remote party hung up.
IncomingCall	The line is ringing.
IncomingWink	An inbound wink has been detected.
Reply(RequestID, Reply)	VoiceHub has replied to an *AsyncHubRequest.*
Request(Request, RequestType, Reply, Broadcast)	VoiceHub is making a *Request.*

Application Generators

An *application generator* is a product which, as its name suggests, produces source code for you from a user-friendly interface. A typical example is Parity Software's CallSuite Wizard. CallSuite Wizard helps you write, test, edit and maintain your telephony source code. As a CallSuite user, you may never need to look at one code statement which CallSuite Wizard generates — you simply assign names to the blocks of code (*Routines*) which CallSuite Wizard generates and then use these names in your program.

Telephony Controls 609

For the most complete and up-to-date information, refer to the on-line CallSuite Wizard User Guide.

CallSuite AppWizard

With Microsoft Visual Basic, Delphi and Visual C++, CallSuite Wizard can generate a complete application to get you jump started.

In Visual Basic, this feature is automatically provided when you select CallSuite Wizard from the Add-Ins menu in a project where CallSuite Wizard has not previously been used. The AppWizard will lead you step by step through the process of creating a new project. When the project has been created, it will be ready to run and either accept incoming calls or dial outgoing calls.

In Visual C++, you will find CallSuite AppWizard in the standard AppWizard list for creating new projects.

In Delphi, you will find the AppWizard in the Tools menu.

CallSuite Wizard

CallSuite Wizard writes and edits computer telephony program source code for you. You can think of CallSuite Wizard as a highly specialized code editor.

The code that CallSuite Wizard creates uses Parity Software CallSuite ActiveX controls to perform computer telephony actions such as touch-tone menus.

You don't have to use CallSuite Wizard to write CallSuite code. If you wish, you can use Basic, C++, Delphi, or another language supporting ActiveXs to write your own code. However, CallSuite Wizard can make it easier for you: you don't have to remember properties or methods, and you can't make a syntax error.

CallSuite Wizard can create a complete skeleton application to get you started, and can create most or all of your telephony code. However, in most cases the Wizard does not create all the code in the application for you. You use your chosen language (Basic, C++, Delphi, etc.) for the overall flow of your application. When you want to invoke computer telephony operations, you use the name of a piece of code which you created in the Wizard. This gives you the most powerful combination for creating your application: the full flexibility of your programming language, without having to memorize or look up the program statements needed by CallSuite controls.

We call the pieces of code created by CallSuite Wizard *Routines*.

A CallSuite project will contain several Routines. With one click, you can generate source code for all these Routines in Basic, C++, Delphi, or another supported language. If you want to change the Routines later, you can call up the project and make modifications as you wish. You can even change the target language from Basic to C++, copy Routines from a Basic project to a C++ project, and so on.

The following screen-shot shows the CallSuite Wizard main window.

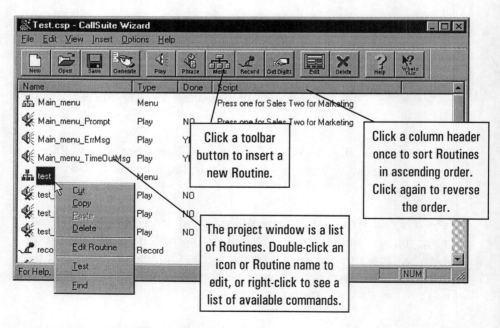

The CallSuite Wizard main window.

Telephony Controls   611

Routines

A Routine is a series of source code statements that can be invoked by a single name. In Basic, a Routine is a procedure. In C++, a Routine is a function. Other languages may have different names for the same concept. (When "Routine" is spelled with a capital R, this denotes a CallSuite Wizard Routine).

CallSuite Wizard defines several types of Routine which cover the operations most commonly required in call processing applications. The most common Routine types are shown in the following table.

Routine Type	Operation Performed When Routine Executes
Play	Play a single voice file to the caller.
Phrase	Speak a phrase composed of several elements, usually including variable information such as numbers or dates, for example "you have 12 new messages", or "the next train leaves at 7:25 pm". The calling program will pass variables to the Phrase Routine which are to be spoken.
Record	Record a voice file. Offers options to the caller to review and re-record the file if desired.
GetDigits	Prompt the caller for one or more touch-tone digits. If the caller does not enter the required number of digits, the GetDigits Routine will play an error message and ask for the digits again, up to a specified number of re-tries. The GetDigits Routine will return the digit(s) entered as a string, or a special string such as "TIMEOUT" to indicate an error.
Menu	Prompt the caller for one touch-tone digit from a pre-defined set, for example 1, 2 or 3. The Menu Routine will return the digit selected as a string, or one of the strings "INVALID" or "TIMEOUT", as appropriate.

Routine Example

A typical example of a Routine you will find in most applications is the main menu, which will of course be a Menu Routine, which we'll call "MainMenu".

The MainMenu Routine will be created by using the Menu Editor dialog in CallSuite Wizard, which can be invoked via the New Menu tool in the toolbar, by Insert | New Menu, or by Ctrl+M.

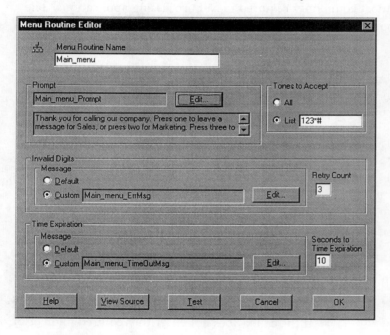

The Menu Routine Editor in CallSuite Wizard.

A Menu Routine must have a Routine name, a prompt, an Error message, and a Timeout message. The prompt is a Play Routine that plays at the start of the Routine; the voice file will presumably explain the valid choices to the caller. The error and timeout messages are Play Routines that play when the caller has made an invalid choice or no choice for a certain amount of time.

Telephony Controls 613

You can use default Play Routines provided with CallSuite Wizard for the Error and Timeout messages, or you can choose Custom, then Edit to open the Play Routine Editor and set up different messages.

A simple Visual Basic event procedure to handle an incoming call could look like this.

```
Private Sub VBocx1_IncomingCall()
    Dim Digit As String
    Call VBocx1.OffHook ' Answer call
    Digit = MainMenu(VBocx1)
    ' process rest of call
End Sub
```

MainMenu is the Routine (Visual Basic function) which CallSuite Wizard writes for you. It prompts the caller to press a digit, gets the digit from the caller, checks that the digit is valid and then either returns (if the digit is acceptable) or plays an error message and re-tries the prompt (if there was no digit or an invalid digit) until Retry Count expires.

The value returned by MainMenu is the digit pressed by the caller, or "TIMEOUT" if invalid digits were pressed or there was an error.

As with all the Routine Editors, you can see the source code which will be produced by pressing the View Source button, you will see something like the following.

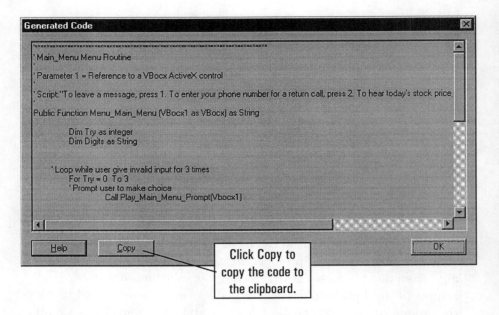

Other types of Routines use Play Routines. For example, GetDigits Routines start by playing a prompt to ask the user for digit(s). A Play Routine is called to play this prompt.

Recording Prompts And Testing Routines

To test Routines and record prompts, you will need either:

- A Wave-compatible sound card and microphone. The Wave card must be capable of playing and recording 6 kHz and 8 kHz Wave data (most, but not all, Wave cards can do this: standard for Wave is 11 kHz and 22 kHz, some cards can use only these rates). To use this method, you'll use SimPhone with CallSuite Wizard.

- A telephony board connected to a telephone via a phone line simulator. CallSuite Wizard will take the Dialogic line off-hook when it starts.

All Routine editors except Play include a Test button (in the case of the Play editor, simply press the "play" button – the black triangle – to hear the prompt). Pressing Test executes the code which CallSuite Wizard has written for you. This allows you to

Telephony Controls 615

hear the prompts and experiment with pressing touch-tone digits etc. to check the flow of the Routine that you have built. For Test to work, you will need either SimPhone or a Dialogic board and phone line simulator for recording prompts.

Using CallSuite Wizard With Development Environments

If you are using Visual Basic, Visual C++, or any of the supported languages, you can choose to run CallSuite in one of three different ways:

- As an add-in or tool that starts from the development environment. For example, you can start CallSuite Wizard from the Visual Basic Add-Ins menu or the Visual C++ Tools menu. This is the recommended method.

- As a stand-alone editor of Visual Basic, Visual C++, or Delphi files.

- Using "Copy / Paste" through the Windows clipboard. Using this method, the Wizard puts source code statements into the clipboard, you can then paste them into your editor.

Visual Basic Add-In

When you use CallSuite Wizard as an Add-In, you can add CallSuite Wizard support to an existing Visual Basic project or create a new project with sample CallSuite Wizard Routine already included. Either way, you'll go through a set of screens to help you set up your project. Follow the directions on the screens, then click CallSuite Wizard from the Add-Ins menu again to start CallSuite Wizard when you need to add, edit or test Routines in your project.

The Basic source code (.bas) file containing the CallSuite source code will be automatically added to your Visual Basic project. Each time you open your project, you will be able to select CallSuite Wizard from the Visual Basic Add-Ins menu to add new Routines or edit previously-created Routines. We usually recommend using the Wizard as an Add-In.

Visual C++

When you create a new project in Visual C++, you can use the CallSuite AppWizard to automatically set up a skeleton telephony project you can edit to meet your application's requirements. After you've set up a project, you can start CallSuite Wizard and add Routines to the project by clicking Tools | CallSuite Wizard. When you exit CallSuite Wizard, Cswizard.cpp is created and Vbocx.cpp and Vbocx.h are altered to contain the code for your Routines. These files are included in your project. If you want to add new Routines or edit any Routines you're created, you can start CallSuite Wizard from the Tools menu at any time.

Delphi

You can use the CallSuite AppWizard to generate a skeleton telephony project you can edit to meet your application's requirements. After you've set up a project, you can start CallSuite Wizard and add Routines to your project by clicking Tools | CallSuite Wizard from the Delphi menu. When you exit CallSuite Wizard, Cswizard.pas is created.

As A Stand-Alone Editor

If you prefer, you can use CallSuite Wizard to create and edit a .bas, .cpp, or .pas file which you manually add to a Visual Basic, Visual C++, or Pascal project. If you do this, you must make sure that CallSuite Wizard and development environment are not using the same project at the same time because this can result in conflicts if an application like Visual Basic has the file opened. For this reason, we usually recommend starting CallSuite Wizard from your development environment. The main use of using CallSuite Wizard as a stand-alone program is to allow you to copy and paste Routines between different projects.

Copy / Paste Through The Clipboard

If you prefer, you can select View Code, Copy from any Routine editor. This will copy the source code in the language of your choice to the Windows Clipboard. You can then switch focus to your development environment code editor and paste the code into your project. The big disadvantage of this method is that once the code has been pasted, it cannot be edited by CallSuite Wizard, only by your development environment.

Chapter 40

VOS

How To Avoid C Programming

So — you're a programmer, and you want to develop a custom voice application. You've read the chapters on the Dialogic APIs, devices, state machines etc., and you're rather depressed. You can do this, you think, but your product isn't going to be finished for a long time. It's going to be a bear to debug, maintain and improve, too.

There is another way — and that's where my own company, Parity Software Development Corporation, Sausalito, CA, http://www.paritysw.com, (415) 332-5656, fax (415) 332-5657, sales@paritysw.com comes in. We've already discussed our CallSuite family of ActiveX controls, including VoiceBocx. Our flagship product, VOS, provides a high-performance run-time engine and applications-oriented language for creating Dialogic-based voice processing applications on Windows, UNIX and MS-DOS without the need to know a low-level language like C, and without the complexity of state-machines. VOS support a very wide range of telephony hardware and features, including almost all the technologies described in this book.

If you're interested in finding out more about Parity's VOS and CallSuite products, there's a form at the very back of this book which you can fill out and mail or fax to Parity Software. I recommend trying our "Power Pack CD" product, which provides extensive documentation of many Parity Software products and working evaluation versions of VOS and VoiceBocx.

What Is VOS?

VOS includes two main elements:

A high-performance, multi-tasking run-time engine.

By allowing multiple programs to execute in parallel under DOS, VOS gives the developer an environment where one program can control one conversation without worrying about what is happening on other lines. A function to play a speech file, for example, will "block" until the play is completed. In other words, the "play" command stops the execution of that particular program until the play finishes, at which point the program resumes execution. VOS takes care of servicing other lines as the play is taking place.

VOS is *not* a pre-emptive multi-tasking system such as UNIX. Pre-emptive systems interrupt an executing task when the PC clock ticks, handing over control to the next task. This involves significant overhead, and — at least in the case of common system like standard UNIX — does not lend itself to real-time response systems like voice processing. The unique proprietary architecture of VOS has no overhead at all associated with switching between tasks.

An application-oriented, structured language.

VOS programs are written in an application-oriented language which offers the usual features expected in a modern structured language:

Arithmetic expressions, like *(A + B)*C.*

Logical conditions, like *(var1 > 1 and var2 < 3).*

Decision-making with *if..else..endif.*

Loops *for .. endfor, while .. endwhile, do .. until.*

Multi-way decision with *switch .. case* (ideal for implementing menus).

Subroutines and functions with named arguments.

The design of a program is thus greatly simplified: the flow of the interaction with a caller can be described by the *if..else,* loops, subroutine calls and other features of a structured programming language.

Why VOS?

If VOS is a programming language which looks something like C, and Dialogic gives you C libraries at no charge, then why use VOS? There must be good reasons why VOS is the development tool chosen by thousands of developers around the world.

Simple, Robust Application Language.

While the VOS language superficially looks something like C, it is in fact much simpler and more robust. Many users with no prior C experience have easily learned to use VOS, so you don't have to hire C "gurus" with high salaries. Users who already know C will be much more productive with VOS because so much of the detail work is handled internally by the VOS engine. The simplicity of the language will cut your development and testing time to a fraction, getting your products to market more quickly.

Wide Range Of Supported Technologies

VOS supports almost all the technologies described in this book.

Uniform Environment

Each board vendor has its own API with its own design philosophy and quirks. Each Dialogic product has a different style of API, and even the same product has very different APIs on different platforms. With VOS, all boards and APIs are presented in a very similar, uniform way, making it easy to incorporate many different technology types into your system.

New Technologies Today

Parity Software works with Dialogic and other major technology vendors to make sure that VOS supports all the latest features. You don't have to worry about the learning curves, poor documentation and other problems with beta and first release products.

Multi-Tasking On MS-DOS

If you want to use DOS – an inexpensive, well-understood and well-supported platform requiring very little memory and very little CPU overhead – then VOS gives you an environment where it is easy to run many phone lines. In fact, Parity Software's biggest single-PC installations are still done in DOS (384 channels in one PC is the latest record as I'm writing this chapter in the spring of 97). And, since VOS gives you cross-platform portability (see next point), you don't lock your development into a platform which may one day be obsolete.

High Performance

The hand-crafted, highly tuned engine in VOS is hard to beat for performance. As we discussed earlier in the chapter on our "Ping Pong" benchmark, VOS for DOS beats *any* C program on NT, and VOS for NT beats, or is close to, all but the most optimized C code on NT.

Cross-Platform Portability

VOS offers the exact same language and Dialogic function calls on Windows, UNIX and DOS. You only invest in learning one way to do your development. With the Dialogic C API, and with most other software development tools, you lock yourself into one platform.

VOS

Proven Engine

With tens of thousands of installations and eight years of field experience, the VOS engine is robust and well-tested. If you create a custom-purpose C application, it may have many hidden problems and you will be very dependent on one or a few talented software engineers to maintain the product.

Greeter Revisited

To see how much simpler VOS makes the programming of voice applications, we can return to the example of the Greeter application described in the State Machines chapter.

A multi-tasking version of the entire Greeter program, which took several pages of obscure C code, can be written in just 12 lines of VOS code as follows:

```
dec
    var line:2;      # Dialogic line number
    var digit:1;     # Digit from caller
enddec

program
    line = arg();
    answer(line);
    sc_play(line, "GREET.VOX");
    sc_getdigits(line, 1, 4, 4);
    digit = sc_digits(line);
    sc_play(line, "MSG" & digit & ".VOX");
    sc_onhook(line);
    restart;
endprogram
```

The *arg()* function returns the argument which was passed to the program when it was started. To take a simple example, suppose we have a four-line system, a program like the following could be used to start four copies of Greeter, one on each line:

```
program
    spawn("GREETER", 1);
    spawn("GREETER", 2);
    spawn("GREETER", 3);
    spawn("GREETER", 4);
endprogram
```

A *for-loop* could also have been used. The *spawn* function starts a new program as a separate task, the current task continues. The first argument to *spawn* is the program file name, the second argument is passed through to the program which is started, where it can be retrieved by the *arg* function. This mechanism gives great flexibility in controlling tasks on different Dialogic lines, which may, for example, all be running different applications.

The *sc_getdigits* function starts an operation to wait for touch-tone digits from the caller. The *sc_digits* function returns the digit buffer for the given channel, which should contain the digit entered by the caller.

The *restart* statement "jumps" back to the beginning of the program. In this example, as in most VOS applications, calling *restart* automatically readies the system for a new incoming call. There's nothing magic about *restart* that performs this task, it happens just because the beginning of the program goes on-hook and waits for a call.

In the following paragraphs, we give a more complete overview of the VOS language.

Dialogic Basics

The following list covers the main capabilities of the Dialogic voice boards, and the corresponding VOS commands:

Audio Processing
Play speech from file	*sc_play*
Record speech to file	*sc_record*
Get touch tones from caller	*sc_getdigits*

Telephone Signaling
Go off-hook	*sc_offhook*
Go on-hook	*sc_onhook*
Dial	*sc_dial*
Dial with call progress	*sc_call*
Wait for incoming call	*sc_wait*
Disconnect processing	*onsignal*

Creating A Touch-Tone Menu

One of the basic tools in automatic call processing is the ability to play a menu and get a response from the caller.

A typical menu is:

"Please press 1 to do this, press 2 to do that, press 3 to do..."

To make the menu, we first play the pre-recorded file containing the voice saying the menu. Then we wait for the caller to enter a touch-tone digit.

The VOS code might look something like this:

```
sc_play(line, "menu.vox");
sc_getdigits(line, 1);
digit = sc_digits(line);
if (digit eq 1)
    do_one();
endif
if (digit eq 2)
    do_two();
endif
if (digit eq 3)
    do_three();
endif
if (digit < 1 or digit > 3)
    bad_input();
endif
```

The *sc_getdigits* command waits for the caller to enter one or more digits. There are two parameters to the command:

1. The Dialogic line number

2. The number of digits to get

Thus, *sc_getdigits(line, 1)* waits for a single digit.

The *sc_digits* command returns the contents of the digit buffer (the place where VOS stores digits dialed by the caller).

The VOS *if.. endif* command executes the commands between the if and the endif only when the "condition" following the if is True.

In this example we used the commands *do_one, do_two, do_three* and *bad_input*. These are "user-defined functions", new commands that the programmer can invent by writing VOS code. User-defined functions are called functions or subroutines in other languages.

An alternative to the *if*-statement for making menus is to use a decision-making syntax called *switch..case*, which is often a more convenient solution.

Answering An Incoming Call

To answer a typical call using VOS, the following steps are made:

1. Wait for ring	sc_wait(line);
2. Go off-hook	sc_offhook(line);
3. Play greeting	sc_play(line, filename);

The variable line is the Dialogic line (also called port or channel) number where the given action is to be performed.

Making An Outgoing Call

To make an outgoing call using VOS, the following steps are usually made:

1. Go off-hook	*sc_offhook(line);*
2. Wait for dial tone	*sleep(20);* or (better, if supported by the voice board) use Perfect Call Progress with an "L" character in dial string.
3. Dial, analyze call progress	*sc_call(line, number);*
4. Get call progress result	*sc_getcar(line);*

Answering a call may require a different sequence of commands if a digital trunk (T-1, E-1, ISDN) is used.

Transferring A Call

To transfer a call using a typical PBX or Centrex system, you must:

1. Make a flash-hook

2. Wait for new dial tone

3. Dial the new extension or phone number

4. Hang up

In VOS, a flash-hook can be "dialed" like a digit by using the ampersand (&) symbol. Numbers are dialed using the *sc_call* or *sc_dial* commands, the difference is that *sc_call* does call progress analysis, *sc_dial* does not. A pause can be "dialed" by using the comma (,) character. So, to transfer to extension 123, you would:

1. Flash-hook, pause, dial	*sc_dial(line, "&,123");*
2. Hang up	*sc_onhook(line);*

This is called a blind transfer because no attempt is made to determine whether the extension is busy or answered. A supervised transfer uses call progress analysis to determine the result of the dialing. This would be done as follows:

1. Flash-hook, pause, dial	*sc_call(line, "&,123");*
2. Get call analysis result	*sc_getcar(line);*
3a. If connected: Hang up to transfer	*sc_onhook(line);*
3b. If not connected: Flash-hook again	*sc_dial(line, "&");*

If the call progress analysis determines that the transferred call was answered, the VOS program will simply hang up using *sc_onhook*, patching the original caller through to the new extension. If the transfer does not succeed, i.e. if call progress analysis determines that the call was not answered, the second flash-hook will return to the original caller, perhaps to offer the option of leaving a voice mail message.

The original caller will be listening to silence or music on hold while the call progress analysis is taking place.

Terminating A Call

To end a call, simply go on-hook using sc_onhook. Most programs will want to start again at the beginning and get ready to receive another call. This is done using the restart command.

1. Go on-hook to hang up	*sc_onhook(line);*
2. Return to start of program	*restart;*

Detecting A Disconnect

Reacting to a hang-up from the caller is a little more complicated. The caller may hang up at any point in the VOS program. To handle this situation, VOS provides a number of options for dealing with disconnects. These are unique features of the VOS language — other development environments have a much harder time dealing with the different challenges posed by disconnects.

The most commonly used mechanism VOS provides is the *onsignal* function. When a disconnect is detected, VOS interrupts the program wherever it is currently executing, and jumps to the onsignal function. The *onsignal* function can contain any VOS commands. The *onsignal* function will usually finish with either a *return* command or a *restart* command. A *return* command sends the program back to the point where it was originally interrupted, and will continue exactly as before (except that variables may be been changed in the *onsignal* function). If a *restart* command is executed, the program goes back to the start of the program and is ready to receive the next call.

A simple example of an *onsignal* function, suitable for most VOS applications, is:

```
onsignal
    restart;
end
```

This reacts to a disconnect by going back to the beginning of the program, which will presumably either wait for the next call to arrive or initiate a new call.

A problem with using onsignal may be that the program is in the middle of doing something important when the disconnect signal arrives. For example, it may have been in the middle of doing some database updates. This situation can be handled using the *sc_sigctl* command, which allows a disconnect signal to be "deferred". Calling *sc_sigctl("(")* says "I am beginning to do something important—don't jump to onsignal even if a disconnect signal does arrive". Calling *sc_sigctl(")")* says "OK, if a disconnect signal did arrive, now jump to *onsignal*, otherwise just carry on as usual."

Another option is to use the *sc_trans* command, which asks "Did a disconnect occur since the last time a *sc_clrtrans* command was issued?".

The *sc_use* command tells VOS "when you detect a hang-up signal on this line, jump to onsignal subject to any deferment specified by *sc_sigctl*". We can add disconnect processing to our example program as follows (new lines are shown in **bold** type):

```
program
    voslog("Prog started");
    trace(1);
    trace(2);
    sc_onhook(1);
    sc_use(1);
    sc_wait(1);
    sc_offhook(1);
    sleep(10);
    sc_record(1, "msg.vox");
    sc_clrdigits(1);
    sc_play(1, "msg.vox");
    restart;
endprogram

onsignal
    restart;
end
```

Notice that we moved the *sc_onhook* command from the end of the program to the beginning. This means that the program will correctly go on-hook to get ready for the next call however the call ended – through onsignal or through restart.

It is a useful technique to clean up in preparation for a new call at the beginning rather than at the end of a VOS program, this enables you to put a restart anywhere in your program and let the start of the program worry about cleaning up.

Robust and Fast

The VOS language is easy to learn, robust and fast. It is specially designed for call processing – for example, there is no dynamic memory allocation, so live systems cannot run out of memory. The VOS language is simpler than Visual Basic and much simpler than C or C++.

Best Features Of C And Pascal, Easier To Learn And Use

VOS was designed to borrow some of the best ideas from the C language, but to avoid many of the problems which make C difficult for the beginner (syntax) and for the expert (data types, dynamic memory allocation).

Beginners will find the VOS language easy to learn, experts who know C will quickly adapt to the improvements which VOS offers.

For example of the improved syntax in VOS, if-statements and loops in C have two types: "simple statements" (an expression followed by a semi-colon), and "compound statements" (one or more statements enclosed in curly brackets { ... }). In C, this can lead to subtle bugs which are very hard to find. Even experienced C programmers can write a statement like this:

```
for (;;);
    DoThis();
```

The extra semi-colon at the end of the first line means that *DoThis()* is executed once and is not controlled by the loop. In VOS, a for loop is terminated by an endfor. If you make the same error in the VOS language:

```
for (;;);
    DoThis();
endfor
```

then it doesn't matter – the extra semi-colon simply creates an empty statement which does nothing, *DoThis()* is of course still inside the loop. For another example, in C, beginners are often make mistakes in using the *switch ... case* statement because a case "falls through" to the next case unless a break statement is inserted, which is easily forgotten. In VOS, a case is ended by the next case, if you want to "fall through", you must write a goto or use some other technique. VOS does provide some of the same short-hands as C, such as $x++$ for "add one to x", but if you don't like the short-hands, you don't need to learn them, there is always another way.

Source Code

VOS is a case-sensitive language. All parts of the language recognize the difference between upper-case letters (*ABC*...) and lower-case letters (*abc*...).

Comments may be included in source code by using the # character (this is the so-called "pound" sign, which may appear as £ or another special character on non-US PCs; it is the ASCII character with code 35 decimal, 23 hex). Comments continue up to the end of a line.

The end-of-line mark has no syntactical significance except that it terminates a comment. Multiple statements may be included in a single line, although this is discouraged because it generally makes the source code harder to read.

Values

All values in VOS are stored as character strings. Character strings are written as a sequence of characters inside double quotes:

    "This is a string".

Numerical Values

Numbers are represented by strings of decimal digits. For example, one hundred and twenty three is represented as the string of three characters *"123"*. The double quotes may be omitted when writing a number, so *"123"* and *123* both represent the same string of three characters.

Logical Values

True and False are also represented as strings. False is represented as an empty string containing no characters, written "", and True is represented as *"1"* (a string containing the decimal digit *1*).

Variables

A variable has a name and contains a character string. All variables are set to empty strings when VOS starts and when a *restart* statement is executed.

A variable has a maximum length, if a string longer than this length is assigned to the variable, it will be truncated to this maximum length. If you are running a Debug version of VOS, a warning message is issued if a string is truncated by assigning it to a variable. (If you want to avoid the warning, assign using a *substr* function).

Why a maximum length for each variable? Why doesn't VOS allocate memory dynamically? This is because call processing systems must often run unattended for days, weeks and months at a time. VOS is designed to avoid dynamic memory allocation in all areas to avoid problems caused by memory fragmentation and running out of memory in a live system. This is one of many reasons that we chose to design a new language for VOS rather than use an existing language.

Variables must be declared (given a name and maximum length) before they can be used. Variables can be declared in two places: before the start of the main program, or inside of a function. An example program with a variable declaration block looks like this:

```
dec
    var x : 2;
    var y : 3;
enddec

program
    x = 12;
    y = 123;
    y = "Too long"; # y becomes "Too"
endprogram
```

Two variables are declared: one named x, with a maximum length of 2 characters, and one named y, with a maximum length of 3 characters.

A variable name must start with a letter and may continue with any number of letters, digits (0..9) and underscore characters (_). The VOS language is always case-sensitive, so the variable names X and x refer to two different variables.

Constants

A constant is a named value, that is a name assigned to a fixed string of characters. Constants are declared inside a dec..enddec block, for example:

```
dec
    var x : 2;
    var y : 3;
    const LINE_NUMBER = 1;
    const MAX_TIME = 60;
enddec
```

The name of a constant may be used anywhere a value is expected in the language. For example, a constant may be used to specify the maximum length of a variable:

```
dec
    const MAX_DIGITS = 7;
    var PhoneNumber : MAX_DIGITS;
enddec
```

Arithmetic Expressions

Values may be combined into expressions. A value, variable or constant is a simple example of an expression. More complex expressions are formed using operators. Operators take one, two or three values and produce a single value as a result. A familiar example is +, the addition operator, which combines its left-hand side and its right-hand side to give a single result, which is the sum of the two values. VOS includes a full set of arithmetic operators:

Operation	Expression
Add	*Left + Right*
Subtract	*Left - Right*
Multiply	*Left * Right*
Divide	*Left / Right*
Change sign	*-Right*

Operators have different "strengths", also called "precedences". For example, multiplication is stronger than addition, so in the expression $A+B*C$, the multiplication $B*C$ is performed first and the result added to A. Parentheses (...) may be used to create groups and force evaluation in the desired order. For example, in the expression *(A+B)*C* the addition will be performed first.

Logical Conditions

VOS includes logical operations (and, or, not) and comparison operators (greater than, less than...) which give logical results (True or False). True is represented as *"1"*, False is represented as *""*. Logical operators which combine two True / False values include:

Operation	Expression
Logical AND	*Left and Right*
Logical OR	*Left or Right*
Logical NOT	*not Right*

Comparison operators which combine values and give a logical result include:

Operation	Expression
Greater than	*Left > Right*
Greater than or equal	*Left >= Right*
Less than	*Left < Right*
Less than or equal	*Left <= Right*
Equal numerically	*Left eq Right*
Not equal numerically	*Left <> Right*
Equal as string	*Left streq Right*
Not equal as string	*Left strneq Right*

There is a subtle difference between being equal as a number and equal as a string. A string is converted to a number by taking all the characters up to the first non-digit. So, "123", "0123", and "123ABC" are all equal numerically but are different when compared as strings.

This is especially important to remember when making menus using the pound or star keys:

```
"*" eq "*"      # This gives True
"#" eq "*"      # This also gives True!
```

because we are comparing zero with zero. Be sure to use *streq*, not *eq*, when comparing non-numeric values.

Logical conditions can include arithmetic operators and parentheses just like other expressions. In fact, VOS makes no distinction between logical conditions and other expressions except to convert the final result to a logical value rather than a numerical value. Logical conditions are used in if-statements and to control for, while and do..until loops.

As an example, this is a logical condition used in an if-statement:

```
if (Year eq 1996 and Month > 6)
```

Assignments

The operator = (assignment) is a special case of an operator. The left-hand side must be a variable (or array element, see the on-line help for more about arrays). The right-hand side is any expression. The following are examples of valid assignments:

```
x = 1;
Month = 3;
MonthName = "March";
TotalSeconds = Hours*3600 + Mins*60 + Secs;
```

The short-hands from C: ++, --, +=, -=, /= and *= are available if you like them; if you don't like them you don't need them.

String Concatenation

The & operator combines two strings by placing the characters of the right-hand side following the characters of the left-hand side. This is called string concatenation. For example, *"A" & "B"* gives *"AB"*. For another example:

```
FirstName = "Joe";
LastName = "Smith";
FullName = FirstName & " " & LastName;
```

This results in *FullName* being assigned *"Joe Smith"*.

Loops

A loop is a way to repeat one or more statements. The VOS language includes three types of loop: for, do..until and while. The choice is mostly a matter of style and taste, any given task that needs a loop can be written using any of the three types.

The syntax is as follows:

```
for (initialize ; test ; increment)
    statements
endfor

do
    statements
until ( test );

while ( test )
    statements
endwhile
```

The text shown in italics is replaced by appropriate code:

statements
One or more statements (which may themselves be loops) which are executed zero or more times as the loop repeats.

initialize
An expression which is executed once before the for loop starts. In the case of the *for*-loop, this may be left blank if not required.

test
A logical condition which is evaluated each time through the loop and determines whether the loop continues executing. In the case of the *for*- and *while*-loop, it is evaluated before each loop, if True the loop continues to run. If *test* is False the first time through, the statements inside the loop are never executed. In the case of the *do..until* loop, the loop continues until the test is True. In the case of the *for*-loop, the test may be left blank, in which case the value is always assumed to be True and the loop repeats for ever, or until exited by means of a *goto, jump* or *return* statement.

increment
An expression which is executed once at the end of each iteration through the *for*-loop. If the *test* is False the first time through the loop, the *increment* is never executed. In the case of a *for*-loop, this may be left blank if not used.

The most common example of a loop is stepping through all values of a variable from 1 to an upper limit, say *Max*. The following loops all compute the sum $1+2+...+Max$ using a variable n:

```
Sum = 0;
for (n = 1; n <= Max; n++)
    Sum = Sum + n;
endfor

Sum = 0;
n = 1;
while (n <= Max)
    Sum = Sum + n;
    n++;
endwhile

Sum = 0;
n = 1;
do
    Sum = Sum + n;
    n++;
until (n > Max);
```

If you don't like the short-hand $n++$, you can use $n = n + 1$ or $n += 1$ instead. Also, $Sum = Sum + n$ could be replaced by $Sum += n$ if you prefer.

A loop which repeats for ever (or until exited by a *goto, jump* or *return*) may be written using a *for*-loop:

```
for (;;)
    # ...
endfor
```

or a while-loop:

```
while (1)
    # ...
endwhile
```

Remember that True is represented as *"1"*, so the while test is always True.

Switch / Case

A *switch/case* block is a way to make a decision between several different options. *If .. else .. endif* is a one- or two-way decision, *switch / case* allows for multiple decisions. A typical example where *switch / case* is useful is for touch-tone menus where the user may press one of several different choices.

The basic idea behind a *switch / case* statement is that an expression is matched against several possible values. When a matching value is found, the following statements are executed. This is a simple example using switch statement:

```
Digit = sc_digits(line);
switch (Digit)
case 1:
    One();
case 2:
    Two();
case 3:
    Three();
case "*":
    Star();
default:
    Invalid();
endswitch
```

One(), *Two()* .. *Invalid()* are functions which are defined elsewhere.

The expression in parentheses following *switch* is evaluated. The value is then compared with the expression following each *case* until a match is found. The first time a match is found, the statements following that *case* are executed until the next *case*, *default* or *endswitch* is encountered. Execution is then transferred to the first statement following *endswitch*. If no matching cases are found, the statements following *default* are executed. If no matching cases are found and there is no *default*, then the entire *switch .. endswitch* block is skipped.

Full expressions, not just constants or variable names, are permitted
following *switch* and following each *case*. As in other VOS language
constructs, *switch .. endswitch* statements may contain further switches,
loops etc. as required (although this is usually bad style—deep nesting
usually results in hard-to-read code; we recommend using functions to
break up such code into easy read pieces).

Goto And Jump

Goto and *jump* statements are used to branch to a given location within
your program.

The target of a *goto* is a label, which is a name followed by a colon. A
label may be applied to any statement. For example:

```
    n = 1;
Repeat:
    if (n > Max)
        goto Done;
    endif
    Sum = Sum + n;
    n = n + 1;
    goto Repeat;
Done:
    vid_write("Sum = ", Sum);
```

shows yet another way to compute a sum. Repeat and Done are labels.
The *goto* statement is written:

```
    goto labelname ;
```

where *labelname* is defined somewhere in the same program section
(main program or given function). When a goto statement is executed,
control is transferred to the statement with that label and execution
continues from there. The label may be before or after a *goto* which
references that label. You can't use a *goto* to branch from one function to
another or from a function to the main program.

A *jump* statement transfers control to a special label in the main
program. This may be done from anywhere in the program. If a *jump* is
executed from within a function, the stack is cleared, which means that

the memory of all function calls up to that point is erased. A jump label is defined using a *label* statement:

    label *labelname* :

which may be applied to any statement in the main program. The *jump* statement looks like this:

    jump *labelname* ;

Jumps are useful for features like "press star to return to the main menu", which can be very awkward to implement using other flow-of-control statements.

Calling VOS Functions

A function is a sequence of statements which has a name and produces a value. A function can have one or more "arguments", also called "parameters". Arguments are values which are copied into the function, inside the function arguments are referred to by names which behave very like variables except that they may not be assigned values.

There are three types of functions in VOS:

- Built-in functions. These are functions which are "hard-coded" into VOS and vlc, and are therefore always available to be used in a program.

- User-defined functions. These are functions which are written in VOS source code.

- RLL functions. These are functions which are written in C or C++. They are loaded from binary files called Runtime Link Libraries (RLL). On DOS, an RLL is a TSR. On Windows, an RLL is a DLL.

The syntax for calling a function is the same for all three types of function. The name of the function is given, followed by a list of arguments in parentheses, separated by commas. If there are no arguments, the parentheses must still be given. Here are some examples of calls to functions:

```
vid_write("Hello");   # Built-in function
MyFunc(); # User-defined
code = Fquery(queue, index);    # RLL
```

Since the syntax is exactly the same, you can't tell from the call to the function whether it is built-in, user-defined or in an RLL.

When a function name is used in an expression, the function is executed (this is known as "calling the function") and the value produced by the function (the "return value") is obtained. The name of the function its list of arguments is thrown away and replaced by the return value.

All functions return values. If no return value is specified, an empty string (string containing zero characters, written as "") is returned. Many functions such as *vid_write* don't return any useful value, *vid_write* always returns an empty string. If the return value is not used in the statement, as in:

```
vid_write("Hello");  # Return value not used
```

then VOS simply ignores the return value. If a useful value is returned, it may be used in an expression. For example, suppose that a function named *MyFunc* requires two arguments and returns a value. Then the following are valid statements:

```
MyFunc(1, 2);  # Ignore return value
x = MyFunc(1, 2); # Store return value in x
# Use return value in expression:
y = x*MyFunc(1, 2) + z;
```

The arguments to a function may be constants, variables or expressions. The following are all valid:

```
MyFunc(1, 2);   # Constants
MyFunc(x, y);   # Variables
MyFunc(z, y+123); # An expression
```

An expression used as a function argument may itself contain function calls. Like most other features of the VOS language, "nesting" or "recursion" to many levels is permitted.

VOS Language Functions

You can define your own function in the VOS language. Function definitions appear following the endprogram statement which ends the main program.

Let's define a simple function named Add2 which adds two numbers (yes, this is not very useful, but it's a good example):

```
func Add2(Arg1, Arg2)
    return Arg1 + Arg2;
endfunc
```

The function definition starts with func, followed by the name of the function, followed by a list of zero or more parameter names separated by commas. The function definition ends with endfunc.

The function is exited by the return statement. There may be any number of return statements which may appear anywhere within the function.

A return statement may appear like this:

```
return;
```

in which case an empty string is returned, or an expression may be given:

```
return expression;
```

in which case the expression is evaluated and the resulting value returned. In our example, we want to add the two arguments Arg1 and Arg2, so Arg1+Arg2 is the expression we need.

If no return statement is given, an empty string is returned when execution reaches the endfunc statement.

Any series of statements which would be valid in the main program can be included in a function (except the label statement).

Optionally, a dec...enddec block may be included immediately following the func statement. This is used to declare variables and constants for use only within the current function. For example,

```
func Add2(Arg1, Arg2)
    dec
        var Sum : 10;
    enddec

    Sum = Arg1 + Arg2;
    return Sum;
endfunc
```

Function arguments are passed "by value". This means that a string value is copied into each argument name (Arg1 and Arg2 in our Add2 example) when the function begins execution. You can't assign a value to an argument:

```
Arg1 = 1; # Illegal!
```

This means that you can't change the value of a variable by passing the name of the variable as a function argument. This is because it is the value stored in the variable, not the variable name, which is copied into the function. If you really do need to change variables passed into a function, you can use the "unary indirection" operators & and *. The expression &v is the internal VOS variable number of v, the expression *v is the variable whose internal number is v. These two can be used to pass variables "by reference" into a function. See the on-line documentation for more details.

Functions Files And Libraries

VOS language functions may appear in separate files. If a function has not been defined in the main source code file, vlc will search for an external file containing the function. If a function named MyFunc has not been defined, vlc will search file a file named MYFUNC.FUN The first 8 characters of the function name are converted to upper case, and the extension .FUN is appended. (On UNIX, conversion is to lower case, and .fun is appended). The environment variable FUNCDIR may be set to specify a search path for function files. Like the DOS PATH variable, it is a list of one or more directories separated by semi-colons.

Function libraries are created using mkvl. Function libraries can be encrypted so that you can distribute library files without disclosing your source code. See the on-line help for mkvl and dmpvl for more details.

Include Files

A source code file may include the complete contents of another file by using the include statement. This is typically used to include standard lists of constant declarations. For example:

```
dec
    include "project.inc"
    var MyVar : 2;
enddec
```

The keyword include is followed by the name of a file in double-quotes. An include keyword may occur anywhere in the source code, and is replaced by the contents of the referenced file. For a silly example, if the file plus.inc contains the single character +, then:

```
x = y include "plus.inc" z;
```

is equivalent to:

```
x = y + z;
```

The environment variable INCDIR may be used to specify a search path for include files.

Chapter 41

Sizing Your System

Introduction

There are several issues to be considered when choosing computer hardware to support a PC-based voice response unit. Among the most important are:

1. How many lines will you need to support in your system.

2. Hard disk size and speed. Size determines the amount of speech data that can be stored as audio files, speed determines the number of lines that can simultaneously be serviced when playing or recording speech data.

3. CPU type and speed. Probably less important than the hard drive selection, however the CPU must be fast enough that the application and disk will not be limited by processor throughput.

4. Number of free slots. As PC-compatible computers become more powerful, and as voice processing applications on one PC support more lines and include more enhancements such as voice recognition and fax requiring extra cards, the number of slots can become a critical resource.

5. RAM size. Extra RAM can be used for several purposes, including RAM-disks for commonly used prompts, storing speech card buffers, and storing application data.

6. Add-in cards such as LAN cards, video cards and so on.

How Many Lines Do You Need?

Often, one of the hardest decisions in building a VRU revolves around the question of how many lines the system should support. Obviously, the number of lines is the same as the number of simultaneous calls the system can support without "busying out", i.e. giving busy tones to new callers. But how many calls will your system receive?

The first point to realize is that many VRU systems have most of their lines idle most of the time. Calls tend to be clustered around certain times — when a TV ad airs, when office workers check their voice mail after lunch, and so on. The system should be dimensioned to handle the *peak load* in an acceptable way, which will probably mean that the system will have significant free capacity the rest of the time.

The most important number to estimate when calculating loads is the *average call length* in minutes or seconds. The longer the average call, the more lines will be needed to support a given number of calls per hour.

Suppose, for example, that a VRU system will be installed to help a call center handle peak loads in response to television advertising. In the eight minutes following the airing of a TV ad, an average of 100 calls "get lost" because they cannot be handled by the team of live operators. It is decided to install a VRU with the ability to greet the caller with an invitation to record their name, address and telephone number so that sales literature can be mailed to them. Trials indicate that this transaction can be completed in an average of two minutes.

If we denote the number of calls per minute during peak load as CPM, and the average call length in minutes by ACL, the minimum number of lines will be required to handle the peak load is:

$$\text{LINES} = \text{ACL} \times \text{CPM}$$

In our example, we have:

$$\text{ACL} = 2 \text{ minutes}$$

$$\text{CPM} = 100/8 = 12.5 \text{ calls/minute}$$

so we derive:

$$\text{LINES} = 2 \times 12.5 = 25 \text{ lines.}$$

Of course, this calculation assumes averages, so it might be advisable to allow a margin for error and install, say, 32 lines.

Hard Disk Size

The hard drive size is probably the easiest design parameter to estimate in advance. The calculation is simple: how many minutes of speech data will be needed stored as audio files? The requirements can be determined depending on the digitization method chosen from the following table:

Bits per Sample	Rate (Hz)	Bytes per minute
4	6053	181,590
4	8000	240,000
8	8000	480,000
8	11025	661,500

Using the Dialogic default of 6KHz 4-bit ADPCM, the lowest quality, most compact storage method, one hour of storage requires 181,590 x 60 = 10,895,400 bytes, so a good rule of thumb to remember is:

> 6 kHz 4-bit ADPCM = *Ten megabytes per hour of storage.*

Using the network standard of 64Kbps (8KHz 8-bit PCM), the requirement rises by more than a factor of two to 480,000 x 60 = 28,800,800 bytes, so remember the handy approximation:

> 64 kbps = *Thirty megabytes per hour of storage*

Remember that a FAT file system wastes an average of half a cluster in storage space in each file, so if speech data is being stored in many small files of a few seconds each, there will be an additional factor to take into account due to wasted space. (A cluster is the unit of disk allocation used by a FAT, for a typical hard drive it may be 8, 16 or 32 kb, giving a wastage of 4, 8 or 16 kb per file).

Hard Disk Speed, Number Of Lines

There is a simple rule associated with hard disk speed: if you want to run many lines, use the drive with the fastest access time that you can afford. The heaviest load on the PC in most voice processing applications is moving speech data to and from the hard drive, which means that the faster the hard drive, the more lines the system will be able to support.

Even based on the assumption that it is disk data input/output that is the bottleneck, it is still hard to estimate the total number of lines that a given PC configuration can support. The disk seek time, average latency, data transfer rate and other factors combine with bus speed, device driver overhead and other hard-to-predict factors to produce the important number: the data throughput in bytes/second that can be achieved from the disk.

By far the best idea is to test out the software on the intended PC platform with as many lines as possible making calls. Of course, this is often not practical, so simplified benchmarks can be designed which can help measure the practical limit on the disk throughput, and hence the maximum possible number of lines which could be supported by a given PC configuration.

The following program is designed to provide a simple benchmark which will provide an upper limit on the number of lines that a given PC configuration will support — without having to go to the trouble of installing speech card hardware and getting 24 close friends to call the system. It is designed using the POSIX-style *open*, *read*, *write* functions for file i/o, and it can therefore be compiled and run on Windows, DOS and UNIX.

The idea is to create a number of files, 16 as configured in the source code, then to perform a repeated sequence of read or write operations, using a random number generator to skip "at random" between the files, reading or writing a buffer of a given size. At given intervals, set by constants in the program, a file will be created. The idea is to simulate the operating system file input/output operations which must be performed by the speech card driver software when many lines are active. The number measured by the program is the effective disk throughput in bytes read/written per second from these files. By using the known transfer rates required by different audio digitization schemes, this final transfer rate is used to give an estimated upper limit on the number of lines that the tested computer could support.

The program was written and tested using Microsoft C version 6.0, but should be compatible with most C compilers on the PC. The reader (at this point, at least) is assumed to be a reasonably experienced C programmer who can read the program and understand how to implement it on his or her own machine. Other readers may skip ahead.

```c
/*
bench: Simple disk I/O benchmark for
voice programs
*/

#include <stdlib.h>
#include <fcntl.h>
#include <io.h>
#include <time.h>

unsigned irand(unsigned);    /* Random number */

#ifndef O_BINARY    /* If nonDOS O/S */
#define O_BINARY    0
#endif

#define BUFFSZ      (16*1024)
    /* Bytes in one buffer */
#define BUFFS       10
    /* Buffers in one file */
#define HANDLES     16
    /* Nr files open */
#define RATIO1      4
    /* Open every RATIO1 reads/writes */
#define RATIO2      4
    /* Do RATIO2 x more reads than writes */

int iters;
    /* Number of buffers to read/write */
char buff[BUFFSZ];
    /* Read/write buffer */
int handle[HANDLES];
    /* File handles */
char filename[HANDLES][4];
    /* File names */
long start_time, stop_time;
    /* Measured start/stop time */
long trate;
    /* Transfer rate */

int n;
int iter;
int buff_nr;
unsigned bytes;
int count1;
int count2;
void main(int argc, char **argv)
    {
    if (argc < 2)
        {
        printf("Use: BENCH n\n");
```

```
        exit(1);
        }
    iters = atoi(argv[1]);

/* Close preopened handles 0, 2   5 */
/* so that they're */
/* available for new files */
    close(0);
    for (n = 2; n <= 5; n++)
        close(n);

/* Construct file names "1", "2" ... */
/* and create file */
    for (n = 0; n < HANDLES; n++)
        {
        sprintf(filename[n], "%d", n);
        printf("Creating '%s'\n",
          filename[n]);
        handle[n] = open(filename[n],
O_RDWR | O_BINARY | O_TRUNC | O_CREAT, 0666);
        if (handle[n] < 0)
            {
            printf("Can't open '%s'\n",
            filename[n]);
            exit(1);
            }
        for (buff_nr = 0; buff_nr < BUFFS;
          buff_nr++)
            if (write(handle[n], buff, BUFFSZ)
            != BUFFSZ)
                {
                printf("Write error\n");
                exit(1);
                }
        }
/* Start the clock */
    start_time = time((long *) 0);

/* .. and start disk operations */
    count1 = 0;
    for (iter = 0; iter < iters; iter++)
        {
    /* Pick file at random */
        n = irand(HANDLES);
        if (count1++ == RATIO1)
            {
            close(handle[n]);
            handle[n] = open(filename[n],
O_RDWR | O_BINARY | O_TRUNC | O_CREAT, 0666);
            if (handle[n] < 0)
                {
```

```c
            printf("Reopen error\n");
            exit(1);
            }
        if (write(handle[n], buff, BUFFSZ)
        != BUFFSZ)
            {
            printf("Write error\n");
            exit(1);
            }
        count1 = 0;
        continue;
        }
    if (count2++ >= RATIO2)
        {
    /* Append to end of file */
        if (lseek(handle[n], 0L, 2) < 0)
            {
            printf("Seek error\n");
            exit(1);
            }
        if (write(handle[n], buff, BUFFSZ)
        != BUFFSZ)
            {
            printf("Write error\n");
            exit(1);
            }
        count2 = 0;
        }
    else
        {
        long size, pos;
    /* Seek to random position in file */
        size = lseek(handle[n], 0L, 2);
        pos =
irand((unsigned) (size/BUFFSZ))*BUFFSZ;
        if (lseek(handle[n], pos, 0) !=
        pos)
            {
            printf("Seek error\n");
            exit(1);
            }
        if (read(handle[n], buff, BUFFSZ)
        != BUFFSZ)
            {
            printf("Read error\n");
            exit(1);
            }
        }
    }
stop_time = time((long *) 0);
printf("Benchmark complete\n");
```

```
        printf("Elapsed time     %ld seconds\n",
            stop_time - start_time);
        printf("Bytes moved  %ld\n",
            (long) iters*BUFFSZ);
        trate = ((long) iters*BUFFSZ)
            /(stop_time - start_time);
        printf("Transfer rate   %ld bytes/sec\n",
            trate);
        printf("Max lines 4bit 6KHz: %ld\n",
            trate/3026);
        printf("Max lines 8bit 8KHz: %ld\n",
            trate/8000);
        exit(0);
        }

/* irand: Generate random number by */
/*    linear congruential method */
unsigned irand(unsigned n)
        {
        static long _ran = 218765432;

        _ran = (_ran*123)%17857;
        return (unsigned) (_ran%n);
        }
```

Be careful to consider the effect of disk cache software and watch for other software which may give a misleadingly good result. When running the benchmark, a cache or buffer system may be able to speed up the operation of the benchmark program more than a real-life application which may be using many more different files. When running the benchmark, disable any cache software you may have. This is because a real system is likely to be using a wider range of data than the benchmark. When running a live system, you should of course take maximum advantage of available buffers and cache software if possible to speed operations and reduce the load on the hard drive.

Remember, of course, that the limit is an upper limit only, the more overhead that is required by the rest of the application apart from speech play and record operations will reduce the number of lines which can be supported.

It has been said that there are lies, damn lies, and benchmarks. Any benchmark should be taken with a dose of skepticism, and should be used as a guide only.

Trying out the benchmark on some of the PCs in my office, the following results were obtained:

PC	HD Size	HD Seek	Xfer Rate	Max Lines 4/6	8/8
P6/200	6000Mb	9ms	915,000	292	111
486/66	500Mb	9ms	819,200	270	102
386/33	120Mb	18ms	139,000	45	17
386/16	60Mb	30ms	134,000	44	16
8086/8	20Mb	75ms	20,300	6	2

The symbol "4/6" refers to 4-bit, 6KHz sampling rate, "8/8" to 8-bit, 8KHz sampling. There is surprisingly little difference between the 386/33MHz computer with a 18ms average seek time on the hard drive and the 386/16MHz with a 30ms hard drive. One possibility is that the 386/33 had a hard drive which was badly fragmented, increasing the average access time to any part of a file. If I had thought of this at the time, I would have run a disk defragmentation utility (found with many packages such as Norton Utilities and PC Tools) to make the results more comparable.

While the benchmark may not produce very accurate numbers for the maximum number of lines, it may indicate the relative importance of the different factors — CPU speed, hard drive seek time, fragmentation, various types of disk cache, and so on. My recommendation is to develop experience with this simple benchmark coupled with rigorous "real-life" testing of your application. With the experience you develop, you may be able to refine the benchmark to give a more accurate prediction relative to the characteristics of your particular application.

Automated Testing

Even if you are able to test out your system with the required number of speech card channels available, it can still be a problem to create realistic loads on all lines in a test situation. You may not have enough phone lines available in your lab, and you don't want to tie up 12 or 24 people in boring, repetitive testing work.

The solution may be to build automated testing features into your software. Even if no phone line is attached, the speech card can be made to go off-hook as if an incoming call had been detected, accept simulated touch-tones and play or record speech files. Whenever touch-tone input is expected, the program could select a pre-assigned digit, or choose from a list of possible digits using a random-number generator. It would be wise to weight the probabilities in favor of selecting valid digits, otherwise you are unlikely to get very far into the menu tree. The tester could call in on one or two lines and evaluate the responsiveness of the system. This type of automated testing procedure can also flush out bugs in the software which cause the system to lock up, corrupt database files, and so on.

Chapter 42

Which Operating System?

Introduction

Windows 95? DOS? UNIX? Something else? Which operating system should you choose for your call processing system? There is, of course, no simple answer. In this chapter we'll explore some pros and cons of the different options.

The material for this chapter originated in an article called "The Hourglass Of Death". This title was chosen to dramatize an important limitation of 16-bit Windows. At the time, there was an explosion of interest in using Windows 3.1, and especially Visual Basic, as a platform for developing call processing systems. However, in many cases, Windows 3.1 is *not* a good choice for a call processing platform, and the article was hopefully intended to provide a reality check. If you see an hourglass icon, your call processing application is dead! Within a second or two at most, messages will stop playing or recording, touch-tones dialed by the caller will not be processed, incoming calls will be ignored. For example, starting Word for Windows 3.1 usually results in an hourglass icon lasting ten seconds or more. If you are running a call processing application when you start Word, the app will simply freeze.

Visual Basic exacerbates the weaknesses of Windows for call processing, so if you're thinking that a VB custom control is the coolest way to create your next application, you may want to read on.

What about my vested interests? Perhaps I'm trying to promote my favorite operating system or trash one I don't like? Not so. Parity Software supports a cross-platform, fully compatible series of VOS engines which run on all the popular operating systems including DOS, Windows 3.1, Windows 95, Windows NT and UNIX (OS/2 is the exception, only because Dialogic does not currently provide the right kind of driver for us). Parity Software is also offering Visual Basic custom controls for customers who prefer to use that environment and can live with its limitations. One benefit of Parity's "Any Application, Any Platform" strategy is that it allows us to offer the best possible advice to our customers when choosing a platform. We can say, "if you want to use platform XYZ, we have the best tools available, but be careful of these pitfalls...".

Platforms

There are at least three platforms to consider: the server, the client and the development tool. All three platforms may be combined in a single computer running a single operating system, or they may be three different platforms running different operating systems. The best combination depends on the application.

The Server

The voice cards are controlled directly by the call processing server. Requests are received by the server from the client software, which are analyzed and converted to API commands to the voice card device driver(s). The software running the server will be handling many simultaneous conversations. Often it will be running unattended in a phone room or basement with no human operator. It needs to be high-performance, reliable and robust. In many cases, it will have no user interface or only a few simple commands (show line status, shut down). In some systems, the client and server software will be combined into a single program running on one computer. A typical auto-attendant, which will be a PC running voice mail software with voice cards installed, is a well-known example.

The Client

The client is the rest of the application software other than the server. In a call center, the client will be an agent station which receives an notification from the server when an incoming call is received, perhaps with the caller's number retrieved via ANI. In an audiotex service bureau, the client software may be used to install and configure the services which run on the server. In an integrated messaging system, the client will be e-mail/fax/voice mail management software. The client is the generally the software that your end-user sees. It therefore probably needs a user-friendly interface, at least menu-driven and perhaps graphical.

The Development Tool

If you are creating a call processing system, you have a choice of development tools ranging from low-level coding using the C language, an applications-oriented language like VOS, forms-based/menu-driven "application generators" and join-the-icons entry-level software for end-users.

It is important to realize that the style of development tool used to create the software has nothing to do with the style of the software which is created. A highly graphical, user-friendly interface can be created with an unfriendly procedural language typed into a character-mode editor (say, Microsoft Visual C/C++ for Windows NT). Remember that even with Visual Basic, you spend most of your time writing old-fashioned procedural code. In most cases, you want to provide user-friendly client software for your agents and system administrators, this does not necessarily mean that your call processing software development tool needs to be user-friendly or Windows-based. On the contrary, a friendly "join-the-icons" type of applications generator may be very inflexible and provide you with no facilities for customizing a Windows-based user interface for the finished product, and a powerful procedural language like VOS may provide you not only with a much more powerful development environment but also with convenient hooks to interface builders like Visual Basic. Developers need powerful and flexible development tools, servers need to be efficient and robust, end-users need friendly, menu-driven and graphical interfaces.

Let's look at some examples to see how the different platforms are used in real systems.

A call center can be built as follows. Each agent station has a DOS PC running Clipper applications for data entry and account queries. Incoming phone lines and agent headsets are connected to an ACD (Automatic Call Distributor, a special type of PBX designed for call centers). A Voice Response Unit (VRU), a PC containing Dialogic cards, is also connected to ACD extensions. All the computers are connected via a LAN. Here, the agent stations are the clients and the server is the VRU. In this article, when I say "server", I mean a call processing server. This may or may not be part of the same computer which contains the disk server, database server etc.

One of Parity Software's customers runs a 180-line audiotex/fax service bureau in Prague, Czechoslovakia. Six digital E1 trunks each carrying 30 conversations terminate on PCs running VOS for DOS. These are connected via a fiber-optic LAN to a Netware server and to a number of other PCs which are used as agent and administrative work-stations, most of them running Visual Basic applications under Windows 3.1.

Which Operating System? 661

A CTI-based call center could have a Netware file server running TSAPI to control a PBX. The agent stations might be running a Windows for Workgroups application for database management. The database could be stored on a dedicated Oracle server. The applications software might have been developed using Visual C++ on Windows NT.

So, what about the different operating systems?

MS-DOS

Pros: inexpensive, well-known, low overhead: smaller and faster than any other O/S. Cons: no native GUI, no multi-tasking provided by the operating system, new APIs may not be accessible, server software is exceptionally hard to write without a good tool.

In many cases, DOS is still the best platform for call processing, and retains a market share in computer telephony which is much higher than in the main-stream desktop software world. DOS takes very little memory and never takes a CPU cycle unless the application software specifically makes a request. Writing call processing server software for DOS is hard using low-level tools like the C language because multiple conversations must be managed by a single program, requiring difficult techniques called "state machine programming". The low overhead of DOS means that you can run more lines with less memory and a slower CPU than any other operating system. The lack of a native GUI may be an issue if the client and server are in the same box. The fact that DOS cannot run more than one program at one time may be an issue for remote maintenance. If the DOS PC is on a LAN, remote control programs such as Co-Session can be found which can access other work-stations; if the PC is stand-alone, this may be a problem. Also, new APIs (Application Programming Interfaces) such as MAPI, Microsoft's Messaging API, or ODBC for database access, are often provided as Windows DLLs and are therefore not accessible directly from DOS. For smaller systems (say, 2 to 16 lines), a DOS server and Windows client can be combined into one box using OS/2, which can provide robust multi-tasking between a DOS session and a Windows session. Many VOS for DOS developers integrate solutions with a graphical interface into a stand-alone box using this configuration.

Windows 3.1, Windows for Workgroups

Pros: Not many, Windows 95 is usually a better choice. Cons: Multi-tasking has serious problems for call processing servers, major voice board vendors (including Dialogic) do not provide Win 3.x device drivers, software is hard to write (even harder than DOS) without a good tool.

Win 3.x is well-suited to support client software, as long as it is in a separate PC from the call server or in a separate session in a robust multi-tasking environment such as OS/2 can provide. If the server software which controls multiple conversations directly runs under Win 3.x, then the cooperative multi-tasking scheme used by Windows becomes a major problem. Win 3.x applications must voluntarily relinquish control in order for another application to run. Well-behaved software will do this at regular intervals, but even if all your software is well-behaved, there are still times when Windows hogs the processor with a single process for a significant length of time, such as when loading a new application from disk. When you see an hourglass icon, this means that Windows or your application is grabbing the CPU for its own exclusive use and does not relinquish control until the hourglass changes to another icon. Even if you never see an hourglass, Windows can still decide to perform time-consuming operations (e.g., consolidate unused memory areas) during which time your call processing application is stopped. In these situations, the call processing server software will be stopped, and your application will die. Ironically, old-style DOS applications, unlike Windows apps, can be pre-emptively multi-tasked under Win 3.1, so for example VOS for DOS may run better in a DOS box under Win 3.x than the native VOS for Windows engine. However, even pre-emptively multi-tasked DOS boxes grind to a halt when you are looking at an hourglass, so this is not a true solution.

In summary: by all means, provide a Win 3.x user interface for your end-users, but don't run the voice boards with a Win 3.x control program.

Visual Basic

Pros: Very widely known and used programming environment (more than one million sold), great for building client software. Cons: Requires a program instance for each conversation (in other words, 24 lines means 24 .EXEs running).

Visual Basic allows the developer to draw a user interface on the screen and to attach Basic code to events caused by interacting with user interface controls such as buttons and scroll bars. For example, the programmer can write Basic code to be executed whenever a given button is pressed.

VB is great for client applications, which are usually run by live users sitting at a PC, but server software interacts with the users via a telephone line, and here VB is much less attractive. There are no user interface elements to press or click, the call proceeds in a linear fashion from beginning (incoming ring or dial out) to end (hang-up), just like programs written in pre-Windows languages. The only way for software tools vendors like Parity Software to extend Visual Basic by adding call processing functionality is by creating a "custom control" (ActiveX). The custom control mechanism is designed for user interface items like buttons; it stretches this interface to its limits to make a custom control for call processing. More natural would be to define new functions and subroutines for the VB language, but there is currently no way to do this.

The call processing control becomes an invisible element in the finished application, actions such as playing messages and waiting for touch-tones have to be done by assigning values to control properties, a mechanism usually used for changing sizes, fonts, colors and so on in a visible control. Touch-tones received and other notifications from the call in progress could be represented as Visual Basic events, but this would result in a complex state-machine programming model and is therefore rarely, if ever, done. The one true asynchronous event is a caller hang-up, which is awkward to deal with in Visual Basic. If a hang-up triggers a special hang-up event procedure which the user writes in Basic, it does not, as would be desirable, interrupt the Basic code which is currently executing. The only way to make your Visual Basic code jump somewhere when an event happens in the middle of a routine is to report an error condition to the Visual Basic run-time and to use the "on error goto" mechanism. This can only happen when your Basic code references the call processing control, so the hang-up may go undetected for some time. Also, any true Basic run-time error will be treated as a hang-up, which makes debugging and testing the application more difficult.

Visual Basic is not designed as a multi-tasking engine: when Basic code is executing, it will not give up control to another program unless a special function (DoEvents) is called, or a reference is made to the call processing control, which may then take the opportunity to a Windows API Yield() call which allows another program to run. (Contrast this with VOS for Windows, where VOS will swap to other tasks without the developer having to worry about how and when this is done). These factors mean that Visual Basic code for a call processing server must be carefully written so that any loops or other time-consuming operations are kept to a minimum or contain embedded calls to DoEvents() as appropriate, otherwise other lines will not be serviced while this code is executing. Calls to APIs such as ODBC may also "hang" other lines unless requests terminate quickly, which they may not. Sending a database request that requires a complex SQL statement may block the application for several seconds. If this proves to be a problem, there is no solution without re-writing the DLL providing the service (not for the faint of heart). Running multiple lines requires that multiple instances of the .EXE be loaded, a situation which Win 3.x does not handle very efficiently.

As a result of these issues, Visual Basic cannot usually be recommended for serious call processing applications except for clients on a separate platform from the voice cards.

Windows NT

Pros: Robust 32-bit multi-tasking engine, Windows GUI, is being adopted by some corporations as preferred choice for server platforms. Cons: big and slow, voice board drivers may not be available or not as mature, not real-time, DOS box cannot run voice board drivers.

Unlike Win 3.x, Win NT uses pre-emptive multi-tasking, which means that an executing program will receive regular attention from the processor even if an hourglass is being displayed by another program or another time-consuming operation is in progress. This avoids the most important weakness of Win 3.x for call processing servers. However, Win NT is big (expect to buy at least 24 MB of RAM even for a small system) and slow (the pre-emptive multi-tasking and interprocess security features eat up a lot of CPU time). Also, to date, Win NT has generally been a low priority for voice card vendors when it comes to providing drivers. This is likely to change over the next couple of years if Win NT gains market share. Despite the pre-emptive multi-tasking, NT is not a real-time system, which means that there is no guaranteed maximum time that will elapse from an event until it receives service by an application — for example, responding to a touch-tone. This same comment applies to all the platforms discussed here except DOS, where it is possible (though difficult) to write real-time software, and QNX, which is a true real-time platform.

Windows 95

Pros: Offers an environment for 32-bit programs with reasonably robust multi-tasking, Windows GUI, smaller and faster than NT, has won wide acceptance on desktop. Cons: less robust than NT, drivers not available for most call processing cards.

Win 95 us quite similar to Win NT in its ability to run 32-bit programs in a pre-emptive multi-tasking environment. Most of the same NT comments apply, except that Win 95 is be smaller and faster since Microsoft has sacrificed some protection and other features for speed and size. Win 95 will also run 16-bit Win 3.x apps, but these are not be pre-emptively multi-tasked and suffer the same weaknesses as running under Win 3.1.

Win 95 has replaced Win 3.1 as the platform of choice for client applications.

OS/2

Pros: Robust, 32-bit, pre-emptive multi-tasking, GUI, DOS box will run voice board drivers. Cons: Long-term future in doubt.

From a developer's point of view, creating native OS/2 apps is very comparable with Win 95 or Win NT 32-bit apps, with generally slightly better performance than NT. OS/2 provides a robust environment for multi-tasking DOS and Win 3.x applications in one box. The question is, will OS/2 survive long-term? Microsoft clearly aims to position Win 95 as an OS/2-killer, and they have proved to be formidable marketers.

UNIX

Pros: Robust, 32-bit, pre-emptive multi-tasking, good connectivity for some specialized areas (Internet TCP/IP, databases such as Informix), a corporate standard for many phone companies. Cons: poor integration with GUI, many different UNIX flavors which are not compatible, high license cost compared with competitors such as NT, some popular UNIX flavors have slow process switching and inter-process communication.

There are many flavors of UNIX currently being used for call processing. SCO currently has the largest market share by most estimates, but SCO is losing ground to leaner and meaner competitors. QNX, for example, is a real-time UNIX variant: probably the only real-time operating system which is a serious option for PC-based call processing. SCO does not yet offer multi-threading (where a single program can have multiple routines executing in parallel), only multi-tasking between programs. This means that process switching and inter-process communication is relatively slow (one benchmark at Parity Software showed that one widely-used vendor's UNIX driver was five times slower to process events under SCO UNIX than under DOS on the same machine). In addition to high license fees, UNIX also has hidden costs as expert UNIX programmers and system administrators often demand high salaries.

Chapter 43

Which Programming Tool?

Hourglass Of Death: The Sequel

Some time ago I wrote a white paper entitled "The Hourglass of Death" which appeared in the May 95 issue of *Computer Telephony*. It talked about the different operating systems and tools developers could use to create C-T systems. A lot has changed since then. Windows 95 and Windows NT 4.0 are here, and 32-bit Windows applications are the fastest-growing segment of the desktop market, a trend also followed by telephony. We at Parity Software thought it was time to revisit the question of programming languages and operating systems so that we could provide concrete advice to our developer customers. Pretty simple, we thought – we'll write a couple of performance benchmarks and run them on different platforms, we'll have numbers in a week or two. Wrong. It proved to be a real engineering challenge.

In this chapter I'll share our experiences in designing and testing performance benchmarks for multi-line voice processing cards. We analyzed MS-DOS and Windows NT. Why these two platforms? DOS is still very widely used by developers, and has the major advantage that the operating system takes no overhead unless an explicit request is made by an application to do something. Hand-tuned code on DOS should have the lowest overhead of any alternative platform running on the same processor. Windows NT is our pick as the strategic platform of the future. It's a robust, pre-emptively multi-tasked, fully 32-bit operating system with the world's favorite graphical user interface. NT is designed from the ground up for server applications.

It would have been interesting to compare with Windows 95, which takes less memory and may be faster than NT for some types of applications, but at the time of writing Win95 drivers are not available for high-density voice cards, and we wanted to test at least 48 lines. Also, we do not believe Windows 95 is well-suited for voice servers because of the less robust design and reliance on 16-bit components.

We did want to include UNIX, but there were two obstacles. There are many flavors of UNIX, we would really have to test all of them to find which performed best (that's a lot of work, and we have other stuff to do like develop quality software tools). In addition, UNIX lacks some of the handy tools which come with NT, such as the Performance Monitor which shows the memory and CPU usage of each running application, making some of the important numbers much harder to obtain. We did some tests on UNIX to help optimize the UNIX version of our VOS kernel, but not comprehensively enough that we felt able to publish the results. Generally, UNIX seemed to have more overhead than NT.

Which Programming Tool? 669

It's Harder Than You Think

How do you make a voice processing performance benchmark? It's not as simple as you might think. Most benchmarks measure the time it takes to run a specially designed application. How fast is a Windows PC? Run a series of Word, Excel and PowerPoint macros which simulate typical office application tasks and see how long it takes from start to finish. That's the idea behind *PC Magazine's* Winstone and other benchmarks. But consider a typical voice application. It plays an opening greeting, plays a main menu, waits for a digit to be dialed, plays a message, and hangs up. Let's assume we press the digit one second after the main menu finishes playing. The time taken is pretty much fixed: the opening greeting is 5 seconds, the main menu is 10 seconds, one second pause, the digit is detected, the informational message takes 20 seconds: total 5+10+1+20 = 36 seconds. If you're running on an IBM XT, perhaps there's a tiny additional delay while the voice card drivers do stuff, so maybe it takes 36.001 seconds on your Pentium 100 system and 36.003 seconds on your XT. The problem is that typical voice applications spend most of their life waiting for a long operation to finish (play, record, get-digits etc.), which means that the vast majority of the time is spent idle and the total time taken is dominated by the fixed length of these operations. The overhead of the operating system and programming language is concentrated in the brief bursts of activity when a message is sent to the application to report the completion of a play or the detection of a digit, and when the application makes a request to the voice driver to begin the next action (by making an "API call"). In the 36 seconds it takes to run our typical app (an eternity for a CPU), this happens only a handful of times and the overhead of message passing and API calls is a vanishingly small fraction of the total time. Another problem is that playing and recording messages involves other components, such as the disk drive, unrelated to the voice card.

Ping Pong

Introducing: Ping Pong. Two voice channels are connected via a phone line. Line 1 sends a touch-tone digit to line 2. When line 2 detects the digit, it responds by dialing a digit back to line 1. Line 1 responds with a new digit, and so on. Dialing one digit is just about the quickest thing a voice card can do, giving us the best shot at pumping many messages per second through the system, and doesn't involve the disk or other components except the voice card and drivers. If two channels exchange a sequence of 100 digits, the overhead of passing messages back and forth will be incurred 100 times. Comparing the time taken for the same test on two different systems is a good comparison of the overhead caused by dispatching messages, making API calls and switching between tasks. If the difference in times is very small, you can increase the number of digits dialed until the spread is a matter of seconds and can be reproduced reliably in repeated runs. By adding more pairs of channels which ping-pong digits to each other, you can see how the overhead increases when many lines are active. Very little code is needed to implement Ping Pong in a given language, so the differing speeds of, for example, compiled C versus interpreted Visual Basic should not make a significant difference.

We believe that Ping Pong is an ideal benchmark for measuring the combined overhead of the operating system, voice card driver and programming environment. A configuration that scores well on Ping Pong will be able to run more lines or more features (database access etc.) than a system that scores badly on Ping Pong.

If You Try This At Home...

If a voice channel is dialing, its digit detection feature is temporarily turned off. This is designed to eliminate recognizing the echo of the digit which may come back through the phone network. A small additional guard time is added to allow for network delay in the echo, so digit detection is not turned on until a few milliseconds after the dial has finished. So, if line 1 is dialing, line 2 may recognize the digit and dial the response digit so quickly that line 1 will fail to detect it because digit detection has not yet been turned back on. (All we knew was that our tests weren't working until we finally figured this one out!). The first solution we tried was to ask the voice channel to report a digit when the digit stops, so-called "trailing edge detection". (The usual method is to report the digit as soon as it is recognized.) This guarantees that the response digit is not sent until after the first digit has stopped. However, the guard time following a digit dial can still cause the recognition to fail, so we needed the additional step of increasing the length of the digit tone so that a sufficiently long part of the reply tone would be "heard" following the end of the guard time. By

Which Programming Tool?

trial and error, we discovered that a tone time of 170 ms (upped from the default of 100 ms), combined with trailing edge detection, resulted in a reliable test with all flavors of Ping Pong. So that meaningful comparisons could be made, we made sure the same 170 ms setting was used in all Ping Pong flavors.

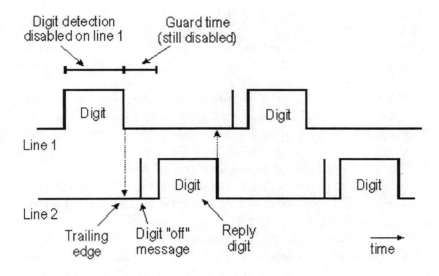

Two lines running Ping Pong exchanging two pairs of digits.

To run Ping Pong on more than one pair of lines, you have to worry about synchronization. You will be running several different programs, threads or tasks, one for each line or pair of lines. You don't want to start timing until all pairs have been started so that the system is fully loaded. Programmers usually solve this kind of problem by communicating between different programs using messages, semaphores, shared memory or other tricks. However, using these methods introduce their own overhead, making them unsuitable for use in a call processing benchmark. We came up with a better solution. Each pair just starts ping-ponging without worrying about whether the other lines are running yet. The test starts with a set number of "warm-up" cycles which are not counted, followed by a number of measured cycles. By trial and error, the number of warm-up cycles is tuned so that we can be sure that all pairs are up and running by the time the warm-up is completed. Using this technique, there is no need for the pairs to communicate in any way.

Programming Models

None of Parity's engineers has yet made the cover of a fashion magazine. No, a programming model is a method used by the application software and driver to structure the flow of a system. The simplest model to understand and program is called "synchronous". If the application wants to play a sound file, it makes an API call which will probably look something like *PlayWaveFile(Channel, WaveFileName)*. *PlayWaveFile* is a subroutine which calls the voice card driver; the call returns several seconds later when the play is finished. The program is suspended while *PlayWaveFile* is in progress, and this means that a separate executable must be running for each channel. The other extreme is called a "state machine" model. An API call like *PlayWaveFile* returns immediately to the application: it starts the play but doesn't wait. Later, the driver sends a message to the application when the play finishes. This allows a single executable to control more than one phone line, and is the only model which works on MS-DOS since DOS only allows one executable to run at a time. The big problem with state machine code is that it is very hard to write – you can't use the conventional commands in a programming language like "if" "then" and "else" statements to describe the flow of an application, you must keep track of the current state of each line in a state table.

There are two methods which may be used for the voice driver to send messages to the application: polling and call-back. In a polling model, the application must regularly call an API subroutine, maybe named *GetMessages()*, to check for messages. In a call-back model, the roles are reversed: the driver calls one of your subroutines each time a message is reported. Think of polling as like a voice mail system which you must call to check for messages, call-back is like having a pager. Everything else being equal, polling may have higher overhead because most calls to *GetMessages()* will be made when there are no messages pending, which is just a waste of time. Dialogic's version of *GetMessages()* on Windows NT, *sr_waitevt(0)*, took a very slow 7 ms on a 100 MHz Pentium with no messages are pending in our tests, making a modest 20 calls per second takes 140 ms/1 sec, which is about 14% of the total power of the CPU just to poll an empty message queue! (We expect much better performance from the native NT drivers from Dialogic, due out in Summer '97). However, the call-back model has its problems because you must be careful to avoid interference between the called subroutine and the rest of your code (programmers call these "critical sections"), and this model may be more difficult to manage when messages from other sub-systems, such as database managers and the keyboard, must also be taken into account. The theoretical efficiency of a given programming model may also be compromised by the implementation of the voice card driver. In the case of Dialogic's NT driver, the call-back model was no more efficient that polling, apparently a background thread created by the Dialogic DLLs is polling the board even when using a call-back model.

Multiple Lines

There are three main options for managing multiple lines: a state machine, multiple executables and multiple threads. (Technical note: when I say "executable", I mean an executing instance of an EXE or equivalent which is running a single thread). A state machine is a single executable, this technique can be used on any operating system. Running multiple executables is as simple as starting a different program (or a separate copy of one program) for each phone line, but requires that the operating system is multi-tasking. Multi-threading is a feature offered by NT and some UNIX flavors which allows a single program to have more than one execution path at the same time. A thread is a light-weight version of a process or task: it shares code and data with other threads in the same program, the operating system multi-tasks between all active threads. The typical use for a thread is to recalculate or print in the background while a user edits a spreadsheet in the foreground. Since threads within one program share code and data, the operating system should be able to switch between them faster than between separate executables with their separate address spaces, and we would therefore expect less overhead using threads. Windows NT version 4.0 offers an even lighter-weight relative of the thread called a "fiber" which may be faster still, we were not able to test this because Dialogic drivers do not (at the time of writing) support this latest NT version. Incidentally, early reports from the computer press indicate that NT v4.0 may be a little slower than v3.52, so there may be reasons to stick with the older release for a while.

The Tests

We wanted to test using a digital trunk to avoid any problems caused by distortion or echo on an analog line, we also wanted as many channels as possible which is much easier in a digital configuration. We designed a test system where we connected two Dialogic D/240SC-T1 cards together in one PC using a "null T-1 cable". This gave up to 48 channels for testing Ping Pong. Our test machine had a 100 MHz Pentium CPU with 96 MB RAM (yes, you read that right: almost one hundred megabytes). We chose this PC despite the relatively slow processor speed because it had the most RAM of the available machines in our lab (you'll see why we needed lots of memory in a moment).

We tested the three most popular options for creating PC-based call processing applications: the C programming language, Parity Software's VOS and Visual Basic using custom controls. C and VOS were tested on both DOS and NT, Visual Basic was of course tested on NT only. We used Parity Software's VoiceBocx ActiveX with Visual Basic. We tested each Ping Pong flavor on 2, 4, 8, 16, 32 and 48 channels (1 to 24 pairs).

On NT, we had four different C programming options. To make a state machine, you can using polling or call-back. Using the easier synchronous model, we could use multi-threading or run multiple executables. We tested all four possibilities. On DOS, the only option for C programming is a polled state machine model, so that's what we tested.

On NT, in addition to measuring the Ping Pong time, we also used NT's Performance Monitor to look at the total memory and CPU usage. A common question from our customers is how much memory and processor speed is needed to run a system of a given size. Our results help provide some recommendations. One final measurement we made was the startup time to get the system running. We found that it took a long time just to start many programs, especially under Visual Basic. That may not matter too much in a production system, but for development and testing a startup time of over a minute (to start a T-1 system based on Visual Basic) is significant compared with about 5 seconds (the same system based on VOS); during that entire minute each time the system is run the developer can't do much except watch the screen. (Shutting down can take time too — in Visual Basic it is awkward to automate a shutdown of a large number of separate programs. We didn't test this because we felt that the shutdown process was much more application-dependent than starting up).

Speed Results

The first table shows the time taken to run Ping Pong for 100 digits.

Nr of Chans	DOS C Poll	DOS VOS	NT C Poll	NT C Callback	NT C EXEs	NT C Threads	NT VOS	NT V Basic
2	10	13	19	13	19	19	14	15
4	10	13	19	13	20	19	14	15
8	10	13	19	13	22	19	14	15
16	10	13	19	14	23	20	15	22
32	11	15	20	16	36	26	17	74
48	11	18	24	24	54	34	25	Failed

Time Taken for Ping Pong, 100 digits per run.

To make the numbers easier to read, the numbers have been scaled so that the fastest version, a state machine polling model on DOS running just a single pair of channels, is 10 units. The actual time taken was 37.2 seconds, so to get times in seconds multiply the entry in the above table by 3.72. It was not possible to measure with Visual Basic on 48 channels since Ping Pong simply failed to run — there were system errors, Dialogic driver time-outs and other failure symptoms on each attempt. As the CPU and memory usage numbers verify, there was simply too much overhead from loading Visual Basic and OLE 48 times for the software to run successfully.

As you can see from the table, the speed rankings vary a little depending on the number of channels being run. Since overhead matters most in larger systems, the following table shows speed rankings based on 32-channel Ping Pong, the largest number of lines that all models were able to run. The rankings are shown in the following table.

	O/S	Model	Time
Best	MS-DOS	C Poll	11
	MS-DOS	VOS	15
	NT	C Callback	16
	NT	VOS	17
	NT	C Poll	20
	NT	C Multi-threads	26
	NT	C Multi-EXEs	36
Worst	NT	Visual Basic	74

Speed rankings based on 32-channel Ping Pong

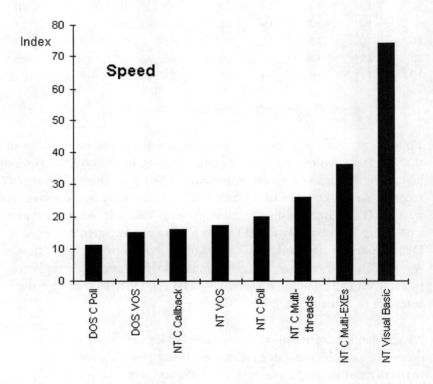

Models compared by speed (smaller values are better).

CPU Load Results

CPU loading was measured by using NT's Performance Monitor utility. Results are accurate to the closest 5%.

Nr of Chans	NT C Poll	NT C Callback	NT C EXEs	NT C Threads	NT VOS	NT V. Basic
2	50%	5%	5%	5%	5%	5%
4	50%	5%	5%	5%	5%	10%
8	50%	10%	15%	5%	10%	25%
16	50%	30%	35%	15%	25%	95%
32	60%	65%	75%	20%	55%	100%
48	65%	65%	90%	20%	60%	(Failed)

CPU usage according to Performance Monitor.

The CPU usage numbers may be distorted by the fact that Ping Pong is designed to pump messages through the system as quickly as possible. The quicker Ping Pong runs, the more messages are passed. What we'd really like to know is how much CPU gets used on each API call and message – we shouldn't penalize more efficient programming models if they use more CPU simply because they're processing more messages. To get the CPU load per message, we should correct the CPU use percentage by dividing by the number of messages per second. The number of messages per second is inversely proportional to the time taken for Ping Pong to process 100 digits. A measure of the CPU load per message is therefore obtained by multiplying the CPU load percentage by the Ping Pong time shown in Table 1. Scaling the results so that the lowest load is 10, we obtain a result we call the "CPU load index" as shown in the following table, based on the 32-channel numbers.

	Model	CPU Load Index
Best	C Multi-threads	10
	VOS	18
	C Callback	20
	C Poll	23
	C Multi-EXEs	51
Worst	Visual Basic	142

CPU loading index ranking (NT only).

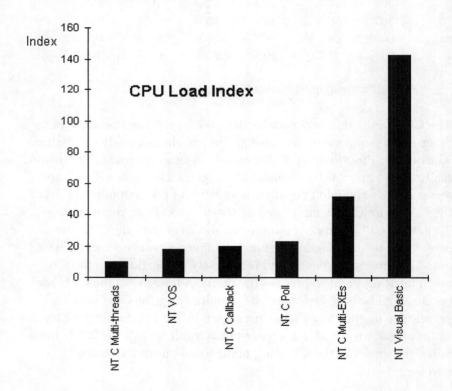

CPU load index (smaller is better).

Note that the large CPU load shown for Visual Basic has very little to do with the fact that VB is an interpreted language. The great majority of the overhead comes from running multiple EXEs and the fact that each API call and message must traverse a lengthy path through

Windows' OLE sub-system. We would expect very similar results from Visual C++ or Delphi, which are compiled languages. Verifying this, VOS is also a p-code interpreted language, and you can see that VOS actually scores better than any of the C programming models.

Memory Usage

NT's Performance Monitor was used to estimate the memory usage of the different Ping Pong flavors. Results are again not precise since they vary slightly from run to run depending on the history of NT's internal dynamic memory allocation. Note that the memory usage is *in addition* to anything already loaded by the operating system. NT generally requires at least 20 MB, so you should add 20 MB plus any additional components such as database managers, remote maintenance programs and so on to the NT numbers below to find the total memory required. Plus, of course, in a real system the voice application code will require significantly more memory because Ping Pong is a very small program. NT does offer virtual memory, so a system may load successfully if the RAM required is more than the physical RAM, but paging increases exponentially and may quickly have a major impact on performance in this case. This does not mean that systems cannot be successfully designed with less memory, but the developer should be careful to run systematic tests with peak loading in the desired configuration.

Nr of Chans	DOS C Poll	DOS VOS	NT C Poll	NT C Callback	NT C EXEs	NT C Threads	NT VOS	NT V. Basic
2	0.1	0.4	1.7	1.7	1.8	1.8	3.5	5.3
4	0.1	0.4	1.7	1.7	3.1	1.9	3.5	9.0
8	0.1	0.4	2.0	1.8	5.9	2.1	3.5	16.5
16	0.1	0.4	2.2	2.0	11.1	2.4	3.7	31.6
32	0.1	0.5	2.7	2.2	21.7	3.0	3.8	62.3
48	0.1	0.6	3.1	2.4	32.2	3.6	3.8	95.0

Memory usage (MB).

	O/S	Model	Memory (MB)
Best	MS-DOS	C Poll	0.1
	MS-DOS	VOS	0.5
	NT	C Callback	2.2
	NT	C Poll	2.7
	NT	C Multi-threads	3.0
	NT	VOS	3.8
	NT	C Multi-EXEs	21.7
Worst	NT	Visual Basic	62.3

Memory usage rankings based on 32-channel Ping Pong.

Visual Basic appears to be a big memory hog. To run 48 lines, corresponding to a typical system with two T-1s, the VB version of Ping Pong requires about 95 MB in addition to 20 MB for NT, making a total of 115 MB. Note that Ping Pong is a very small application, and the only control which is loaded onto a VB form is the VoiceBocx ActiveX to control the Dialogic board. Each additional control is likely to add something in the range of 0.4 to 4 MB per channel. As a quick test, we ran a modified version of the Visual Basic Ping Pong where we opened a database using Visual Basic's built-in database control. The memory requirements shot up, as shown in Table 7. A whopping 155.3 MB was needed in addition to NT's 20 MB or so. In a real-life system, the Visual Basic application could have other custom controls loaded and there would be other software also loaded under NT (database server, remote maintenance software etc.), so the total memory requirement for a 48-line system written in Visual Basic under NT can easily reach into the 150 MB to 200 MB range.

We also repeated the test using a minimal ActiveX control (which we called "Empty") which has no code other than the framework generated by Visual C++'s Control Wizard. This verifies that VoiceBocx is not wasting memory, the difference between the memory requirement with VoiceBocx and the memory requirement with Empty, about 1 MB per channel, is explained by the fact the Dialogic DLLs must be loaded into each copy of the EXE, it is not VoiceBocx which is wasting memory.

Nr. instances	"Empty" control	VoiceBocx	VoiceBocx + database ctrl.
2	3.0	5.3	8.3
4	5.0	9.0	13.9
8	9.2	16.5	26.5
16	17.6	31.6	50.5
32	33.7	62.3	107.8
48	66.6	95.0	155.3

Visual Basic memory usage, values shown in MB.

We've had skeptical reactions from some people when we've quoted these numbers. Why does VB need so much memory? One important factor is the way Windows NT organizes memory. Each process (running EXE) has its own memory space which is kept completely separate from every other process. If a program requires 1 MB of data, running two copies of the program will require 2 MB. Every program which runs loads a complete copy of the Windows system DLLs (KERNEL32.DLL, USER32.DLL, GDI32.DLL) into memory. (This is quite different from 16-bit Windows, where a DLL is loaded only once and shared between all processes). Visual Basic adds its own run-time DLL (VBRUNxx.DLL) and loading an ActiveX requires all the OLE DLLs in addition. Any Dialogic application must also load the Dialogic DLLs, which will include at least LIBDXX.DLL and LIBSRL.DLL. A one-line C program which simply opens a Dialogic channel requires a couple of megabytes to load for each copy.

Fortunately for Parity Software's credibility, you can try your own Visual Basic memory usage test in a matter of minutes. Create an empty Visual Basic project, load a couple of controls onto the form (load external ActiveX/OCX controls rather than built-in controls), build the EXE. Start Performance Monitor and configure it to monitor memory usage. Start the EXE, and you will see the meter jump up. Start further copies, and it will make a big jump for each copy. You can quickly demonstrate for yourself that running 48 copies of a realistic Visual Basic application can require a *lot* of memory.

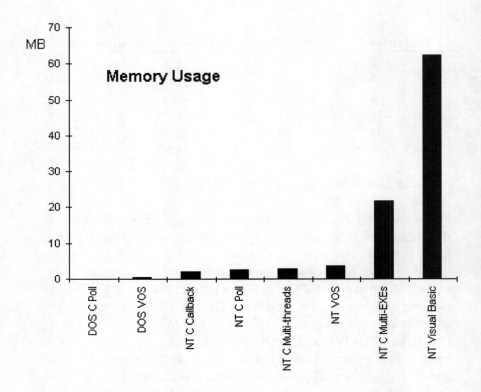

Memory usage (smaller is better).

Startup Times

We had not originally planned to include the startup time in our testing. However, we got quite frustrated waiting up to a minute or two for the Visual Basic test to get started, this can be a significant damper on productivity for a developer. It might also be a factor if a system must be re-started quickly after being taken down for updating, maintenance or after a crash.

Which Programming Tool?

Nr Chans	NT C Poll	NT C Callback	NT C EXEs	NT C Threads	NT VOS	NT Visual Basic
2	2	2	2	2	2	15
4	2	2	2	2	2	18
8	2	2	3	3	2	24
16	4	4	5	4	5	45
32	7	7	10	8	8	110
48	10	10	24	10	10	(Failed)

Startup times (seconds), smaller is better.

Again basing our ranking on the 32-channel test, we arrive at the following result.

	Model	Time (secs.)
Best	C Poll	7
	C Callback	7
	VOS	8
	C Multi-threads	8
	C Multi-EXEs	10
Worst	Visual Basic	110

Startup time ranking (NT only).

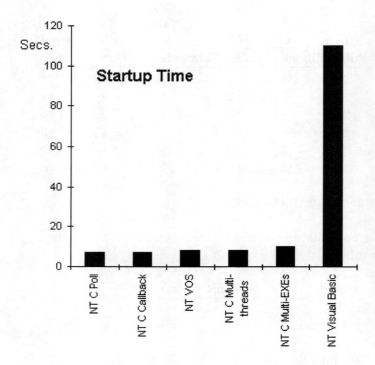

Startup time (smaller is better).

Interpretation

There are lies, damn lies and benchmarks. What do all these numbers really mean?

We feel that the memory usage and startup time values give clear guides to what will happen in a real-world application. Since Ping Pong is a very simple application, real systems will take more memory and be slower to start. However, this will be very application-dependent. The numbers shown by Ping Pong give an excellent idea of the relative performance penalties extracted by the different development models.

The CPU usage and speed index measurements are more open to discussion. Ping Pong was designed to measure the overhead of making API calls and processing messages. A real application will, on average, probably process significantly fewer messages per second than Ping Pong, and will be loading other components of the system, such as the

Which Programming Tool? 685

disk drive. However, Ping Pong should be effective in measuring the *relative* performance of two different programming models. If Ping Pong takes 95% of the CPU capacity running a 32-line test, it is a fair bet that you would not be very successful running a system of this size or larger on that particular computer. A peak load in a typical application will resemble the loading caused by Ping Pong, and even if this only happens for a short length of time the system must still be able to cope.

One question we did not investigate was the dependence on the Dialogic drivers in our tests. The driver architecture is clearly very inefficient in certain respects (7 ms on a 100 MHz Pentium to poll an empty event queue!), using drivers from other vendors might have had a significant impact on the results. We would anticipate that a more efficient driver architecture would give a larger advantage to state machine models on NT, especially the call-back model.

There are no doubt other benchmark designs which would shed light on other aspects of real-life computer telephony systems. We'd love to hear from any of you who have comments on our approach to Ping Pong, suggestions for alternative benchmarks, details of your own tests. E-mail us at tech@paritysw.com.

Conclusions

DOS beats NT across the board for performance.

MS-DOS is more efficient than NT because DOS has no overhead unless an application makes an API call asking DOS to do something. NT has additional overhead due to task-switching and other operations, and requires a lot more RAM.

VOS for DOS beats all C programming models on NT.

Even with a full-blown application generator engine which interprets multi-threaded p-code and manages other devices such as the screen, keyboard, serial port etc., VOS for DOS shows better performance than any of the C programming models on NT. Parity Software has many years' experience in refining the VOS kernel on DOS, and as a result even today VOS still provides a very competitive and stable environment which is especially well-suited to high-density systems.

The biggest systems created so far in a single PC by Parity Software customers have been built using VOS for DOS.

Visual Basic comes dead last, by a wide margin, in every performance category.

It is not surprising that Visual Basic performs poorly with many lines since it was not designed for real-time applications, for running many applications simultaneously or for meeting the special demands of telephony. VB is a great option for entry-level computer telephony developers who are already familiar with the Basic language and want to create smaller systems, however with the current architecture it will always be soundly beaten by development tools which are optimized for higher-density systems. The limitations of VB are due to running multiple EXEs and loading large driver DLLs and/or custom controls, so we expect the same performance limitations with Delphi, Visual C++, Optima++, PowerBuilder and other tools.

C programmers: Write polled state machines on DOS, on NT the choice is less clear.

If you are a C programmer, DOS is the most efficient platform, and a polled state machine your only model. Writing multiple EXEs is the easiest C programming option, but has slower performance and takes a lot more memory so in most cases it appears to be well worth taking the extra trouble to write state machine or multi-threaded code. For the best possible performance on NT, a C programmer will need to create a sophisticated, load-balanced state machine architecture like the one we created for the VOS NT kernel.

VB is fat and slow, there is no benefit in further tuning of the VoiceBocx ActiveX.

As you might imagine, one of our objectives with creating benchmarks was to enable us to optimize the performance of our VOS and CallSuite development tools. In the case of VoiceBocx, the CallSuite telephony control for Visual Basic and other environments, it turned out that the overhead was almost entirely in the host environment, so there was no point in us trying to optimize VoiceBocx further, the benefit in a real application would be negligible.

VOS has similar performance to C both on DOS and NT.

The results show that VOS is generally more efficient than some C programming options, a little less efficient than others. However, there is no one C programming model that beats VOS in all areas, so we feel that VOS has an excellent compromise design. Did we optimize VOS specifically to look good on Ping Pong? Absolutely not. We used other benchmarks (which we don't discuss here) to optimize typical applications with less telephony messaging and also p-code intensive applications. We were able to optimize VOS so that a pure p-code application, making no telephony API calls, executes almost as fast as on DOS, and still is able to process messages in a highly efficient manner. We created a hybrid state machine model which dynamically adapts to the message through-put from different system components (telephony driver, p-code interpreter, keyboard etc.) and assigns more processor power to the different elements as required. VOS provides a programming language which is simpler than Visual Basic combined with a very simple programming model which anyone with a little programming experience can understand. Combined with the very low memory requirements of VOS (a 48-line system can easily be run in a 4 MB machine on DOS or a 32 MB machine running NT), VOS provides an excellent combination of high performance and simple, robust programming language for telephony developers.

Glossary

...with thanks to Harry Newton for many of the following definitions. Harry is the author of "Newton's Telecom Dictionary" published by Telecom Library Inc. 212-691-8215, 800-LIBRARY or 800-999-0345.

A & B Bits

Bits used in digital environments to convey signaling information. A bit equal to one generally corresponds to loop current flowing in an analog environment, A bit zero corresponds to no loop current, i.e. to no connection. Other signals are made by changing bit values: for example, a flash-hook is sent by briefly setting the A bit to zero. See A & B Signaling.

A & B Leads

Additional leads used typically with a channel bank two-wire E&M interface to certain types of PBXs (also used to return talk battery to the PBX).

A & B Signaling

Procedure used in most T-1 transmission links where one bit, robbed from each of the 24 subchannels in every sixth frame, is used for carrying dialing and controlling information. A type of in-band signaling used in T-1 transmission.

A-Law Encoding

Digital telephony equipment usually stores and transmits audio signals in 8-bit samples using PCM encoding. The value of the 8-bit sample corresponds to the amplitude (volume, loudness) of the audio signal at the time the sample was taken. There is a translation table which is used to determine the amplitude given the sample value. There are two different tables in common use: the A-law table, which is widely used in Europe and Asia on E-1 trunks, and Mu-law (m-law) encoding used in the US, Canada and a few other countries.

ABCD Bits

Signaling bits used on E-1 digital trunks to convey information about the state of a call.

For example, ABCD bit values may be used to signal an incoming call, disconnect, seize and so on. They are closely analogous to the A and B bits commonly used on T-1 digital trunks. E-1 is a digital trunk standard used in many countries outside of North America. There are 32 channels (time-slots) in contrast to 24 channels on T-1, hence the faster bit rate of 2.048 MHz versus 1.544 MHz for T-1. Channels 0 and 16 are used to carry the ABCD bits and synchronization (framing) bits, hence only 30 channels are available for audio conversations. Thus, ABCD bits are carried out-of-band in contrast to the in-band robbed-bit scheme used by T-1.

Unfortunately, the conventions used for the meanings of bit values vary widely from country to country and even from region to region in different countries. For example, in Greece, the "idle" line condition is signaled by ABCD=1010, a seize is signaled by changing to ABCD=0010. In Italy, idle is ABCD=0101 and seize is ABCD=1001. Differences in ABCD and interregister signaling is a major headache for equipment vendors to different E-1 implementations.

ACD

See Automatic Call Distributor.

ActiveX

A marketing term used by Microsoft to refer to many different types of OLE objects in Windows. Microsoft wants you to pronounce ActiveX as "active" with a silent X, but nobody does. In telephony, most often encountered in reference to ActiveX controls, which are custom controls which can be loaded into "visual" tools like Visual Basic and Delphi.

Adaptive Differential Pulse Code Modulation

See ADPCM.

ADPCM

Adaptive Differential Pulse Code Modulation. A speech coding method which calculates the difference between two consecutive speech samples in standard PCM-coded telecom voice signals. This calculation is encoded using an adaptive filter and therefore is transmitted at a lower rate than the standard 64 Kbps technique. Typically, ADPCM allows an analog voice conversation to be carried within a 32k-kbit/s digital channel; 3 or 4 bits are used to describe each sample, which represents the difference between two adjacent samples. Sampling is generally done 8,000 times a second (8 KHz). The Dialogic default ADPCM format, however, uses 4-bit sampling at 6 KHz.

ADSI

See Analog Display Services Interface.

AEB

Analog Expansion Bus. The analog voice processing bus designed by Dialogic for interfacing DTI/124, D/4x, AMX and other voice response component boards which fit in an AT-expansion slot of a PC. See also PEB, which is the more modern digital PCM expansion bus.

AMIS

See Audio Messaging Interchange Specification.

Amplitude

The distance between high or low points of a waveform or signal. Also referred to as the wave "height." When applied to audio signals, the amplitude is a measure of the loudness of volume of the sound.

Amplitude Distortion

The difference between the output wave shape and the input wave shape.

Amplitude Equalizer

A corrective network that is designed to modify the amplitude characteristics of a circuit or system over a desired frequency range. Such devices may be fixed, manually adjustable, or automatic.

Amplitude Modulation

Also called AM, it's a method of adding information to an electronic signal in which the signal is varied by its height to impose information on it. "Modulation" is the term given to imposing information on an electrical signal. The information being carried causes the amplitude (height of the sine wave) to vary. In the case of LANs, the change in the signal is registered by the receiving device as a 1 or a 0. A combination of these conveys different information, such as words, numbers or punctuation marks.

Analog

Comes from the word "analogous," which means "similar to." In telephone transmission, the signal being transmitted — voice, video, or image — is "analogous" to the original signal. In other words, if you speak into a microphone and see your voice on an oscilloscope and you take the same voice as it is transmitted on the phone line and ran that signal into the oscilloscope, the two signals would look essentially the same. The only difference is that the electrically transmitted signal (the one over the phone line) is at a higher frequency. In correct English usage, "analog" is meaningless as a word by itself. But in telecommunications, analog means telephone transmission and/or switching which is not digital. See Analog Transmission.

Analog / Digital Converter

An A/D Converter. Pronounced: "A to D Converter." A device which converts an analog signal to a digital signal.

Analog Display Services Interface

ADSI. A protocol for driving special analog telephones with small LED screens. Used for example in enhanced banking-by-phone applications where the caller can see menus or account information on the display.

Analog Expansion Bus

See AEB.

Analog Transmission

A way of sending signals — voice, video, data — in which the transmitted signal is analogous to the original signal. In other words, if you spoke into a microphone and saw your voice on an oscilloscope and you took the same voice as it was transmitted on the phone line and threw that signal onto the oscilloscope, the two signals would look essentially the same. The only difference would be that the electrically transmitted signal would be at a higher frequency.

ANI

Automatic Number Identification. A phone call arrives at your home or office. At the front of the phone call is a series of digits which tell you, the phone number of the phone calling you. These digits may arrive in analog or digital form. They may arrive as touchtone digits inside the phone call or in a digital form on the same circuit or on a separate circuit. You will need some equipment to decipher the digits AND do "something" with them. That "something" might be throwing them into a database and bringing your customer's record up on a screen in front of your telephone agent as he answers the call. "Good morning, Mr. Smith."

Anti-Aliasing

A computer imaging term. A blending effect that smoothes sharp contrasts between two regions of different colors. Properly done, this eliminates the jagged edges of text or colored objects. Used in voice processing, anti-aliasing usually refers to the process of removing spurious frequencies from waveforms produced by converting digital signals back to analog.

API

See Applications Programming Interface.

Applications Programming Interface

A set of functions provided by an operating system or device driver. Usually the functions are provided as subroutine calls which can be used by programs written in C, or sometimes other programming languages.

Application Generator

A program to generate actual programming code. An applications generator will let you produce software quickly, but it will not allow

Glossary 695

you the same flexibility had you programmed it from scratch. Voice processing "application generators", despite the name, often do not generate programming code. Instead they are self-contained environments which allow a user to define and execute applications.

ARU

See Audio Response Unit.

ASCII

American Standard Code for Information Interchange. A character set which gives a numerical value from 32 to 127 to commonly used letters, numbers and symbols. For example, an upper-case A is assigned the value 64. The IBM PC extended this standard to 255 characters which included symbols required in countries other than the USA such as accented letters. This extended character set is sometimes called the 8-bit ASCII character set to distinguish it from the original standard, which is then called the 7-bit character set since values from 0 to 127 can be represented using 7 bits, and can therefore be transmitted over modem lines using 7 rather than 8 data bits, for example. Computer files stored using these characters are called ASCII text, or sometimes just text files. These files almost always use one byte (an 8-bit binary unit which can take a value from 00000000 = decimal zero to 11111111 = decimal 255 when interpreted as a number) for each character. ASCII files are simpler than word processing files, which have complex codes embedded within them, and have the advantage that most programs can read and write in this format. However, there are only ASCII codes for the most rudimentary formatting information: tab, end of line and end of page markers. All other information such as font size, tab positions and so on is lost when a file is stored as ASCII.

ASR

Automatic Speech Recognition.

Assisted Telephony

A simplified set of functions for TAPI which makes it easy for applications to do basic things like dialing a call.

ASVD

Analog SVD. See SVD.

Asynchronous Mode

A mode for calling driver functions where a function call to perform a time-consuming action like playing a file returns immediately to the application once the operation is started, and later notifies the application of the completion of the operation by sending a message or event.

Audio

Sound you hear which may be converted to electrical signals for transmission. A human being who hasn't had his or her ears blown by listening to a Sony Walkman or a ghetto blaster can hear sounds from about 15 to 20,000 hertz.

Audio Frequencies

Those frequencies which the human ear can detect (usually in the range of 20 to 20,000 hertz). Only those from 300 to 3,000 hertz are transmitted through the phone, which is why the phone doesn't sound "Hi-Fi."

Audio Menu

Options spoken by a voice processing system. The user can choose what she or he wants done by simply choosing a menu option by hitting a touchtone on his phone or speaking a word or two. Computer or voice processing software can be organized in two basic ways — menu-driven and non-menu driven. Menu-driven programs are easier for users to use, but they can only present as many options as can be reasonably spoken in a few seconds. Audio menus are typically played to callers in automated attendant/voice messaging, voice response and transaction processing applications. See also Menu and Prompts.

Audio Messaging Interchange Specification

AMIS. Issued in February 1990, AMIS is a series of standards aimed at addressing the problem of how voice messaging systems produced by different vendors can network or inter-network. Before AMIS, systems from different vendors could not exchange voice messages. AMIS deals only with the interaction between two systems for the purpose of exchanging voice messages. It does not describe the user interface to a voice messaging system, specify how to implement AMIS in a particular systems or limit the features a vendor may implement.

AMIS is really two specifications. One, called AMIS-Digital, is based on completely digital interaction between two voice messaging systems. All the control information and the voice message itself, is conveyed between systems in digital form. By contrast, the AMIS-Analog specification calls for the use of DTMF tones to convey control information and transmission of the message itself is in analog form. AMIS was discussed in detail in the October 1990 issue of *Business Communications Review*. AMIS specifications are available from Hartfield Associates, Boulder CO. 303-442-5395.

Audio Response Unit

A device which translates computer output into spoken voice. Let's say you dial a computer and it says, "If you want the weather in Chicago, push 123," then it gives you the weather. But that weather would is "spoken" by an audio response unit. Here's a slightly more technical explanation: An audio response unit is a device that provides synthesized voice responses to dual-tone multi-frequency signaling input. These devices process calls based on the caller's input, information received from a host data base, and information carried with the incoming call (e.g., time of day). ARUs are used to increase the number of information calls handled and to provide consistent quality in information retrieval.

Audiotex

A generic term for interactive voice response equipment and services. Audiotex is to voice what on line data processing is to data terminals. The idea is you call a phone number. A machine answers, presenting you with several options, "For information on Plays, push 1; for information on movies, push 2; for information on Museums, push 3." If you push 2, the machine may come back, "For movies on the south side of town, push 1; for movies on the north side of town, push 2, etc." See also Information Center Mailbox.

Audiotext

A different, and less preferred, spelling of Audiotex. See Audiotex.

Automated Attendant

A device which is connected to a PBX. When a call comes in, this device answers it and says something like, "Thanks for calling the ABC Company. If you know the extension number you'd like, press that extension now and you'll be transferred. If you don't know it, press "0" (zero) and the live operator will come on. Or, wait a few seconds and the operator will come on anyway." Sometimes the automated attendant might give you other options, such as, "dial 3" for a directory. Automated attendants are very new. They are connected also to voice mail systems ("I'm not here. Leave a message for me."). Some people react well to automated attendants. Others don't. A good rule to remember is before you spring an automated attendant on your people/customers/subscribers, etc., let them know. Train them a little. Ease them into it. They'll probably react more favorably than if it comes as a complete surprise. The first impression is rarely forgotten, so try to make it a good experience for the caller. See also Dial By Name.

Automatic Call Distributor

ACD. A specialized phone system used for handling many incoming calls. Once used only by airlines, rent-a-car companies, mail order companies, hotels, etc., it is now used for any company that has many incoming calls (e.g. order taking, dispatching of service technicians, taxis, answering technical questions, etc.). An ACD performs four functions. 1. It will recognize and answer an incoming call. 2. It will look in its database for instructions on what to do with that call. 3. Based on these instructions, it will send the call to a recording that "somebody will be with you soon, please don't hang up!" or to a voice response unit (VRU). 4. It will send the call to an operator within a group of operators as soon as that operator has completed their previous call, and/or the caller has heard the canned message.

The term Automatic Call Distributor comes from distributing the incoming calls in some logical pattern to a group of operators. That pattern might be Uniform (to distribute the work uniformly) or it may be Top-down (the same agents in the same order get the calls and are kept busy. The ones on the top are kept busier than the ones on the bottom). Distributing calls logically is the function most people associate with an ACD, though it's not the most important. The management information which the ACD produces is much more valuable. This information is of two sorts: 1. The arrival of incoming calls (when, how many, which lines, from where, etc.) and 2. How many callers were put on hold, asked to wait and didn't. This is called information on Abandoned Calls. This information is very important for staffing, buying lines from the phone company, figuring what level of service to provide to the customer and what different levels of service (how long for people to answer the phone) might cost.

Automatic Call Sequencer

ACS. A device for handling incoming calls. Typically it performs three functions. 1. It answers an incoming call, gives the caller a message, and puts them on "Hold." 2. It signals the agent (the person who will answer the call) which call on which line to answer. Typically, the call which it signals to be answered is the call which has been on "hold" the longest. 3. It provides management information, such as how many abandoned calls there were, how long the longest person was kept on hold, how long the average "on hold" was, etc.

Automatic Call Unit

ACU. A device that places a telephone call on behalf of a computer.

Automatic Callback

When a caller dials another internal extension and finds it busy, the caller dials some digits on his phone or presses a special "automatic callback" button. When the person he's calling hangs up, the phone system rings his number and the number of the original caller and the phone system automatically connects the two together. This feature saves a lot of time by automatically retrying the call until the extension is free.

Automatic Circuit Assurance

ACA is a PBX feature that helps you find bad trunks. The PBX keeps records of calls of very short and very long duration. If these calls exceed a certain parameter, the attendant is notified. The logic is that a lot of very short calls or one very long call may suggest that a trunk is hung, broken or out of order. The attendant can then physically dial into that trunk and check it.

Automatic Clock Fallback

A fault-tolerant feature of a system with several components which all access a trunk or bus. One component (the "clock master") is responsible for setting the clock for the bus or trunk, other components take their clock from this master clock. In the event that the master device fails, Automatic Clock Fallback immediately assigns another device to be the master so that the system can continue to function.

B-Channel

An ISDN channel used to convey audio or data rather than signaling information. See D-Channel. "B" stands for "bearer", as in the bearer of good (or bad) tidings, and "D" stands for "data".

Baby Bell

A term used for an RBOC, one of the Regional Bell Operating Companies responsible for local calls in the US following the break-up of AT&T.

Backward Signal

A tone signal used in compelled signaling, usually on E-1 networks. A backward signal is essentially a tone sent to the originator from the receiver of a call. For example, if a device is dialing a number, the device

Glossary 701

is the originator and the central office may send backward tones to the device to request information about the call routing.

Basic Rate Interface

BRI. An ISDN connection with just one B-channel (for audio) and one D-channel (for data and routing information). When you finally get an ISDN phone at home, it will probably be based on a BRI service. See also PRI (Primary Rate Interface).

Battery

All telephone systems work on DC (direct current). DC power is what you use to talk on. Often the DC power is called "talking battery." Most key systems and many PBXs plug directly into an AC on the wall, but that AC power is converted by a built-in power supply to the DC power the phone system needs.

All central offices (public exchanges) used rechargeable lead acid batteries to drive them. These batteries perform several functions: 1. They provide the necessary power. 2. They serve as a filter to smooth out fluctuations in the commercial power and remove the "noise" that power often carries. 3. They provide necessary backup power should commercial power stop, as in a "blackout" or should it get weak, as in a "brownout."

In short, "battery" is the term used to reference the DC power source of a telephone system. 2. Storage battery used with central office switching systems and PBXs serving locations which cannot tolerate outages. Batteries serve the following purposes: Act as a filter across the generator or power rectifier output to smooth out the current and reduce noise; provide a cushion against periodic overloads exceeding the generator/rectifier capacity; supply emergency power for a limited time in event of commercial power failure.

Board Device

In the Dialogic API, some function calls reference board devices to set parameters or perform operations which affect all channels on that board. A board device may be a "virtual board", a single physical board may include more than one virtual board and hence more than one board device.

Board Locator Technology

See BLT.

BLT

Board Locator Technology. Some Dialogic boards have an ID switch which can be set to 0, 1, 2... These are known as BLT boards, and can share a single IRQ and shared address range in memory.

Also, Bacon, Lettuce and Tomato sandwich, a favorite of telephone engineers everywhere.

Bong

A tone that long distance carriers and value added carriers make in order to signal you that they now require additional action on your part — usually dialing more digits.

BRI

See Basic Rate Interface.

Busy

In use. "Off-hook". There are slow busies and fast busies. Slow busies are when the phone at the other end is busy or off-hook. They happen 60 times a minute. Fast busies (120 times a minute) occur when the network is congested with too many calls. Your distant party may or may not be busy, but you'll never know because you never got that far.

Cadence

In voice processing, cadence is used to refer to the pattern of tones and silence intervals generated by a given audio signal. Examples are busy and ringing tones. A typical cadence pattern is the US ringing tone, which is one second of tone followed by three seconds of silence. Some other countries, such as the UK, use a double ring, which is two short tones within about a second, followed by a little over two seconds of silence.

Call Center

A place where calls are answered and calls are made. A call center will typically have lots of people (also called agents), an automatic call distributor, a computer for order-entry and lookup on customers' orders. A Call Center could also have a predictive dialer for making lots of calls quickly.

Call Completion

This is industry jargon for "putting the call through". When a call has been completed, there is an unbroken ("complete") circuit made between the caller and recipient of the call. This circuit is known as the talk path.

Call Progress Analysis

The automated determination by a piece of telecommunications equipment as to the result of dialing a number. For example, the result of the analysis might be a busy tone, ringing at the other end but no answer after a pre-set number of rings, an answered call and so on. The analysis involves detecting the various call progress tones which will be generated by the telephone network as the call is put through.

Call Progress Monitoring

Closely analogous to call progress analysis, call progress monitoring may be active during the entire length of a conversation. For example, when a call is placed across a PBX or in a country which does not provide for loop current drop disconnect supervision, it may be necessary for equipment to monitor for a "re-order" or dial tone to determine that the caller hung up. This would be classified as call progress monitoring since it must take place during the entire call, not just when a number is dialed or a transfer is initiated.

Call Progress Tone

A tone sent from the telephone switch to tell the caller of the progress of the call. Examples of the common ones are dial tone, busy tone, ringback tone, error tone, re-order, etc. Some phone systems provide additional tones, such as confirmation, splash tone, or a reminder tone to indicate that a feature is in use, such as confirmation, hold reminder, hold, intercept tones.

Call Status Transition Events

Dialogic jargon for unsolicited notifications which are generated by a call, such as a disconnect.

Call Supervision

Another term for call progress analysis or call progress monitoring.

Called Subscriber Identification

See CSI.

Caller ID

A name for a service which displays the calling party's telephone number on a special display device.

Calling Tone

See CNG.

CAS

1. Communicating Applications Specification. A high-level API (application programming interface) developed by Intel and DCA that was introduced in 1988. CAS enables software developers to integrate fax capability and high-speed, error-corrected file transfer into their applications.

2. Channel Associated Signaling. A form of out-of-band signaling used on digital trunks. Similar in some ways to Common Channel Signaling (CCS) except that some bits (typically 4 out of the 8-bit value in the

Glossary

time-slot) are reserved to provide a fixed framing pattern for synchronization purposes. Time-slots 0 and 16 are CAS signaling time-slots on most E-1 protocols.

CCM

A type of Dialogic board, including some DTI, DMX and MSI board models.

CCS

Common Channel Signaling. A form of out-of-band signaling used on digital trunks. One time-slot (the "common channel") is used to convey signaling information. An example of a CCS implementation is the D channel of ISDN or the CCITT number 7 protocol which is used in CO to CO communication on DS-1 level trunks. The full band-width of the common channel is available for signaling, framing is accomplished by other means, generally by a bit pattern appended to each frame following the set of time-slot samples.

CCITT

Comite Consultatif Internationale de Telegraphique et Telephonique (International Telephone and Telegraph Consultative Committee). An international standards body. Renamed on March 1, 1993 to the Telecommunications Standard Sector (TSS).

CED

Called Station Identification. A tone used in the handshaking used to set up a fax call: the response from a fax machine to the called machines CNG tone.

Central Office

See CO.

Centrex

Centrex is a business telephone service offered by a local telephone company from a local central office. Centrex is basically single line telephone service delivered to individual desks (the same as you get at your house) with features, i.e. "bells and whistles," added. Those "bells and whistles" include intercom, call forwarding, call transfer, toll restrict, least cost routing and call hold (on single line phones).

Centrex is known by many names among operating phone companies, including Centron and Cenpac. Centrex comes in two variations — CO and CU. CO means the Centrex service is provided by the Central Office. CU means the central office is on the customer's premises.

CFR

Confirmation To Receive frame. May be sent as part of Phase B of the Group 3 fax protocol.

Channel

A path of communication, either electrical or electromagnetic, between two or more points. Also called a circuit, facility, line, link or path.

Channel Device

In the Dialogic API, most function calls address channel devices. A channel device typically processes the audio from a single telephone connection.

Channel Bank

A multiplexer. A device which puts many slow-speed voice or data conversations onto one high-speed link and controls the flow of those "conversations." Typically the device that sits between a digital circuit — say a T-1 — and a couple of dozen voice grade lines coming out of a PBX. One side of the channel bank will be connections for terminating two pairs of wires or a coaxial cable — those bringing the T-1 carrier in. On the other side are connections for terminating multiple tip and ring single line analog phone lines or several digital data streams. Sometimes you need channel banks. Sometimes, you don't. For example, if you're shipping a bundle of voice conversations from one digital PBX to another across town in a T-1 format — and both PBXs recognize the signal — then you will probably not need a channel bank. You'll need a Channel Service Unit (CSU). If one, or both, of the PBXs is analog, then you will need a channel bank at the end of the transmission path whose PBX won't take a digital signal. See Channel Service Unit and T-1.

Channel Service Unit

CSU. A device used to connect a digital phone line (T-1 or less) coming in from the phone company to either a multiplexer, channel bank or directly to another device producing a digital signal, e.g. a digital PBX, or data communications device. A CSU performs certain line-conditioning, and equalization functions, and responds to loopback commands sent from the central office. A CSU regenerates digital signals. It monitors them for problems. And it provides a way of testing your digital circuit. You can buy your own CSU or rent one from your local or long distance phone company. See also CSU and DSU.

Clear-Back

A signal from the CO to telecommunication equipment which usually indicates that the party at the other end of the line has hung up. Largely synonymous with "disconnect".

Clear-Forward

A signal from telecommunications equipment to the CO that a disconnect is desired: in other words, that the equipment wishes to terminate the call.

CNG

Calling Tone. The piercing "whistle" tone (1,100 Hz) of a fax machine to inform the caller that it is ready to receive a transmission.

CO

Central Office. In North America, a CO is that location which houses a switch to serve local telephone subscribers. Sometimes the words "central office" are confused with the switch itself. In Europe and abroad, the words "central office" are not known. The more common words are "public exchange." But those words tend to refer more to the switch itself, rather than the site, as in North America. CO was the name of a magazine published by Telecom Library Inc, the publisher of this book. See also Central Office or Public Exchange.

CO Lines

These are the lines connecting your office to your local telephone company's Central Office which in turn connects you to the nationwide telephone system.

CO Simulator

A desktop device which pretends to act like a mini-central office. The smallest version will consist of two lines and two RJ-11 jacks. Plug a phone into both jacks. Pick up one phone. You hear dial tone. Dial or touchtone two or three digits. Bingo, the second phone rings. You pick up the second phone. You can have a conversation with yourself or with a machine — like a voice processing system. Most central office simulators can simulate normal on-hook, off-hook, dialing, answering, speaking, etc. Some now can simulate caller ID features — including number of person calling.

Co-Articulation

When speaking a pair of words such as "seven nine" or "test tube" people will generally omit the consonant which starts the second word, and will in fact say something like "seven'ine" or "test'ube". This phenomenon is called co-articulation, and was invented by people to make sentences easier to say and to make life difficult for developers of voice recognition systems. If your voice recognizer can cope with co-articulation, you have a continuous recognizer, and you can laugh at your competitors who merely have connected or discrete recognizers.

Codec

Coder/decoder. A device used to convert analog to digital or to compress and decompress audio.

Common Channel Signaling

See CCS.

Communicating Applications Specification

See CAS.

Compelled Signaling

Compelled signaling is a method of conveying numerical and status information between two pieces of equipment. It is often used on E-1 trunks for dialing numbers, getting DDI digits and so on. Single and dual tones may be used, depending on the specifics of the protocol. To send one item of information, such as one digit, both the sender and the receiver of the information play one tone. The sequence of events is like this:
1. The sender begins transmitting a tone.
2. The receiver detects the tone. In response, the receiver begins transmitting a response tone.
3. When the sender detects the response, it turns off the original tone.
4. When the receiver detects the end of the original tone, it stops sending the response tone.

It sounds complicated, and it is. R2/MF signaling is based on the compelled signaling model, and comes in many variations in different regions.

Conference

A configuration where three or more callers are able to talk to each other. Some parties may be in a "listen-only" mode where they can hear but not speak to other parties.

Compression

Changing the storage or transmission scheme for information so that less space (fewer bits) are required to represent the same information. Compressing data means that less space is required for storage and less time for the transmission of the same amount of data. Comes in two flavors: lossless compression, where the original information can be reconstructed precisely, and lossy compression, where something close to the original can be reconstructed but some details may differ.

Connected Speech

A technical term used to describe speech made of a series of utterances which come in relatively quick succession without co-articulation. See Co-Articulation. Connected speech is intermediate between discrete speech and continuous speech. Usually applied to the capability of a voice recognizer to recognize words from this type of speech.

Continuous Speech

A technical term used to describe speech made of a series of utterances which come in relatively quick succession with co-articulation. See Co-Articulation. Usually applied to the capability of a voice recognizer to recognize words from this type of speech.

Co-operative Multitasking

The type of multi-tasking used in 16-bit Windows where a single application hogs the CPU until it voluntarily gives up control, at which point Windows may give control to another application.

CSI

Called Subscriber Identifier. The "name" of a fax device, transmitted to the fax device at the other end in the course of establishing a fax call. Typically a telephone number and/or company name.

CSU

1. Channel Service Unit. Also called a Data Service Unit. A device to terminate a digital channel on a customer's premises. It performs certain line-conditioning and equalization functions, and responds to loopback commands sent from the central office. A CSU sits between the digital line coming in from the central office and devices such as channel banks or data communications devices.

2. Channel Sharing Unit. Line bridging devices that allow several inputs to share one output. CSUs exist to handle any input/output combination of sync or async terminals, computer ports, or modems and thus these units are variously called modem sharing units, digital bridges, port sharing units, digital sharing devices, modem contention

Glossary

units, multiple access units, control signal activated electronic switches or data-activated electronic switches.

CTI

Computer Telephony Integration. Sometimes used as a general term for any system where a computer interacts with a telephone line. Usually, CTI is used to refer to the ability to control a switch (PBX or ACD) via a link to a computer.

Custom Control

A user interface element like a button or scroll-bar, made by a third-party vendor. Most custom controls have a graphical user interface, but some, including most of those used in telephony, may be invisible in the running application.

Cut-Through

See DTMF Cut-Through.

CVSD

Continuously Variable Slope Differential Modulation. A method used to compress voice data.

D-Channel

The Data channel on an ISDN trunk. On a Primary Rate Interface ISDN trunk, there are 24 channels (time-slots) just as on a usual T-1 trunk. One or more channels, the D-channels, are used to convey information such as dialed digits, ANI and DNIS information, routing and billing codes etc. depending on the type of ISDN service used. Therefore there are only 23 or fewer audio channels (called B-channels) available. This is a form of out-of-band signaling protocol which can give faster responses and more flexible services than the in-band DTMF and robbed-bit technology in wide use today.

DCS

Digital Command Signal. A packet of information sent from a fax device to another fax device specifying the modem speed, image width, compression method and page length of the transmission to follow.

DDI

Another way of spelling DID.

Decadic Signaling

A fancy way of referring to pulse dialing.

Dial String

A string of text characters used to represent a dialing process. The characters will represent the digits to dial, other characters may have special meanings. For example, P often means switch to pulse dialing, comma (,) often means pause.

Dial Tone

The sound you hear when you pick up a telephone. Dial tone is a signal (350 + 440 Hz) from your local telephone company that it is alive and ready to receive the number you dial. If you have a PBX, dial tone will typically be provided by the PBX. Dial tone does not come from God or the telephone instrument on your desk. It comes from the switch to which your phone is connected to.

Dialogic

Dialogic Corporation, Parsippany, NJ, is one of the leading manufacturers of interactive voice processing equipment and software. They sell equipment through value added resellers, dealers and distributors. Many of their dealers "add value" to the Dialogic components by doing their own specialized software programming, tailoring Dialogic products to particular specialized (and useful) applications.

Differential Pulse Code Modulation

A method of compressing voice data.

DID

Direct Inward Dialing. You can dial inside a company directly without going through the attendant. This feature used to be an exclusive feature of Centrex but it can now be provided by virtually all modern PBXs and some modern hybrids. Sometimes spelled DDI, especially in the UK.

Digital Command Signal

See DCS.

Digital Signal Processor

See DSP.

Digital Trunk

Generic name for a telephone connection which uses digital rather than analog transmission technology. Common examples are T-1 in the US and E-1 in Europe.

Digitization

The process of converting an analog signal to a digital signal by measuring the value of the analog signal at regular intervals and converting this to a numerical value (sample).

Diphone

A pair of sounds. It is sometimes useful to represent speech as a series of diphones. For example, in phonetic spelling the word "voice" is written \vóis\, and is composed of the diphone pair \vó\ and \is\.

Direct Dialing In

Also known as DDI. See DID.

Direct Inward Dial

See DID.

Disconnect

Term for a signal that the called party hung up, or that your equipment wished to terminate a call.

Disyllable

Another word for diphone. See Diphone.

DNIS

Dialed Number Identification Service. DNIS is a feature of 800 and 900 lines. Let's say you subscribe to several 800 numbers. You use one line for testing your advertisements on TV stations in Phoenix; another line for testing your advertisements on TV stations in Chicago; and yet another for Milwaukee. Now you get an automatic call distributor and you terminate all the lines in one group on your ACD. You do that because it's cheaper to man and run one group of incoming lines. One queue is more efficient than several small ones, etc. You have all your people answering all the calls. You now need to know which calls are coming from where. So your long distance carrier sends you the call's DNIS — the numbers the person dialed to reach you. Those DNIS digits might come to you in many ways, depending on the technical arrangement you have with your long distance company. In-band or out-of-band. ISDN or data channel, etc. Make sure you understand the difference between DNIS and ANI. DNIS tells you the number your caller called. ANI is the number your caller called from.

DS-0

Digital Signal, level Zero. Pronounced "D-S Zero." DS0 is 64,000 bits per second. It is equal to one voice conversation digitized under PCM. Twenty-four DS0s (24x64 Kbps) equal one DS1, which is T-1 or 1.544 million bits per second.

DS-1

Digital Service, level 1. It is 1.544 Mbps in North America, 2.048 Mbps elsewhere. Why there's no consistency is one of those wonderful, unanswered, questions. The 1.544 standard is an old Bell System standard. The 2.048 standard is a CCITT standard. Standard for 1.544 Mbps is 24 voice conversations each encoded at 64 Kbps. Standard for 2.048 megabits is 30 conversations.

Drop And Insert

A drop and insert system has the ability to perform some form of voice processing on a call (when it is "dropped" to a device such as a voice card) and also to pass the call through the system without performing any processing (when the call is said to be "inserted" into the outgoing trunk).

DSP

Digital Signal Processor. A Digital Signal Processor is a specialized computer chip designed to perform speedy and complex operations on digitized waveforms. Useful in real-time processing of sound and video.

DSVD

Digital Simultaneous Voice Data protocol. See SVD.

DTMF

Dual Tone Multi-Frequency. A fancy term describing push button or Touchtone dialing. (Touchtone was a registered trademark of AT&T.) In DTMF, when you touch a button on a pushbutton pad, it makes a tone, actually a combination of two tones, one high frequency and one low frequency. Thus the name Dual Tone Multi Frequency. In U.S. telephony, there are actually two types of "tone" signaling, one used on normal business or home pushbutton/touchtone phones, and one used for signaling within the telephone network itself. When you go into a central office, look for the testboard. There you'll see what looks like a standard touchtone pad. Next to the pad there'll be a small toggle switch that allows you to choose the sounds the touchtone pad will make — either normal touchtone dialing (DTMF) or the network version (MF).

The eight possible tones that comprise the DTMF signaling system were specially selected to easily pass through the telephone network without attenuation and with minimum interaction with each other. Since these tones fall within the frequency range of the human voice, additional considerations were added to prevent the human voice from inadvertently imitating or "falsing" DTMF signaling digits. One way this was done to break the tones into two groups, a high frequency group and a low frequency group. A valid DTMF tone has only one tone in each group. Here is a table of the DTMF digits with their respective frequencies. One Hertz (abbreviated Hz.) is one cycle per second of frequency.

Digit	Low frequency	High frequency
1	697	1209 Hz.
2	697	1336
3	697	1477
4	770	1209
5	770	1336
6	770	1477
7	852	1209
8	852	1336
9	852	1477
0	941	1336
*	941	1209
#	941	1477

There are four other digits defined in the DTMF system and usable for specialized applications that cannot be generated by standard telephones. They are:

A	697	1633 Hz.
B	770	1633
C	852	1633
D	941	1633

Normal telephones (yours and mine) have 12 buttons, thus 12 combinations. Government Autovon (Automatic Voice Network) telephones have 16 combinations, the extra four (those above) being used for "precedence," which in Federal government parlance is a designation assigned to a phone call by the caller to indicate to communications personnel the relative urgency (therefore the order of handling) of the call and to the called person the order in which the message is to be noted.

DTMF Cut-Through

The ability of equipment to respond immediately to a received touch-tone even when a voice prompt is being played.

Dual Tone

A tone composed of two frequency components. The best known examples are touch tones. Other examples are dial tone, busy tone, R2 and so on.

Dumb Switch

A slang word for a telecommunications switch that contains only basic switching software and relies on instructions sent it by an outside computer. Those instructions are typically fed the "dumb" switch through a cable from the computer to one or more RS-232 serial ports which the dumb switch sports. The switch makes no demands on what type of computer it talks to, but simply insists that it be able to feed the computer questions and promptly receive responses in a form that it (the switch) can understand.

For example, the dumb switch might signal the computer, "A call is coming in on port 23, what do I do now?" The computer might reply "Answer it and transfer it to extension 23." Or it might say "answer it and put it on hold," or "answer it, put it on hold and play recording number three." In essence, a dumb switch is anything but. It is in reality an empty cage containing whatever network interface cards the user has chosen. Each of these network interface cards is designed to "talk" to one type of telephone line. That line might be a T-1 line. It might be a normal tip and ring loop start line. It might be a tie trunk with E&M signaling. The card may handle one or many lines, but always of the same type. The card knows how to answer a call or pulse out a call on that particular type of line. It has all the telephony smarts. What it lacks is the intelligence of what to do with the calls. That is provided by the outside computer. Well, almost. Most "dumb" switches do contain rudimentary intelligence — a small computer and some memory. That computer is usually programmed to handle "default" calls — and to handle calls should the link to the outside computer fail, or the outside computer itself fail. Dumb switches come in flavors all the way from residing in their own cabinet to being printed circuit cards which reside in one or more of the personal computer's slots. Dumb switches are programmed to do "specialized" telecom applications, for example emergency 911, added value 800 services, cellular switching, automatic call distributors, predictive dialers, etc. They can, of course, be programmed to be "normal" PBXs. The question increasingly being asked is, "If I want to program a specialized telecom application should I use a dumb switch or should I use an open PBX?" And the answer is, "It depends." Depends on what you want to do. Depends on what software is available, etc.

E & M Leads

The pair of wires carrying signals between trunk equipment and a separate signaling equipment unit. The "M" lead transmits a ground or battery conditions to the signaling equipment. The "E" lead receives open or ground signals from the signaling equipment. These leads are also known as Ear and Mouth Leads. The Ear lead typically means to receive and the Mouth lead typically means to transmit. Changes of voltage on these leads convey such information as seizure of circuit, recognition of seizure, release of circuit, dialed digits, etc. In the old days it was the PBX operators who originated trunk calls by asking the long distance carrier for free trunks using their mouth or M lead. If the carrier had a free trunk, the PBX heard about it through its ear or E lead.

See also E & M Signaling.

E & M Signaling

In telephony, an arrangement that uses separate leads, called respectively the "E" lead and "M" lead, for signaling and supervisory purposes. The near end signals the far end by applying -48 volts dc (vdc) to the "M" lead, which results in a ground being applied to the far end's "E" lead. When -48 vdc is applied to the far end "M" lead, the near-end "E" lead is grounded. The "E" originally stood for "ear," i.e., when the near-end "E" lead was grounded, the far end was calling and "wanted your ear." The "M" originally stood for "mouth," because when the near-end wanted to call (i.e., speak to) the far end, -48 vdc was applied to that lead.

When a PBX wishes to connect to another PBX directly or to a remote PBX or extension telephone over a leased voice grade line, a channel on T-1, the PBX uses a special line interface which is quite different from that which it uses to interface to the phones it's attached directly to (i.e. with in-building wires). The basic reason for the difference between a normal extension interface and the long distance interface is that the signaling requirements differ — even if the voice signal parameters such as level and two-wire, 4-wire remain the same. When dealing with tie lines or trunks it is costly, inefficient and too slow for a PBX to do what an extension telephone would do, i.e. go off hook, wait for dial tone, dial, wait for ringing to stop, etc. The E&M tie trunk interface device is the closest thing there is to a standard that exists in the PBX, T-1 multiplexer, voice digitizer telco world. But even then it comes in at least five different flavors.

E-1

E-1 is a digital trunk standard similar to T-1 used in many countries outside of North America. There are 32 channels (time-slots) in contrast to 24 channels on T-1, hence the faster bit rate of 2.048 MHz versus 1.544 MHz for T-1. Channels 0 and 16 are used to carry the ABCD bits and synchronization (framing) bits, hence only 30 channels are available for audio conversations. Thus, ABCD bits are carried out-of-band in contrast to the in-band robbed-bit scheme used by T-1. Unfortunately, the conventions used for the meanings of bit values vary widely from country to country and even from region to region in different countries. For example, in Greece, the "idle" line condition is signaled by ABCD=1010, a seize is signaled by changing to ABCD=0010. In Italy, idle is ABCD=0101 and seize is ABCD=1001. Differences in ABCD and interregister signaling is a major headache for equipment vendors to different E-1 implementations.

ECM

Error Correction Mode. Used in the Group 3 fax protocol.

ECMA

European Computer Manufacturers Association.

ECTF

Enterprise Computer Telephony Forum. Standards body for computer telephony. Look them up at www.ectf.org.

Engaged

British English for Busy.

EOM

End Of Message. Signal used in the Group 3 fax protocol. Basically means end of page.

EOP

End Of Page. Signal used in Group 3 fax protocol. Basically means end of document.

Event

An unsolicited communication from a hardware device to a computer operating system, application, or driver. Events are generally attention-getting messages, allowing a process to know when a task is complete or when an external event occurs.

Exception Dictionary

In text-to-speech, a list of words which violate the usual pronunciation rules. For example, the London theatre district "Leicester Square" would be pronounced "Lie-Sester Skware" by a person familiar with English spelling but not with that particular name. An exception dictionary would store "Leicester" as a special case with an attached pronunciation as "Lesster".

Fax Broadcast

Fax broadcast systems send a copy of one document to several phone numbers. For example, a company with many offices in the US and in Europe might use a fax broadcast to save on telephone toll charges. If an office in London wanted to send a fax to all regional offices in the US it would be less expensive to send one fax to the New York office and then have New York broadcast the document to all other US offices. Other applications of fax broadcasting are in distributing information such as press releases, questionnaires or promotions. Some e-mail services such as MCI mail offer fax broadcast options.

Fax Mail

Analogous to, and perhaps a feature of, voice mail. Fax mail allows a caller to fax a message rather than speaking a message. Fax messages may be retrieved by the mailbox owner from a fax machine or desktop PC which is able to access the stored file and display it as an image on the computer screen. Some voice mail systems allow fax messages to be incorporated into mailboxes.

Fax On Demand

A typical use for fax on demand is to provide product information to potential customers. A caller dials a voice processing unit and selects one or more documents of interest using touch tone menus or voice recognition. If the caller is calling from a fax machine, transmission can being immediately (this is called one-call or same-call faxing). If the caller is using a telephone rather than a fax machine, a fax number can be entered in response to a menu prompt and the fax on demand system will make a later call to that number to deliver the document.

Fax Store And Forward

This refers to the ability of a computer to store a received fax document as a file stored on a hard drive and re-transmit the document in a subsequent call. Analogous to voice store and forward, which simply means "record" and "play back".

Fax Synthesis

The ability of a computer to create a fax document from stored ASCII text, word processing, database, spreadsheet or other information.

Fast Busy

A busy signal which sounds at twice the normal rate (120 interruptions/minute vs. 60/minute). A "fast busy" signal indicates all trunks are busy.

Fiber

A strand of wire, often used when referring to optical cable.

Also, a new type of execution context used in Windows NT, a fiber is something like a high-performance thread that uses co-operative multi-tasking.

Flash-Hook

A brief on-hook period. A common use of the flash-hook is in the domestic "call waiting" feature. A new call comes in while the called party is holding a conversation. Instead of producing a busy signal, the new caller hears ringing, and the called party hears a special "beep" tone. If he or she chooses, the called party can depress the hook-switch on the telephone briefly, sending a flash-hook. The original caller is put on hold, and the new caller gets through. Repeated flash-hooks will swap between the two callers. On a PBX, a flash-hook will produce a new dial-tone, allowing a three-way conference or call transfer.

Forward Signal

A tone signal used in compelled signaling, usually on E-1 networks. A forward signal is essentially a tone sent from the originator to the receiver of a call. For example, if a device is dialing a number, the device is the originator and may send forward signals to the central office which specify the number to dial.

Fourth Column

The digits A, B, C, D which are DTMF tones found on military and some European telephones.

Frame

In Time Division Multiplexing, several conversations are sent together as a rapid succession of samples, one from each channel, in serial on a communications line. A "frame" is one sample from each channel, plus possible synchronization (framing) bits. Framing bits provide a repeating pattern so that two pieces of equipment can establish where in the stream of bits being passed back and forth a frame starts and ends.

FTT

Failure To Train. Sent as part of the Group 3 fax protocol to indicate that the device couldn't handle the transmission parameters. Basically a request to send slower.

Glare

Imagine that you want to make a phone call, and pick up the handset in order to being dialing — but someone was calling you at that exact moment, so you picked up the phone just before it began ringing. Instead of hearing dial tone, you hear the caller saying "Hello, Hello?". This is called glare. If an automated system both can accept incoming calls and make outgoing calls on a given line, glare is a potential problem.

Ground Start

A way of signaling on subscriber trunks in which one side of the two-wire trunk (typically the "Ring" conductor of the Tip and Ring) is momentarily grounded (often to a cold water pipe) to get dial tone. There are two types of switched trunks typically for lease by a local phone company — ground start and loop start. PBXs work best on ground start trunks, though many will work — albeit intermittently — on both types. Normal single line phones and key systems typically work on loop start lines. You must be careful to order the correct type of trunk from your local phone company and correctly install your telephone system at your end — so that they both match. In technical language, a ground start trunk initiates an out-going trunk seizure by applying a maximum local resistance of 550 ohms to the tip conductor. See Loop Start.

Group 1

Analog fax equipment, according to Recommendation T.2 of the CCITT. It sends a US letter (8½ by 11") or A4 page in about six minutes over a voice-grade telephone line using frequency modulation with 1.3 kHz corresponding to while at 2.1 kHz to black. North American six-minute equipment uses a different modulation scheme, and is therefore not compatible.

Group 2

Analog fax equipment, according to Recommendation T.3 of the CCITT. It sends a page in about three minutes over a voice grade telephone line using 2.1 kHz AM-PM-VSB modulation.

Group 3

A digital fax standard that allows high-speed, reliable transmission over voice grade phone lines. All modern fax devices use Group 3, which is based on CCITT Recommendation T.4.

Group 4

A CCITT fax standard primarily designed to work with ISDN. It is considered difficult to implement, and is not in widespread use owing to the low penetration of ISDN (Group 4 cannot work on non-ISDN lines).

Grunt Detection

A crude form of voice recognition which simply responds to sound or silence in response to a prompt. Typically a prompt such as this is used: "For selection one, please say 'Yes' now.....for selection two, please say 'Yes' now...." and so on. Some voice processing cards are able to report whether or not sound is present on the line, grunt detection exploits this ability. Cheaper than paying for true voice recognition capability, but generally not very friendly or reliable.

HCV

High Capacity Voice. A type of voice data compression.

Homologation

The process of obtaining approval from the local regulatory authorities to attach a device to the public telecommunications network. See also PTT.

Hunt

Refers to the progress of a call reaching a group of lines. The call will try the first line of the group. If that line is busy, it will try the second line, then it will hunt to the third, etc. See also Hunt Group.

Hunt Group

A series of telephone lines organized in such a way that if the first line is busy the next line is checked ("hunted") and so on until a free line is found. Often this arrangement is used on a group of incoming lines. Hunt groups may start with one trunk and hunt downwards. They may start randomly and hunt in clockwise circles. They may start randomly and hunt in counter-clockwise circles. Inter-Tel uses the terms "Linear, Distributed and Terminal" to refer to different types of hunt groups. In data communications, a hunt group is a set of links which provides a common resource and which is assigned a single hunt group designation. A user requesting that designation may then be connected to any member of the hunt group. Hunt group members may also receive calls by station address.

In-Band Signaling

Signaling made up of information passed in the channel used to convey the audio signal of a conversation. A touch tone is an in-band signal because it uses sound and can be heard on the audio circuit. Robbed-bit signaling on T-1 is in-band because it used bits which are interpreted as sound. The more modern form of signaling is out-of-band. Several local and long distance companies provide ANI (Automatic Number Identification) via in-band signaling. Some long distance companies provide it out-of-band, using the D-channel in a PRI ISDN trunk.

Inflection

The change in pitch and emphasis in the spoken voice. A typical example of inflection is the difference in the three spoken words "Hello?", "Hello!" and "Hello." where the tone of voice informs the listener that there is a question, exclamation or statement. Natural-sounding inflection is a desirable characteristic of a machine-generated voice, but is difficult to achieve.

Inserted Signaling

When Dialogic DTI digital trunk devices are in inserted signaling mode, it is the DTI device that sets the signaling bits sent to the T-1 or E-1 trunk. In transparent mode, the signaling bit is taken from the PEB bus and passed through the DTI without change. SC bus versions of the DTI bus work in inserted signaling mode only.

Interactive Voice Response

IVR. Think of Interactive Voice Response as a voice computer. Where a computer has a keyboard for entering information, an IVR uses remote touchtone telephones. Where a computer has a screen for showing the results, an IVR uses a digitized synthesized voice to "read" the screen to the distant caller. Whatever a computer can do, an IVR can too, from looking up train timetables to moving calls around an automatic call distributor (ACD). The only limitation on an IVR is that you can't present as many alternatives on a phone as you can on a screen. The caller's brain simply won't remember more than a few. With IVR, you have to present the menus in smaller chunks.

Some people use IVR as a synonym for voice processing, or all computer-telephone integration technology involving spoken responses from the computer.

In more modern usage, IVR has come to refer to computer telephony access to databases, as in banking-by-phone applications.

Information Provider

A business or person providing information to the public for money. The information is typically selected by the caller through touchtones, delivered using voice processing equipment and transmitted over tariffed phone lines, e.g., 900, 976, 970. Typically, billing for information providers' services is done by a local or long distance phone company. Sometimes the revenues for the service are split by the information provider and the phone company. Sometimes the phone company simply bills a per minute or flat charge. A typical "information provider" is American Express, which provides a service — 1-900-WEATHER. By dialing that number you can touchtone in city names and find out temperatures, weather forecasts, etc. Calling 1-900-WEATHER costs several dollars a minute.

Key System

A type of phone system. You can recognize a key system because each phone has a row of push-buttons which are pressed to access a given phone line. Key systems don't integrate well with VRUs because there is usually no flash-hook transfer or hunting feature provided by the phone system.

LATA

Local Access Transport Area. LATAs are geographic areas. A call from two phones in the same LATA is a local call. There are 161 LATAs in the US.

LCR

Least Cost Routing. For example, you may have a choice of long-distance carriers which you can reach by dialing different access codes. If you know the rates charged by each carrier for each region at each time of day, you can program a switch to choose the cheapest carrier for each call that you make.

Local Access Transport Area

See LATA.

Local Loop

The wire which passes between your home phone and the phone company. Generally a length of good, old-fashioned copper wire.

Loop

1. Typically a complete electrical circuit.
2. The loop is also the pair of wires that winds its way from the central office to the telephone set or system at the customer's office, home or factory, i.e. "premises" in telephones.
3. In computer software. A loop repeats a series of instructions many times until some prestated event has happened or until some test has been passed.

Loop Current

The current that flows when there is a live connection through an analog phone line.

Loop Start

You "start" (seize) a phone line or trunk by giving it a supervisory signal. That signal is typically taking your phone off hook. There are two ways you can do that — ground start or loop start. With loop start, you seize a line by bridging through a resistance the tip and ring (both wires) of your telephone line.

Mailbox

A set of stored messages belonging to a single owner. Typically, these will be recorded voice messages, but increasingly mailboxes also include e-mail and fax documents.

MCA

Micro Channel Architecture. The new internal bus inside IBM's new line of PS/2 MCA machines.

MCF

Message Confirmation Frame. Data sent as part of the Group 3 fax protocol.

Media Stream

Jargon used by Microsoft for referring to the message passed along a communications channel. For example, in a fax phone call, the media stream would be the graphics image, in a data call, the media stream might be a transferred file, in a voice call, the media stream is the audio waveform describing the sound.

Menu

Options displayed on a computer terminal screen or spoken by a voice processing system. The user can choose what he or she wants done by simply choosing a menu option — either typing it on the computer keyboard, hitting a touchtone on his phone or speaking a word or two. There are basically two ways of organizing computer or voice processing software — menu-driven and non-menu driven. Menu-driven programs are easier to use but they can only present as many options as can be reasonably crammed on a screen or spoken in a few seconds. Non-menu driven systems may allow more alternatives but are much more complex and frightening. It's the difference between receiving a bland "A" or "C" prompt on the screen — as in MS-DOS — and receiving a menu of "Press A if you want Word Processing," "Press B if you want Spread Sheet," etc. See also Audio Menus and Prompts.

Mezzanine Bus

Another term for a voice bus used to communicate audio and signaling data between different voice processing components. Examples of mezzanine buses are PEB, MVIP and the SC bus.

MF

See Multi-Frequency Signaling.

Micro Channel

A proprietary bus developed by IBM for its PS/2 family of computers' internal expansion cards. Also offered by Tandy and other vendors. Compare with EISA bus.

MMR

Modified Modified Read. A compression scheme used in the Group 3 fax protocol.

MPS

Multi-Page Signal. A tone sent as part of the Group 3 fax protocol.

MR

Modified Read. A compression scheme used in the Group 3 fax protocol.

Mu-Law Encoding

Digital telephony equipment usually stores and transmits audio signals in 8-bit samples using PCM encoding. The value of the 8-bit sample corresponds to the amplitude (volume, loudness) of the audio signal at the time the sample was taken. There is a translation table which is used to determine the amplitude given the sample value. There are two different tables in common use: the A-law table, which is widely used in Europe and Asia on E-1 trunks, and Mu-law (m-law) encoding used in the US, Canada and a few other countries.

Multi-Frequency Tones

Tones which have two or more frequency components. DTMF, MF and R2 tones are examples of dual tones, i.e. multi-frequency tones which have two components. Two components is good because it guards against false positive detections, three or more is getting complicated and is not used in any standard signaling system which this author is familiar with.

Multi-Tasking

Doing several different tasks at the same time on one computer. This should not be confused with Task Switching. In task switching, the computer jumps from one task to another, typically in response to a command from you, the user. For example, in task switching, you might temporarily stop your word processing, jump into your communications package, dial up a database, grab some information, then jump back into

Glossary

your word processor and put that new information into the document you're word processing. However, in true multi-tasking, you could have told your computer to dial the database, grab the information and alert you when it had grabbed the material. At that point you could have included it in your document. But in the meantime, you could have been happily doing word processing.

MVIP

Multi Vendor Integration Protocol. MVIP usually refers to Mitel's voice bus, which has been adopted by some non-Dialogic compatible vendors as an alternative to Dialogic's PEB and SC buses.

Network Module

Dialogic-speak for a voice processing component which connects to a phone line, whether it is an plain only analog line, digital trunk, DID circuit or whatever.

NFAS

An ISDN protocol which allows more than one trunk to share a D channel.

NSF

Non-Standard Facilities frame. Information sent by one fax device to another indicating vendor-specific facilities beyond the standard Group 3 requirements.

NXX

Refers to the local exchange part of a telephone number. A US telephone number is AAA-NXX-LLLL, where AAA is the area code, NXX indicates the local exchange number (given an NXX number, you know which CO switch is attached to the phone), and LLLL is the index of the phone in that local switch.

OCX

The file extension of an ActiveX control.

Off-Hook

When the handset is lifted from its cradle it's Off-Hook. Lifting the hook switch alerts the central office that the user wants the phone to do something like dial a call. A dial tone is a sign saying "Give me an order." The term "off-hook" originated when the early handsets were actually suspended from a metal hook on the phone. When the handset is removed from its hook or its cradle (in modern phones), it completes the electrical loop, thus signaling the central office that it wishes dial tone. Some leased line channels work by lifting the handset, signaling the central office at the other end which rings the phone at the other end. Some phones have autodialers in them. Lifting the phone signals the phone to dial that one number. An example is a phone without a dial at an airport, which automatically dials the local taxi company. All this by simply lifting the handset at one end — going "off-hook."

OLE Control

Custom controls using OLE interfaces are called ActiveX or OCX controls.

On-Hook

When the phone handset is resting in its cradle. The phone is not connected to any particular line. Only the bell is active, i.e. it will ring if a call comes in. See On-Hook Dialing and Off-Hook.

On-Hook Dialing

Allows a caller to dial a call without lifting his handset. After dialing, the caller can listen to the progress of the call over the phone's built-in speaker. When you hear the called person answer, you can pick up the handset and speak or you can talk hands-free in the direction of your phone, if it's a speakerphone. Critical: Many phones have speakers for hands-free listening. Not all phones with speakers are speakerphones — i.e. have microphones, which allow you to speak, also.

One-Call Fax

The caller dials an automated service from a fax machine, selects a document through touch tone or voice recognition and then hits the

Start button on the fax machine to send or receive a fax. This is "one-call" or "same-call" fax, as opposed to "two-call" fax where the caller enters a PIN code or fax phone number, the computer then calls back in a later call with the document.

Operator Intercept

When an invalid number is dialed or an error condition occurs on the network, an operator intercept may occur. In the US, SIT tones are heard followed by a recorded message explaining the problem.

Out-Of-Band Signaling

Signaling that is separated from the channel carrying the information — the voice, data, video, etc. Typically the separation is accomplished by a filter. The signaling includes dialing and other supervisory signals.

Outgoing Register

In compelled signaling protocols, this refers to the device which is transmitting a number to another device. For example, in an outgoing call, the number to be reached is sent from the outgoing register (the calling equipment) to the incoming register (the phone company switch).

PABX

Private Automatic Branch eXchange. Originally, PBX was the word for a switch inside a private business (as against one serving the public). PBX means a Private Branch Exchange. Such a "PBX" was typically a manual device, requiring operator assistance to complete a call. Then the PBX went "modern" (i.e. automatic) and no operator was needed any longer to complete outgoing calls. You could dial "9." Thus it became a "PABX." Now all PABXs are modern. And a PABX is now commonly referred to as a "PBX." Some manufacturers have tried to make their PBX appear different by calling it something else. Rolm/IBM calls theirs the "CBX" (Computerized Branch Exchange). Some others call theirs the "EPABX" (the Electronic Private Automatic Branch Exchange. Then there are the special ones, like Wang's WBX, SRX's SRX, NEC's IMS (Information Management System), etc. See PBX.

PAMD

Positive Answering Machine Detection. A feature of Dialogic's Perfect Call progress analysis firmware which attempts to recognize when an answering machine or voice mail picks up a call, for example by recognizing the "beep" tone which typically precedes the greeting. Not 100% reliable, but can be useful.

Pay-Per-Call

Some phone calls have an added charge which is levied by the phone company and sent to an information provider. Services such as weather, sports scores, stock prices etc. may be provided. They are generally reached by dialing numbers with a special area code or prefix. In the US, national numbers with the 900 area code or local calls with the 976 prefix have added charges. Similar services are now available in many countries. These are termed "pay-per-call" or "premium rate" services.

PBX

Private Branch eXchange. A private (i.e. you, as against the phone company, own it), branch (meaning it is a small phone company central office), exchange (a central office was originally called a public exchange, or simply an exchange). In other words, a PBX is a small version of the phone company's larger central switching office. A PBX is also called a Private Automatic Branch Exchange, though that has now become an obsolete term. In the very old days, you called the operator to make an external call. Then later someone made a phone system that you simply dialed nine (or another digit — in Europe it's often zero), got a second dial tone and dialed some more digits to dial out, locally or long distance. So, the early name of Private Branch Exchange (which needed an operator) became Private AUTOMATIC Branch Exchange (which didn't need an operator). Now, all PBXs are automatic. And now they're all called PBXs, except overseas where they still have PBXs that are not automatic.

PCM

Pulse Code Modulation. The most common method of encoding an analog voice signal and encoding it into a digital bit stream. First, the amplitude of the voice conversation is sampled. This is called PAM,

Pulse Amplitude Modulation. This PAM sample is then coded (quantized) into a binary (digital) number. This digital number consists of zeros and ones. The voice signal can then be switched, transmitted and stored digitally. There are three basic advantages to PCM voice. They are the three basic advantages of digital switching and transmission. First, it is less expensive to switch and transmit a digital signal. Second, by making an analog voice signal into a digital signal, you can interleave it with other digital signals — such as those from computers or facsimile machines. Third, a voice signal which is switched and transmitted end-to-end in a digital format will usually come through "cleaner," i.e. have less noise, than one transmitted and switched in analog. The reason is simple: An electrical signal loses strength over a distance. It must then be amplified. In analog transmission, everything is amplified, including the noise and static the signal has collected along the way. In digital transmission, the signal is "regenerated," i.e. put back together again, by comparing the incoming signal to a logical question: Is it a one or a zero? Then, the signal is regenerated, amplified and sent along its way.

PCM refers to a technique of digitization. It does not refer to a universally accepted standard of digitizing voice. The most common PCM method is to sample a voice conversation at 8000 times a second. The theory is that if the sampling is at least twice the highest frequency on the channel, then the result sounds OK. (See NYQUIST THEOREM.) Thus, the highest frequency on a voice phone line is 4,000 Hertz. So one must sample it at 8,000 times a second. Many PCM digital voice conversations are typically put on one communications channel. In North America, the most typical channel is called the T-1 (also spelled T1). It places 24 voice conversations on two pairs of copper wires (one for receiving and one for transmitting). It contains 8000 frames each of 8 bits of 24 voice channels plus one framing (synchronizing bit) bit which equals 1.544 Mbps, i.e. 8000 x (8 x 24 + 1) equals 1.544 megabits.

Countries outside of the United States and North America use a different scheme for multiplexing voice conversations. It is based not on 24 voice channels, but on 32. This scheme keeps two of the 32 channels for control, actually transmitting 30 voice conversations at a data rate of 2.048 Mbps. The European system is calculated as 8 bits x 32 channels x 8000 frames per second. European PCM multiplexing is not compatible with North American multiplexing. The two systems cannot be directly connected. Some PBXs in the U.S. conform to the U.S. standard only. Some (very few) conform to both. Both the European and North American T-1 "standards" have now been accepted as ISDN "standards." In addition to PCM, there are many other ways of digitally encoding voice. PCM remains the most common. See T-1 and Compression.

PCM-30

Short name of international 2.048 Mbps T-1 (also known as E-1) service derived from the fact that 30 channels are available for 64 Kbps digitized voice each using pulse code modulation (PCM).

PEB

The Dialogic equivalent of MVIP. A voice bus for connecting different voice processing components. Stands for PCM Expansion Bus.

Plain Old Telephone Service

See POTS.

Phoneme

Phonemes are the smallest units of identifiably different sounds.

Phase A, B, C1, C2, D, and E

The stages in a fax transmission:
    Phase A: Establishment.
    Phase B: Pre-message procedure.
    Phases C1 and C2: In-message procedure, data transmission.
    Phase D: Post-message procedure.
    Phase E: Call release.

Play-Off

"False positive" detection of a DTMF digit caused by frequencies in an audio file being played by a voice board matching the frequencies of a touch tone.

Port

Industry jargon for the place where you plug in a phone line. If something has X ports, it can support X phone lines.

Port Density

Jargon for "the number of ports, i.e. phone lines, supported by a system" or "the number of ports per voice processing board". "High density" means that a lot of ports are handled by one or a few boards.

POTS

Plain Old Telephone Service. The basic service supplying standard single line telephones, telephone lines and access to the public switched network. Nothing fancy. No added features. Just receive and place calls. Nothing like Call Waiting or Call Forwarding. They are not POTS services. Pronounced like in "pots and pans".

Positive Answering Machine Detection

See PAMD.

Positive Voice Detection

The ability of a call progress algorithm to distinguish the characteristic frequencies and cadence of the human voice, giving a method of rapidly detecting the "hello" or greeting of a person answering a phone call.

Power Dialer

Another name for a predictive dialer.

Predictive Dialing

An automated method of making many outbound calls without people and then passing answered calls to a person as the calls are answered. Here's the story: Imagine a bunch of operators having to call a bunch of people. Those calls may be for collections. They may be for employee call-ups. They may be for alumnae fund raising. When it's done manually, here's how it works: Before each call operators spend time reviewing paper records or computer terminal screens, selecting the person to be called, finding the phone number, dialing the numbers, listening to rings, listening to phone company intercepts, busy signals and answering machines. Operators also spend time updating the records after each call. Predictive dialing automates this process, with the computer choosing the person to be called and dialing the number and only passing it to an operator when a real live human being answers. There are enormous productivity gains made by screening out answering machines, busy signals, network busy signals, non-completed calls, operator intercepts etc. The result is productivity increases of 300% to 600%. According to generally accepted industry lore, a well-run manual dialing center can get its people talking on the phone for 25 minutes an hour. With a predictive dialer you can get them on the phone making sales, collecting money, etc. for 55 minutes an hour. It's a major productivity gain.

True predictive dialing should not be confused with automated dialing. True predictive dialing has complex mathematical algorithms that consider, in real time, the number of available telephone lines, the number of available operators, the probability of getting no answer, a busy signal, a disconnected number, operator intercept or an answering machine, the time between calls required for maximum operator efficiency, the length of an average conversation and the average length of time the operators need to enter the relevant data. Some predictive dialing systems constantly adjust the dialing rate by monitoring changes in all these factors.

Some people don't like the term "predictive dialing," since they think it's getting "a bad rap" in Washington, DC by being associated with junk phone calls. As a result some people would prefer to call it Computer Aided Dialing.

Pre-Emptive Multi-Tasking

A method of doing multi-tasking where the running process is interrupted by the operating system at regular intervals and control is passed to another process.

PRI

See Primary Rate Interface.

Primary Rate Interface

PRI. The ISDN equivalent of a T-1 circuit. The Primary Rate Interface (that which is delivered to the customer's premises) provides 23B+D (in North America) or 30B+D (in Europe) running at 1.544 megabits per second and 2.048 megabits per second, respectively. There is another ISDN interface. It's called the Basic Rate Interface. It delivers 2B+D over either one or two pairs. In ISDN, the "B" stands for Bearer, which is 64,000 bits per second, which can carry PCM-digitized voice or data. See ISDN.

Private Branch Exchange

A business phone system, often abbreviated to PBX or PABX ("A" for Automatic).

Prosody

Intonation. In text to speech, prosody refers to how natural it sounds — the ups and downs of the sentence.

PSTN

See Public Switched Telephone Network.

PTT

The Post Telephone and Telegraph (PTT) administrations, usually controlled by their governments, provide telephone and telecommunications services in most countries where these services are not privately owned. In CCITT documents, these are the entities referred to as "Operating Administrations."

It is not a simple thing to obtain approval from the PTTs to sell and use telecommunication equipment of any kind in their countries. The world is far from being one in the field of telecommunications. Meeting international requirements typically means providing hardware and software modifications to the product, unique to each country, and then going through an extremely rigorous approval process that can average between six to nine months. Products are required to meet both safety and compatibility requirements.

Public Switched Telephone Network

Usually refers to the worldwide voice telephone network accessible to all those with telephones and access privileges (i.e. In the U.S., it was formerly called the Bell System network or the AT&T long distance network).

Pulse Code Modulation

PCM. The most common and most important method a telephone system in North America can use to sample a voice signal and convert that sample into an equivalent digital code. PCM is a digital modulation method that encodes a Pulse Amplitude Modulated (PAM) signal into a PCM signal. See PCM and T-1.

Pulse Dialing

One or two types of dialing that uses rotary pulses to generate the telephone number.

Pulse To Tone Converter

A device which recognizes the "clicks" made by a rotary dial phone and converts them to DTMF tones. Not always a reliable technology — consider, for example, the problem faced by the device in distinguishing the click made by dialing a "1" digit and a static click caused by lightning or other interference on the line.

PVD

See Positive Voice Detection.

Quantization Noise

Signal errors which result from the process of digitizing (and therefore ascribing finite quantities to) a continuously variable signal.

R1/MF Tone

Often abbreviated to just MF tone. A system of representing digits and a couple of special characters similar to DTMF but using different frequencies. Uses two frequencies for each tone.

R2/MF Tone

Similar to an R1/MF tone, but used in compelled signaling protocols. There are usually two sets of tones, one for forward signals and one for backward signals. These sets are selected from non-overlapping sets of frequencies to avoid ambiguity in the direction of the tone.

RBOC

See Regional Bell Operating Company.

Regional Bell Operating Company

A local telephone company formed following the break-up of AT&T. Also called a Baby Bell.

Register

Sender or receiver of digits in a compelled signaling protocol.

Resource Module

Dialogic-speak for all the voice processing components in a system other than the network module: for example, DTMF detection and generation, voice recognition, text-to-speech and other components which do some useful processing on the channel.

Ring

1. As in Tip and Ring. One of the two wires (the two are Tip and Ring) needed to set up a telephone connection.

2. Also a reference to the ringing of the telephone set.

3. The design of a Local Area Network (LAN) in which the wiring loops from one workstation to another, forming a circle (thus, the term "ring"). In a ring LAN, data is sent from workstation to workstation around the loop in the same direction. Each workstation (which is usually a PC) acts as a repeater by re-sending messages to the next PC in the ring. The more PC's, the slower the LAN. Network control is distributed in a ring network. Since the message passes through each

PC, loss of one PC may disable the entire network. However, most ring LANs recover very quickly should one PC die or be turned off. If it dies, you can remove it physically from the network. If it's off, the network senses that and the token ignores that machine. In some token LANs, the LAN will close around a dead workstation and join the two workstations on either side together. If you lose the PC doing the control functions, another PC will jump in and take over. This is how the IBM Token-Passing Ring works.

RJ-11

RJ-11 is a six conductor modular jack that is typically wired for four conductors (i.e. four wires). Occasionally it is wired for only two conductors — especially if you're only wiring up for tip and ring. The RJ-11 jack (also called plug) is the most common telephone jack in the world. The RJ-11 is typically used for connecting telephone instruments, modems and fax machines to a female RJ-22 jack on the wall or in the floor. That jack in turn is connected to twisted wire coming in from "the network" — which might be a PBX or the local telephone company central office. RJ-22 wiring is typically flat. None of its conductors (i.e. wires) are twisted. You cannot use flat cable for high-speed data communications, like local area networks.

RJ-14

A jack that looks and is exactly like the standard RJ-11 that you see on every single line telephone. Whereas the RJ-11 defines one line — with the two center, red and green, conductors being tip and ring, the RJ-14 defines two phone lines. One of the lines is the "normal" RJ-11 line — the red and green center conductors. The second line is the second set of conductors — black and yellow — on the outside.

Rotary Dial

The circular telephone dial. As it returns to its normal position (after being turned) it opens and closes the electrical loop sent by the central office. Thus it generates pulses for each digit dialed. You can hear the "clicks". The number "seven," for example consists of seven "opens and closes," or seven clicks. You can dial on a rotary phone without using the rotary dial. Simply depress the switch hook quickly, allowing pauses in between to signify that you're about to send a new digit. It's a good party trick.

Same-Call Fax

See One-Call Fax.

Sampling Rate

The number of times per second that an analog signal is measured and converted to a binary number in a digitization process. See Digitization.

SC Bus

The next generation bus now under development by Dialogic and others. See SCSA. Will play a role analogous to the PEB.

Script

1. A written document specifying the wording of menus and informational messages to be recorded when designing a voice response application.

2. The flow-chart or other description specifying the way that a voice response system interacts with a caller.

SCSA

SCSA (pronounced "scuzza") stands for Signal Computing System Architecture. SCSA is a standard for all levels of design of voice processing components from the voice bus chip level to the applications programming interface (API).

SCSA is an open standard. A consortium of leading telecommunications and computing technology players, led by Dialogic and including companies such as IBM and Seimens, are, at the time of writing, cooperating in development of the SCSA specification. The specification documents will be available to anyone who wants them for a nominal fee. There will be no technology license fees charged to developers who wish to create SCSA-compatible products.

The most important components of SCSA for the voice processing developer are a bus and a uniform API.

The primary bus is the SC bus. SC bus is a high capacity bus which is designed to be the "next generation PEB". Where the PEB has up to 32 time-slots, the SCbus will have up to 2,048 time-slots: enough bandwidth for high-fidelity audio, full-motion video and other demanding applications of the future. The standardized API should make SCSA components such as voice processing, voice recognition, speech synthesis, video boards and others accessible independent of the component manufacturer.

SCX Bus

An "out-of-box" voice bus extender which will allow different voice boards using the SC Bus in different computers to be connected together. The computers must be connected via a LAN.

SDK

Software Development Kit. The set of libraries and documentation needed for a C programmer (or other programmer) to create an application using an operating system, driver or other system-level product.

Seize

The process of getting access to a phone line prior to making a call. On a home phone, the action of taking the phone off-hook by lifting the handset is a seize.

SIT Tones

1. Standard Information Tones. These are tones sent out by a central office to a pay phone to indicate that the dialed call has been answered by the distant phone, etc.

2. Special Information Tones. These are tones for identifying network provided announcements. Here's Bellcore's explanation: Automated detection devices cannot distinguish recorded voice from live voice answer unless a machine-detectable signal is included with the recorded

Glossary

announcement. The CCITT, which specifies signals that may be applied to international circuits, has defined Special Information Tones for identifying network provided announcement. The SIT used to precede machine-generated announcements also alerts the calling customer that a machine-generated announcement follows. Since SIT consists of a sequence of three precisely defined tones, SIT can be machine-detected, and therefore machine-generated announcements preceded by a SIT can be classified. At least four SIT encodings have been defined: Vacant Code (VC), Intercept (IC), Reorder (RO) and No Circuit (NC). With the exception of some small stored Program Control Systems (SPCSs) and some customer negotiated announcements, Bell operating companies in North America now precede appropriate announcements with encoded SITs to detect and classify announcements.

Socotel Signaling

A type of R2/MF compelled signaling used on E-1 trunks in France and Spain. The French and Spanish variations are not compatible, a typical problem of E-1 technology.

Span

1. Refers to that portion of a high speed digital system than connects a C.O. (Central Office) to C.O. or terminal office to terminal office. 2. Also called a T-Span Line. A repeatered outside plant four-wire, two twisted-pair transmission line.

Span Line

A T-1 link.

Speech-To-Text

Another name for voice recognition.

State Machine Programming

To control multiple telephone lines in a single voice processing program, a new program structure is required. Dialogic calls this technique "state machine programming". Computer Science called state machines "Deterministic Finite State Automata."

Station Card

An expansion card for a PBX which adds the capacity for more telephone extensions.

Station Set

The phone on your desk. Typically, PBXs use proprietary phones with extra buttons and lights. These are called station sets.

Store And Forward

In communications systems, when a message is transmitted to some intermediate relay point and stored temporarily. Later the message is sent the rest of the way. Not very convenient for voice conversations, but useful for telex type, and other one-way transmission of messages. Telephone answering machines, as well as voice mailboxes are considered forms of Store-and-Forward message switching.

Supervised Transfer

A call transfer made by an automatic device such as a voice response unit which attempts to determine the result of the transfer — answered, busy, ring but no answer — by analyzing call progress tones on the line.

SVD

Simultaneous Voice Data. SVD protocols allow applications to send voice and data over the same modem connection.

Switch

A mechanical, electrical or electronic device which opens or closes circuits, completes or breaks an electrical path, or selects paths or circuits.

T-1

Also spelled T1. A digital transmission link with a capacity of 1.544 Mbps (1,544,000 bits per second). T-1 uses two pairs of normal twisted wires, the same as you'd find in your house. T-1 normally can handle 24 voice conversations, each one digitized at 64 Kbps. T-1 is a standard for digital transmission in North America. It is usually provided by the phone company and used for connecting networks across remote distances. Bridges and routers are used to connect LANs over T-1 networks. In Europe a similar but incompatible service is called E-1 or E1. See A & B Bits for details of signaling.

For a full explanation of T1 see Bill Flanagan's book The Guide to T-1 Networking. (Call 1-800-LIBRARY for your copy.)

Talk-Off

"False positive" detection of a DTMF digit caused by frequencies in the human voice being mistaken for a dialed digit. When recording a voice mail message, it can happen that the system suddenly jumps to a new menu selection: this is talk-off. Happens more often with female voices because they tend to have stronger frequency components around DTMF pitch.

Talking Yellow Pages

Popular audiotex application which allows the caller to locate a particular type of business through menus. The talking yellow pages service may have an option to connect the caller to the desired business. May be done on a pay-per-call basis or financed by the businesses who are listed. May include advertisements for some of the businesses.

TAPI

Telephony Applications Programming Interface. Microsoft's telephony API for Windows.

TDM

See Time Division Multiplexing.

Terminating Event

Dialogic jargon for a message which indicates the completion of a time-consuming operation such as playing a message.

Text Normalization

The process of analyzing ambiguous text prior to speaking it. For example, should "Dr." be pronounced "Drive" or "Doctor"?

Text-To-Speech

Technology for converting speech in the form of ASCII or other text to a synthesized voice.

TIFF

TIFF provides a way of storing and exchanging digital image data. Aldus Corp., Microsoft Corp., and major scanner vendors developed TIFF to help link scanned images with the popular desktop publishing applications. It is now used for many different types of software applications ranging from medical imagery to fax modem data transfers, CAD programs, and 3D graphic packages. The current TIFF specification supports three main types of image data: Black and white data, halftones or dithered data, and grayscale data. A special variant of TIFF, called TIFF/F, has been defined specifically for storing fax images. Note that most standard PC graphics software, at least at the time of writing, doesn't support TIFF/F even it does support other flavors of TIFF file.

Time Division Multiplexing

Time Division Multiplexing, or TDM, is the process used by digital trunks such as T-1 or E-1 and digital voice buses such as MVIP, PEB or SC bus to transmit audio and signaling information. The data is transmitted as a series of frames. Each frame is divided into 24 (T-1) or 32 (E-1) different time-slots, which are 8 bits long. A complete frame is then 24 x 8 or 32 x 8 bits long, together with a one or more extra bits for synchronization. The 24 or 32 channels are combined at one end of the trunk and split apart at the other end. This process happens so quickly that each conversation is seamless.

In telephony applications, each "slot" is usually an 8-bit sample of the amplitude of the sound in the conversation, encoded using PCM (Pulse Code Modulation). There are 8,000 samples transmitted per second. The number represented by the 8 bit sample is a direct measure of the loudness (amplitude) of the sound, converted on a non-linear scale. Unfortunately, E-1 and T-1 use different scales to do the conversion: the scale usually used by T-1 is called the Mu-law scale, E-1 generally uses A-law.

Time Slice

The amount of time a single process is allowed to run uninterrupted by pre-emptive multi-tasking.

Time-Slot

One channel in a TDM transmission.

Tip

1. The first wire in a pair of wires. The second wire is called the "ring" wire.

2. A conductor in a telephone cable pair which is usually connected to positive side of a battery at the telephone company's central office. It is the phone industry's equivalent of Ground in a normal electrical circuit.

Tone Dial

A pushbutton telephone dial that makes a different sound (in fact, a combination of two tones) for each number pushed. The correct name for tone dial is "Dual Tone Multi Frequency" (DTMF). This is because each button generates two tones, one from a "high" group of frequencies — 1209, 1136, 1477 and 1633 Hz — and one from a "low" group of frequencies — 697, 770, 852 and 841 Hz. The frequencies and the keyboard, or tone dial, layout have been internationally standardized, but the tolerances on individual frequencies vary between countries. This makes it more difficult to take a touch-tone phone overseas than a rotary phone.

You can "dial" a number faster on a tone dial than on a rotary dial, but you make more mistakes on a tone dial and have to redial more often. Some people actually find rotary dials to be, on average, faster for them. The design of all tone dials is stupid. Deliberately so. They were deliberately designed to be the exact opposite (i.e. upside down) of the standard calculator pad, now incorporated into virtually all computer keyboards. The reason for the dumb phone design was to slow the user's dialing down to the speed Bell central offices of early touch tone vintage could take. Today, central offices can accept tone dialing at high speed. But sadly, no one in North America makes a phone with a sensible, calculator pad or computer keyboard dial. On some telephone/computer work-stations you can dial using the calculator pad on the keyboard. This is a breakthrough. It's a lot faster to use this pad. The keys are larger, more sensibly laid out and can actually be touch-typed (like touch-typing on a keyboard.) Nobody, but nobody can "touch-type" a conventional telephone tone pad. A tone dial on a telephone can provide access to various special services and features — from ordering your groceries over the phone to inquiring into the prices of your (hopefully) rising stocks.

Touch Tone

A former trademark once owned by AT&T for a tone used to dial numbers. For a full explanation of touchtone, see DTMF.

Training Sequence

Part of the handshake used in establishing a fax call where the two devices can adjust to prevailing line conditions.

Transition Probabilities

Probabilities of moving from one state to another in a state table.

Transparent Signaling

When Dialogic DTI digital trunk devices are in inserted signaling mode, it is the DTI device that sets the signaling bits sent to the T-1 or E-1 trunk. In transparent mode, the signaling bit is taken from the PEB bus and passed through the DTI without change. SC bus versions of the DTI bus work in inserted signaling mode only.

TSI

Transmitter Subscriber Information. Data which may be sent as part of the Group 3 fax protocol.

Glossary

TSS
See CCITT.

TTS
1. Text To Speech. A term used in voice processing. See Text-To-Speech.

2. Transaction Tracking System. A Novell NetWare feature that protects database applications from corruption by backing out incomplete transactions that result from a failure in a network component. When a transaction is backed out, data and index information in the database are returned to the state they were in before the transaction began.

Twist
The ratio between the amplitudes of the two component frequencies in a dual tone such as a touch-tone. If one component is much louder than the other, then there is a high or low twist, if the two amplitudes are the same, the twist is equal to one. The two frequencies in a touch-tone (DTMF tone), for example, should have the same amplitudes: restricting the twist values which are accepted by a DTMF detector can help eliminate talk-off and play-off. See also Talk-Off, Play-Off.

Twisted Pair
Two insulated copper wires twisted around each other to reduce induction (thus interference) from one wire to the other. The twists, or lays, are varied in length to reduce the potential for signal interference between pairs. Several sets of twisted-pair wires may be enclosed in a single cable. In cables greater than 25 pairs, the twisted pairs are grouped and bound together in a common cable sheath. Twisted pair cable is the most common type of transmission media. It is the normal cabling from a central office to your home or office, or from your PBX to your office phone. Twisted pair wiring comes in various thicknesses. As a general rule, the thicker the cable is, the better the quality of the conversation and the longer cable can be and still get acceptable conversation quality. However, the thicker it is, the more it costs.

Two-Wire Circuit

A transmission circuit composed of two wires — signal and ground — used to both send and receive information. In contrast, a four-wire circuit consists of two pairs. One pair is used to send. One pair is used to receive. All trunk circuits — long distance circuits — are four wire. A four-wire circuit costs more but delivers better reception. All local loop circuits — those coming from a Class 5 central office to the subscriber's phone system — are two wire, unless you ask for a four-wire circuit and pay a little more. Twisted-pair is a widely-used type of two-wire circuit. See also Twisted Pair.

VBX

An obsolete type of custom control for Windows used by 16-bit versions of Visual Basic.

Vocabulary

1. A set of pre-recorded audio files used to synthesize phrases such as "you have one hundred twenty five new messages". The vocabulary for this message would probably be the six fragments "You have" / "one" / "hundred" / "twenty" / "five" / "New messages".

2. The set of words or phrases which can be recognized by a voice recognition system from a single utterance by the caller. A typical voice recognition vocabulary would be the digits "one" through "nine", "zero", "oh", "yes" and "no". Specialized vocabularies might consist of the twelve astrological signs, twelve months or specific commands for a menu such as "Stop", "Rewind", "Help", "Next", "Previous".

Voice Board

Also called a voice card or speech card. A Voice Board is a computer add-in card which can perform voice processing functions. A voice board has several important characteristics: It has a computer bus connection. It has a telephone line interface. It typically has a voice bus connection. And it supports one of several operating systems, e.g. MS-DOS, UNIX. At a minimum, a voice board will usually include support for going on and off-hook (answering, initiating and terminating a call);

notification of call termination (hang-up detection); sending flash hook; and dialing digits (touchtone and rotary). See Voice Response Unit.

Voice Cut-Through

The ability of a voice processing system to respond immediately to the spoken voice of the caller. Generally applied to voice recognition where the caller has the option of speaking his or her selection before the menu has finished playing. May simply be a "barge", where the caller says something, gets a "beep" and then speaks the command, or, better, where the caller may speak the desired word which both interrupts the menu and is recognized by the system.

Voice Mail

You call a number. A machine answers. "Sorry. I'm not in. Leave me a message and I'll call you back." It could be a $50 answering machine. Or it could be a $200,000 voice mail system. The primary purpose is the same — to leave someone a message. After that, the differences become profound. a voice mail system lets you handle a voice message as you would a paper message. You can copy it, store it, send it to one or many people, with or without your own comments. When voice mail helps business, it has enormous benefits. When it's abused — such as when people "hide" behind it and never return their messages — it's useless. Some people hate voice mail. Some people love it. It's clearly here to stay.

Voice Mail Jail

What happens when you reach a voice mail message and you try and reach a human by punching "0" (zero) and you get transferred to another voice mail box and you try again by punching "0" or some other number you're told to punch...and you never reach a human. You're stuck forever inside the bowels of this voice mail machine. You're in voice mail jail.

Voice Mail System

A device to record, store and retrieve voice messages. There are two types of voice mail devices — those which are "stand alone" and those which profess some integration with the user's phone system. A stand alone voice mail is not dissimilar to a collection of single-person answering machines, with several added features. You can instruct the machines (voice mail boxes) to forward messages among themselves. You can organize to allocate your friends and business acquaintances their own mail boxes so they can dial, leave messages, pick up messages from you, pass messages to you, etc. You can also edit messages, add comments and deliver messages to a mailbox at a pre-arranged time. Messages can be tagged "urgent" or "non-urgent" or stored for future listening. The range of voice mail options varies among manufacturers.

Voice Messaging

Recording, storing, playing back and distributing phone messages. New York Telephone has an interesting way of looking at voice messaging. NYTel sees it as four distinct areas: 1. Voice Mail, where messages can be retrieved and played back at any time from a user's "voice mailbox"; 2. Call Answering, which routes calls made to a busy/no answer extension into a voice mailbox; 3. Call Processing, which lets callers route themselves among destinations via their touch-tone phones; and 4. Information Mailbox, which stores general recorded information for callers to hear.

Voice Recognition

The ability of a machine to understand human speech. When applied to telephony environments, the limited bandwidth (range of frequencies transmitted by a telephone connection) and other factors such as background noise and the poor quality of most telephone microphones severely limits the ability of current technology to recognize spoken words. Typical systems are able to recognize standard vocabularies of 16 or so words, such as the digits, yes, no and stop.

Voice Response Unit

VRU. Think of a Voice Response Unit (also called Interactive Voice Response Unit) as a voice computer. Where a computer has a keyboard for entering information, an IVR uses remote touchtone telephones. Where a computer has a screen for showing the results, an IVR uses a digitized synthesized voice to "read" the screen to the distant caller. An IVR can do whatever a computer can, from looking up train timetables to moving calls around an automatic call distributor (ACD). The only limitation on an IVR is that you can't present as many alternatives on a phone as you can on a screen. The caller's brain simply won't remember more than a few. With IVR, you have to present the menus in smaller chunks. See Voice Board.

Voice Store And Forward

Voice mail. A PBX service that allows voice messages to be stored digitally in secondary storage and retrieved remotely by dialing access and identification codes. See Voice Mail System.

VQC

Vector Quantizing Code. A method of voice data compression.

VRU

See Voice Response Unit.

Win32

A generic term referring to common APIs provided by 32-bit Windows operating systems such as Windows 95 and Windows NT.

Wink

A signal sent between two telecommunications devices as part of a "handshaking" protocol. On a digital connection such as a T-1 circuit, a wink is signaled by a brief change in the A and B signaling bits from off to on and back to off (the reverse of a flash-hook). On an analog line, a wink is signaled by a change in polarity (electrical + and -) on the line.

Word Spotting

The ability of an automated voice recognition device to pick out certain selected words from "background" conversation. For example, the recognizer might pick out "No" from "No thanks", or "five" from "give me five please" or the word "nuclear" or "assassinate" from a conversation being monitored by the intelligence agencies.

X.400

X.400 is a store-and-forward electronic messaging protocol defining a framework for distributing data from one network to several others. It allows end-users and application processes to send and receive messages which it transfers in a store-and-forward manner. An X.400 message consists of a message envelope and message content.

The message envelope carries addressing, routing, and control certification information. The message content part involves methods of encoding simple ASCII messages, as well as more complex data. It may contain fax, graphics, text, voice, or binary data structures. X.400 is being increasingly viewed as a delivery platform for a variety of services, including e-mail, electronic data interchange (EDI), and others.

Zero Training

Some pulse digit recognition systems require that the first digit dialed to the system be a zero so that the recognizer can learn the individual characteristics of the caller's telephone in order to improve the recognition accuracy. This process is called zero training.

Index

A

A & B leads, 694
A & B signaling, 694
A and B bits, 97, 116, 198, 201, 211, 454, 694
ABCD bits, 211, 694
ACD, 37, 500, 664, 695. *See also* Automatic Call Distributor
Actions, 437
ActiveX, 598, 695
  controls, 591
Aculab, 300
Adaptive Differential Pulse Code Modulation, 695
Adaptive vocabularies, 168
ADPCM, 351, 353, 356, 695
ADSI, 696
Advanced call managers, 510
AEB, 233, 240, 262, 696
AEB Switching, 335
A-law, 194, 351, 378, 539
A-law encoding, 694
AMIS, 696
Amplitude, 346, 696
Amplitude Distortion, 696
Amplitude Equalizer, 696
Amplitude Modulation, 697
AMX/81, 240
AMX/8x, 264, 335
Analog, 90, 697
Analog / Digital converter, 697
Analog Display Services Interface, 224, 515, 697
Analog Expansion Bus, 233, 240, 262, 698
Analog flash-hook, 120
Analog Interface (LSI) devices, 396
Analog seize, 117
Analog transmission, 698
ANI, 30, 93, 99, 698
Answering a call, 628
Antares board models, 295
Antares boards, 293
Antares firmware, 295
Anti-aliasing, 347, 698
API, 312, 365, 512, 698
Application generator, 448, 612, 699
Application Programming Interface, 365
Applications Programming Interface, 512, 698
Area code, 105
Arithmetic operators, 636
ARU, 699
ASCII, 111, 149, 158, 177, 182, 239, 272, 289, 699
ASR, 160, 700
Assignment, 638
Assisted telephony, 700
Assisted Telephony, 523
ASVD, 223, 700
Asynchronous, 372
Asynchronous mode, 408, 410, 700
Audio, 24, 700
Audio frequencies, 700
Audio menu, 700

Audio Messaging Interchange Specification, 701
Audio Response Unit, 701
Audiotex, 702
Audiotext, 28, 702. *See also* Audiotex
Automated attendant, 27, 702
Automated order entry, 29
Automatic Call Distribution, 37
Automatic Call Distributor, 703
Automatic Call Sequencer, 703
Automatic Call Unit, 704
Automatic callback, 704
Automatic Circuit Assurance, 704
Automatic clock fall-back, 314, 704
Automatic Number Identification, 93, 99
Automatic Speech Recognition, 160
Average call length, 651

B

Baby Bell, 104, 705
Backward signals, 202, 211, 705
Basic multi-tasking, 409
Basic Rate Interface, 98, 705
Battery, 90, 705
B-channel, 98, 705
Bearer channel, 98
Bearer mode, 521
Benchmarks, 671
BeSTspeech, 179
Bi-directional, 313
Binary large object, 381
Blind transfer, 630
BLOB, 381
BLT, 400, 706
Board device, 390, 706
Board devices, 390

Board Locator Technology, 400, 706
Bong, 113, 706
BRI, 98, 707
Built-in functions, 644
Bus, 312
Busy, 95, 96, 707

C

Cadence, 127, 707
Call center, 29, 707
Call completion, 82, 707
Call handles, 374, 521
Call processing, 24
Call progress, 94, 629
Call progress analysis, 26, 95, 125
Call Progress Analysis, 707
Call progress monitoring, 126, 708
Call progress tones, 126, 708
Call status transition events, 411, 422, 708
Call supervision, 125, 708
Called Station Identification, 140
Called Subscriber Identification, 140, 708
Caller ID, 100, 709
Caller Identification, 93
CallerHangUp event, 603
Calling tone, 139, 709
CallSuite Wizard, 612
CAS, 152, 709
CCITT, 137, 188, 196, 206, 214, 351, 709, 710
CCITT Standard R2/MF, 196
CCM, 709
CCS, 709
CED, 140, 710
Centers, 104
Central Office, 89, 710

Index

Central Processing Unit, 101
Centrex, 97, 710
CFR, 141, 710
Channel, 421, 490, 710
Channel bank, 34, 711
Channel Bank, 33
Channel device, 390, 711
Channel devices, 390
Channel numbers
    fax, 404
Channel Service Unit, 33, 711
Clear channel data transport, 314
Clear-back, 212, 712
Clear-forward, 212, 712
Client, 500
Client-server, 500
CNG, 139, 712
CO, 89, 712
CO lines, 101, 712
CO simulator, 348, 712
CO switch, 90
Co-articulation, 167, 183, 713
Codec, 346, 713
Comments, 634
Common Channel Signaling, 713
Communicating Applications
    Specification, 152, 713
Compelled signaling, 194, 199, 713
Compression, 714
Computer Supported
    Telecommunications
    Application, 503
Computer Telephone Integration, 499
Computer Telephony, 31
Computer Telephony Integration, 32
Concatenation, 639
Conference, 714

Confidence level, 163
Confirmation to Receive, 141
Connect, 96
Connected properties, 167
Connected speech, 714
Constants, 636
Context, 409
Context sensitive, 568
Context-dependent, 355
Continuous properties, 167
Continuous speech, 714
Continuously Variable Slope
    Differential Modulation, 352
Controls, 591
Conventional Memory, 251
Co-operative multitasking, 715
Cooperative multi-tasking, 410
CPE, 34
CPU, 101
CSI, 140, 715
CSTA, 503
CSU, 33, 715
CTI, 32, 499, 715
CTI link, 31
CTI Link, 503
CTI Link Hardware, 503
Custom control, 598, 715
Custom tones, 543
Customer Premise Equipment, 34
Cut-through, 175, 237, 716
Cut-Through, 167
CVSD, 352, 716

D

D/12x, 270, 301, 302, 306
D/160SC-LS, 284
D/240SC, 284
D/240SC-T1, 285
D/2x, 263

D/300SC-E1, 286
D/320SC, 285
D/41B, 240
D/41D, 240
D/42, 264
D/4x, 239, 261, 339
D/81A, 280
Data, 24
Data channel, 98
Data Circuit Terminating
  Equipment, 226
Data compression, 362
Data Service Unit, 715
Date/Time stamps, 576
Dates, 576
DCB, 281
DCE, 226
D-channel, 98, 716
D-channels, 260
DCN, 142
DCS, 140, 716
DDI, 99, 716
dec, 635
Decadic signaling, 716
Decadic Signaling. *See also* Pulse
  Dialing
DECTalk PC board, 180
Deterministic Finite State
  Automata, 435
Device, 389
Device handle, 372
Device identifier, 373
Device non-specific handlers, 415
Device Programming Interface,
  318
Device-specific handlers, 414
Dial, 116
Dial cut-through, 175
Dial string, 110, 716

Dial tone, 91, 629, 717
Dialed Number Identification
  Service, 99
Dialing a call, 628
Dialing In, 99
Dialogic, 257, 717
  Antares boards, 293
  high density boards, 282
  product line, 257
Dialogic Devices, 389
Dialogic Drivers, 365
  Operating Systems, 366
Dianatel EA24, 300
DID, 99, 717
DID/120, 270
DID/40, 267
Differential Pulse Code
  Modulation, 351, 717
Differential scheme, 350
Digit and alpha-numeric strings,
  576
Digit Capture, 25
Digital Command Signal, 140, 717
Digital Multiplexer, 307, 343
Digital Signal Processing, 352
Digital Signal Processor, 366, 717
Digital trunk, 97, 718
  T-1, 97
Digital trunks
  E-1, 98
  ISDN, 98
Digitization, 346, 718
Digitization methods, 350
Diphone, 184, 718
Direct Dialing In, 718
Direct Inward Dial, 99, 718
Direct Memory Access, 255
Disconnect, 116, 118, 142, 718
Disconnect supervision, 121

Index **761**

Disconnecting a call, 630
Disyllable, 184, 718
DMA Channels, 255
DMX, 276, 307, 343
DNA, 315
DNIS, 99, 718
do..until, 639
Dongle, 295
DPCM, 351
DPI, 318
Drivers, 366
Drop and insert, 36, 268, 719
Drop and insert configurations, 305
Dropped, 268
DS-0, 97, 719
DS-1, 97, 719
DSP, 352, 366, 719
DSVD, 223, 720
DTI channel devices, 398
DTI/100, 269
DTI/124, 267
DTI/211, 278, 306
DTI/212, 279
DTI/xx, 300
DTMF, 93, 720
DTMF cut-through, 237, 721
Dual tone, 721
Dual Tone Multi Frequency, 93
Dual tones, 195
Dumb switch, 39, 509, 721
Dynamic Node Architecture, 315

E

E & M leads, 722
E & M signaling, 723
E&M, 97
E-1, 724
E-1 digital trunks, 98

E-1 signaling, 193
E-1 signaling bits, 198
EA24, 300
ECM, 141, 724
ECMA, 503, 724
ECTF, 316, 724
Element, 586
EMS, 252
End Of Message, 141
End Of Procedure, 141
End office, 104
enddec, 635
endfunc, 645
endprogram, 645
Engaged, 724
Enterprise Computer Telefony Forum, 316
EOM, 141, 724
EOP, 141, 725
eq, 637
Equal access code, 113
Error Correction Mode, 141
Error Popups, 605
European Computer Manufacturers Association, 503, 724
Event, 370, 411, 422, 595, 725
  fired, 595
  triggered, 595
Event Control File, 155
Event handle, 155
Event non-specific handlers, 415
Event procedure, 596
Event queue, 422
Event-specific handlers, 415
Exception, 603
Exception dictionary, 182, 725
Expanded Memory, 252
Expressions, 636

Extended Memory, 252

F

Failure-to-Train, 141
Fast busy, 95, 96, 128, 726
Fax broadcast, 136, 725
Fax channel devices, 398
Fax channel numbers, 404
Fax mail, 136, 725
Fax on demand, 136, 726
Fax store and forward, 26, 726
Fax synthesis, 27, 726
FAX/120, 272, 301
Feature phones, 101
Fiber, 410, 726
Fibers, 409
Fill, 141
Firmware, 366
Flash-hook, 96, 116, 120, 629, 726
for, 639
Forward signals, 202, 211, 727
Fourth column, 93, 727
Frame, 199, 727
Frames, 458
FTT, 141, 727
func, 645
FUNCDIR, 647
Functions, 643

G

Gateway, 45
GetDigits Routine, 615
Glare, 117, 727
Global Tone Detection, 132
Global Tone Generation, 132
Goto statements, 642
Greeting, 565
Ground start, 91, 728

Group 1, 728
Group 2, 728
Group 3, 729
Group 4, 729
Grunt detection, 163, 729
GTD, 132

H

Hang up, 630
Hangup, 118
Hang-up, 116
Hang-up detection, 121, 631
Hardware Interrupts, 247
Harmonics, 346
HCV, 352, 729
HD, 282
High Capacity Voice, 352
High Density, 282
High Memory Area, 253
HLLAPI, 44
HMA, 253
Homologation, 188, 729
Hook switch, 91
Host, 39
Hunt, 99, 729
Hunt group, 730
Hunt group., 102

I

I/O Port Addresses, 250
In-band, 116, 126
In-band signaling, 730
INCDIR, 648
Include files, 647
Incoming register, 202
Indexed Prompt File, 360
Industry Standard Architecture, 232

Index **763**

Inflection, 184, 730
Information provider, 29, 109, 731
Inserted, 268
Inserted signaling, 473, 474, 730
Installing voice card hardware and software, 245
Integers, 576
Integrated messaging, 137
Interactive Voice Response, 28, 41, 731
Inter-LATA service, 105
Intra-LATA service, 105
Invisible controls, 597
IPF, 360
IRQs, 247
ISA, 232
ISA bus, 240
ISDN digital trunks, 98
IVR, 28, 41

J

Jump statements, 642

K

Key system, 732
Key systems, 101

L

Labels, 642
LAN., 42
Languages supported, 613
LATA, 105, 732
LCR, 114, 732
Learning vocabularies, 168
Least-cost routing, 114
Line conditions tolerance, 166
Line device, 373

Line devices, 518
Line source, 258
Linear PCM, 350
LIST INFO chunk, 554
Listen, 490
Listen channel devices, 395
Local Access Transport Area, 105, 732
Local loop, 91, 103, 732
Logical operations, 637
Logical values, 634
Loop, 639, 732
Loop current, 91, 733
Loop start, 91, 733
Lower Memory, 251
LSI devices, 396
LSI/120, 269, 300, 302
LSI/80, 280

M

Mailbox, 27, 733
Main menu, 565
MCA, 232, 733
MCF, 141, 733
Media control, 502
Media mode, 514
Media rate, 521
Media stream, 514, 733
Menu, 733
Menu Routine, 615
Menu-driven, 558
Menus, 29, 627
Message Confirmation Frame, 141
Message delivery, 28
Message handler function, 596
Messages, 501
Method, 596
Mezzanine bus, 734
MF, 93, 734

MF/40, 240, 266
Micro channel, 232, 734
MMR, 144, 734
Modified Huffman run-length encoding, 142
Modified Modified READ, 144
Modified READ, 143
Modular Station Interface, 300, 343
MPS, 141, 734
MR, 143, 734
MSI, 274, 300, 301, 306, 343
MSI Station devices, 399
MSI/C, 275
MSI/SC, 276
Mu-law, 194, 350, 378, 539
Mu-law encoding, 735
Multi Frequency, 93
Multi Vendor Integration Protocol, 735
Multi-frequency tones, 735
Multi-master, 314
Multi-Page Signal, 141
Multi-point, 223
Multi-tasking, 735
Multi-tasking functions, 420
Multi-threading, 409
Multi-Vendor Integration Protocol, 233
MVIP, 233, 735, 736

N

Netware Loadable Module, 501
Network module, 736
Network modules, 300
NFAS, 98, 736
NLM, 501
No ring-back, 96
Non-multi-tasking functions, 420
Non-Standard Facilities, 140
Non-Standard facilities Setup, 140
NSF, 140, 736
NSS, 140
Numbers, 634
NXX, 109, 563, 736

O

OA&M, 504
OCR, 146
OCX, 598, 736
Off-hook, 91, 736
Oki, 353
OLE control, 598, 737
On Error Goto label, 603
One-call fax, 136
One-Call Fax, 737
One-way drop and insert, 305
On-hook, 92, 737
On-hook dialing, 737
Operator intercept, 95, 96, 737
Operators, 636
Optical Character Recognition, 146
Ordinal numbers., 575
Ordinals, 580
Outgoing register, 202, 738
Out-of-band, 116, 126
Out-of-band signaling, 738

P

PABX, 90, 738
PABX, 101
PAMD, 130, 738
Pay-per-call, 29, 739
PBX, 90, 101, 739
PBX extensions, 101
PCM, 350, 356, 739

Index

PCM Expansion Bus, 233
PCM-30, 741
Peak load, 650
PEB, 233, 299, 741
Perfect Call Progress, 629
Phase A, 139
Phase A, B, C1, C2, D, and E, 741
Phase B, 139
Phase C1, 139
Phase C2, 139
Phase D, 139
Phase E, 139
Phone devices., 518
Phone numbers, 576
Phoneme, 184, 741
Phonetic description, 168
Phonetic modification, 183
Phonetic spelling, 183
Phrase Routine, 615
Plain Old Telephone Service, 90, 741
Play Routine, 615
Play-off, 237, 238, 502, 741
Point-to-point, 222
Port, 742
Port density, 742
Positive Answering Machine Detection, 130, 742
Positive Voice Detection, 130, 742
Post Telephone and Telegraph, 188
POTS, 90, 742
Power dialer, 742
Power dialers, 72
Precedences of VOS operators, 637
Predictive dialer, 30
Predictive dialers, 72
Predictive dialing, 37, 742
Pre-empted, 409

Pre-emptive multi-tasking, 409, 743
PRI, 98, 744
Primary centers, 104
Primary Rate Interface, 744
Printer drivers, 511
Private Branch Exchange, 101, 744
Private Branch Exchanges, 90
Programming models, 408, 676
Proline/2V, 240
Properties, 593
Prosody, 744
PSTN, 90, 744
PTT, 188, 744
Public Switched Telephone Network, 90, 745
Pulse Code Modulation, 350, 745
Pulse dialing, 93, 745
Pulse Recognition, 172
Pulse recognizers, 94
Pulse to tone converter, 173, 745
Pulse to tone converters, 94
PVD, 130, 745

##

Quantization noise, 347, 746

R

R1/MF tone, 746
R1/MF tones, 194
R2/MF signaling, 199
R2/MF tone, 746
Raising an exception, 603
Rate, 521
RBOC, 104, 746
Real Memory, 252
Receive bits, 211
Receive channel devices, 395

Record Routine, 615
Regional Bell Operating Company, 104, 746
Regional centers, 104
Register, 746
Registers, 202
Relative Element Address Differentiation, 143
Request ID, 524
Request identifier, 375
Resource module, 455, 746
return, 645
Ring, 91, 116, 119, 747
Ring no answer, 96
Ring voltage, 92
Ring-back, 94
RJ-11 jacks, 152, 232, 263, 266, 270, 275, 281, 337, 348, 747
RJ-14 jacks, 232, 263, 748
RLL functions, 644
Robbed-bit, 201
Rotary dial, 748
Rotary dialing, 93
Rotary digits, 238
Rotary phones, 160, 172, 190
Routines, 614
RS-232 Serial., 42
Run-length, 142

S

Same-call fax, 136
Same-Call Fax, 748
Sample, 346
Sampling rate, 346
Sampling Rate, 748
SC Bus, 233, 312, 313, 748
SC expansion bus, 315
Score, 163
Screen-pop, 505

Script, 748
SCSA, 311, 749
    Server API, 318
SCx Bus, 315
SCX bus, 321
SCX Bus, 749
SCX/160, 322
SDK, 294, 749
Segment address, 251
Segment index chunk, 551
Segmented Wave file format, 550
Segments, 548, 549
Seize, 116, 117, 750
Seizing, 91
Server, 500
Service provider, 518
Services Programming Interface, 512
Shared Memory, 251, 253
Sign, 353
Signal Computing System Architecture, 311
Signaling bits, 97
    E-1, 198
SimPhone, 618
SIT, 128
SIT tones, 128, 750
SmartBridge 96, 330
SmartSwitch, 329
Socotel signaling, 197, 750
Software Development Kit, 294
Software Interrupts, 249
Sound files, 345
Source code, 634
Span, 97, 751
Span line, 751
Speaker dependent, 26
Speaker independent, 26
Speaker verification, 161

Index 767

Special Information Tones, 128
Speech card, 231
Speech to Text, 26, 751
SPI, 512
Spider, 281
SPOX, 294
SpringBoards, 271
SpringWare, 271
Stand-alone, 44
State machine, 435, 676
State machine programming, 751
State machine programming model., 408
State table, 408, 438
State transition, 437
States, 437
Station, 400
Station card, 751
Station set, 90, 751
Step size, 351
Store and forward, 751
streq, 637
Strings, 634
strneq, 637
Subaddress, 112
Super-state, 447
Supervised transfer, 26, 630, 752
SVD, 752
Switch, 90, 752
Switch cards, 327
Switch Driver, 503
Switch driver interface, 504
Switch/Case, 641
SWV file format, 550
Synchronous, 676
Synchronous mode, 410
Synchronous model, 408

T

T-1, 752
T-1 digital trunk, 97
T-1 flash-hook., 121
T-1 seize, 117
T-1 span, 97
Tagged Image File Format, 158
Talk-down, 229
Talking yellow pages, 268, 753
Talk-off, 174, 229, 237, 502, 752
Talk-off suppression, 175
TAPI, 507, 509, 529, 753
 SDK, 516
TCF, 141
TDM, 199, 458, 753
Telephony
 application ideas, 27, 47
Telephony board, 618
Telephony server, 504
Telephony Services API, 504
Telephony Services NLM, 504
Terminate on, 99
Terminating configurations, 304
Terminating event, 371, 411, 422, 753
Test button, 618
Text normalization, 182, 753
Text to Speech, 25
Text-to-speech, 177, 753
Thread, 409
Threads, 409
Throwing an exception, 603
TIFF, 158, 753
Time Division Multiplexing, 458, 754
Time slice, 409, 754
Time-division-multiplexing, 199
Time-out, 563
Times, 576

Time-slot, 97, 199, 453, 754
Time-slot bundling, 314
Tip, 91, 754
Tolerance of line conditions, 166
Toll center, 104
Tone dial, 755
Tone dialing, 93
Touch tone, 93, 756
Touch-tone menus, 627
Touch-tone type-ahead, 237
Trained, 161
Training Check Frame, 141
Training sequence, 140, 756
Transferring a call, 629
Transition events, 411
Transition probabilities, 756
Transitions, 437
Transmit bits, 211
Transmit channel devices, 395
Transmitter Subscriber Information, 140
Transparent signaling, 398, 473, 474, 756
True and false, 634
Trunk, 101, 258, 259
TSAPI, 499
TSAPI client library, 504
TSAPI library, 504
TSI, 140, 756
TSS, 756. *See* Telecommunications Standard Sector
TTS, 25, 177, 756
Twist, 238, 544, 757
Twist ratios, 133
Twisted pair, 97, 103, 260, 757
Two wire analog station card, 90
Two-way drop and insert, 305
Two-wire, 90
Two-wire circuit, 757

Type-ahead. *See also* Cut-Through

U

UMB, 252
Upper Memory, 252
Upper Memory Block, 252
User i/o, 381
User-defined functions, 644

V

Variables, 635
VBX, 598, 757
Vector Quantizing Code, 352
vid_write, 642, 644
View Source button, 617
Virtual boards, 390
Virtual channels, 314
Virtual time-slots, 314
Visual Basic, 619
Visual Basic Extension, 598
Visual C++, 619
Vocabularies
    learning, 168
Vocabulary, 25, 166, 758
Voice, 24
Voice board, 231, 758
Voice bus, 232
Voice card, 31, 32
Voice channel devices, 396
Voice cut-through, 167, 758
Voice generation, 184
Voice mail, 24, 27, 758
Voice Mail jail, 759
Voice Mail system, 759
Voice messaging, 28, 759
Voice recognition, 159, 760
Voice Recognition resource devices, 400

Index

Voice response unit, 25, 32, 760
Voice store and forward, 25, 760
VoiceBocx, 600, 601, 606
VOS, 621
VOX, 353, 356
VQC, 352, 760
VR/121, 273, 301
VR/160p, 304
VR/40, 240, 267
VR/xxp, 274
VRU, 25, 32, 761

W

Wave, 529
while, 639
Whole numbers., 575
Win32, 371, 761

Windows Open Services Architecture, 511
Windows Telephony., 510
Wink, 99, 100, 119, 121, 761
Word spotting, 167, 761
WOSA, 511

X

X.400, 149, 761

Z

Zero training, 174, 762

μ

μ-law, 194

Get Started Today
with Parity Software's Telephony Developer
Power Pack CD

Save $20!

What's the best way to choose a telephony tool? Try before you buy! Get to know the development tool and company you will be working with. The Parity Software Telephony Developer Power Pack CD is the perfect way to get started with your computer telephony project. Get printed User Guides for VOS and CallSuite. Work with our technical support, see how responsive we are to our developers.

VOS™ Evaluation Versions
Included are Evaluation Versions of VOS for NT, 95, UNIX and MS-DOS: complete, fully-functioning VOS packages with these few restrictions: the language translator will accept up to 100 lines of source code, the run-time engine supports only one phone line and shuts down after 15 minutes of continuous operation.

CallSuite™ Evaluation Versions
Experience the award-winning CallSuite Wizard™. Try VoiceBocx® in TAPI, Dialogic or PRI flavors with CallSuite Wizard and SimPhone. The Eval versions are fully-functional with a few restrictions: the Wizard is limited to 10 routines, VoiceBocx supports one phone line only and allows only 15 minutes continuous run-time operation.

Docs And More
Also on the CD is extensive on-line reference information, including an off-line copy of the most popular parts of Parity's Web site, help files for most of Parity's products and demos of several third-party telephony technologies.

For more details, check out Parity Software's web site at:

www.paritysw.com

Regular price $89, special "PC Telephony" price: $69. See order form on next page.

Power Pack CD Order Form

To request <u>more information</u> on Parity Software products, or to <u>order a Power Pack CD</u>, please fill out the following form and fax to:

☎→ **(415) 332-5657**

Please send me:
- ☐ More information about Parity products (n/c)
- ☐ Telephony Developer's Power Pack CD ($69) _____
    - \+ sales tax (in CA add $5.00) _____
    - \+ shipping (US add $14, int'l add $26) _____
- Total _____

Name: _____

Company: _____

Address: _____

City: _____

State: _____ Zip: _____ Country: _____

Phone: _____ Fax: _____

Fax: _____

E-mail: _____

☐ Mastercard ☐ Visa ☐ Amex ☐ COD (US only)

Credit card number: _____

Expires: _____ Cardholder: _____

Signature: _____ ✍

Parity Software

Phone: (415) 332-5656, Europe: +45 39 40 88 03
Fax: (415) 332-5657, Europe: +45 39 40 78 03
sales@paritysw.com
http://www.paritysw.com